Optimization Theory

Applied Optimization

Volume 59

Series Editors:

The titles published in this series are listed at the end of this volume.

Optimization Theory

Recent Developments from Mátraháza

Edited by

Franco Giannessi

Department of Mathematics,
University of Pisa,
Pisa, Italy

Panos Pardalos

Department of Industrial and
Systems Engineering,
University of Florida,
Gainesville, Florida, U.S.A.

and

Tamás Rapcsák

Computer and Automation Institute,
Hungarian Academy of Sciences,
Budapest, Hungary

KLUWER ACADEMIC PUBLISHERS
DORDRECHT / BOSTON / LONDON

A C.I.P. Catalogue record for this book is available from the Library of Congress.

ISBN-13:978-1-4613-7983-6 e-ISBN-13:978-1-4613-0295-7
DOI:10.1007/978-1-4613-0295-7

Published by Kluwer Academic Publishers,
P.O. Box 17, 3300 AA Dordrecht, The Netherlands.

Sold and distributed in North, Central and South America
by Kluwer Academic Publishers,
101 Philip Drive, Norwell, MA 02061, U.S.A.

In all other countries, sold and distributed
by Kluwer Academic Publishers,
P.O. Box 322, 3300 AH Dordrecht, The Netherlands.

Printed on acid-free paper

Softcover reprint of the hardcover 1st edition 2001

CONTENTS

Preface vii

Heuristics for the Process Network Synthesis Problem
Z. Blázsik, Cs. Holló, Cs. Imreh and Z. Kovács 1

Heuristics for Simplified Process Network Synthesis (PNS) Problems with a Blossom-Type Algorithm for the Edge Covering Problem
Z. Blázsik, K. Keserű and Z. Kovács 19

Lower and Upper Bounds on the Probability of the Union of Some Events with Applications
J. Bukszár 33

The Linear Complementarity Problem and the Class of Generalized Positive Subdefinite Matrices
J.-P. Crouzeix and S. Komlósi 45

Computer Experiences with Successive Regression Approximations for Solving Equations
I. Deák 65

A Bundle of Smooth Lagrangians in Mathematical Programming
V. F. Demyanov 81

A Nontopological Two-function Minimax Theorem with Monotone Transformations of the Functional Values
F. Forgó 99

Non-smooth Optimization with Randomization
L. Gerencsér and Zs. Vágó 111

The Sherman-Morrison Formula for the Determinant and its Application for Optimizing Quadratic Functions on Condition Sets Given by Extreme Generators
G. Kéri 119

Duality For D.C. Optimization Over Compact Sets
Juan-Enrique Martinez-Legaz and Michel Volle 139

Characterization of Monotone Operators by Using a Special Class of Preinvex Functions
S. Z. Németh 147

Two Approaches for Parallelizing the UEGO Algorithm
P. M. Ortigosa, I. García and M. Jelásity 159

Convexification of Functions by Nonlinear Coordinate Transformations
T. Rapcsák 179

Packing Equal Circles in a Square I. — Problem Setting and Bounds for Optimal Solutions
P. G. Szabó, T. Csendes, L. G. Casado and I. García 191

Packing Equal Circles in a Square II. — New Results for up to 100 Circles Using the TAMSASS-PECS Algorithm
L. G. Casado, I. García, P. G. Szabó and T. Csendes 207

On Duality for Quasi-convex Supremization and Reverse Convex Infimization
I. Singer 225

A Minimization Problem in Function Spaces
B. Uhrin 255

PREFACE

This volume contains refereed papers based on the lectures presented at the XIV International Conference on Mathematical Programming held at Mátraháza, Hungary, between 27-31 March 1999. This conference was organized by the Laboratory of Operations Research and Decision Systems at the Computer and Automation Institute, Hungarian Academy of Sciences. The editors hope this volume will contribute to the theory and applications of mathematical programming.

As a tradition of these events, the main purpose of the conference was to review and discuss recent advances and promising research trends concerning theory, algorithms and applications in different fields of Optimization Theory and related areas such as Convex Analysis, Complementarity Systems and Variational Inequalities.

The conference is traditionally held in the Mátra Mountains, and housed by the resort house of the Hungarian Academy of Sciences. This was the 14th event of the long lasting series of conferences started in 1973.

The organizers wish to express their thanks to the authors for their contributions in this volume, and the anonymous referees for their valuable comments. Special thanks are directed to our sponsors, the Hungarian Academy of Sciences, the National Committee for Technological Development, the Hungarian National Science Foundation, and last but not least, the Hungarian Operational Research Society.

We would like to thank John Martindale from Kluwer Academic Publishers for helping us produce this volume, Éva Nóra Nagy for corrections and proof-readings, and Péter Dombi for his excellent work on typesetting and editing the manuscript.

January 2001

The Editors

Heuristics for the Process Network Synthesis Problem

Z. Blázsik, Cs. Holló, Cs. Imreh, Z. Kovács
Department of Informatics,
University of Szeged, Hungary

Abstract. A manufacturing system consists of operating units converting materials of different properties into further materials. In a design problem we are to find a suitable network of operating units which produces the desired products from the given raw materials. By considering this network design from a structural point of view, we obtain a combinatorial optimization problem called the Process Network Synthesis or (PNS) problem. It is known that the PNS problem is NP-complete. In this work two heuristic algorithms are presented for solving this problem. Moreover, we establish some worst-case bound theorems for the algorithms under consideration, and we present an empirical analysis on randomly generated PNS problems.

Key words: Heuristics, Process Design

1. Introduction

In a manufacturing system materials of different properties are consumed through various mechanical, physical and chemical transformations to yield desired products. Devices in which these transformations are carried out are called operating units, e.g., a lathe or a chemical reactor. Thus, a manufacturing system can be considered as a network of operating units, which is called process network. The process design problem in general, and flowsheeting in particular, is to construct a manufacturing system. A design problem is defined from a structural point of view by the raw materials, the desired products, and the available operating units. This triplet determines the structure of the problem as a process graph containing the corresponding interconnections among the operating units. Consequently, the appropriate process networks can be described by some subgraphs of the process graph belonging to the design problem under consideration. Naturally, the cost minimization of a process network is indeed essential.

The importance of process network synthesis (PNS) arises from the fact that networks like this are ubiquitous in the chemical and related industries. The foundations of PNS and the background of the combinatorial model studied here can be found in [5], [6], [7], and [8]. Therefore,

[1] This work has been supported by the Ministry of Culture and Education of Hungary, Grant FKFP 0008/1999, and by the Hungarian National Foundation for Scientific Research, Grant T030074.

F. Giannessi et al. (eds.), Optimization Theory, 1–17.

we recall only the definitions in Section 2 where the necessary notions and notation are introduced.

It has recently been proven (see [2], [4], [11]) that the PNS problem is NP-complete. For NP-hard problems, the construction and analysis of heuristic algorithms is a fastly developing area. By heuristic algorithms we mean fast, (polynomial time) algorithms which do not guarantee an optimal solution in general, but always result in a feasible solution which is close to the optimal solution in some sense. Heuristic algorithms are important for several reasons. The feasible solutions found by these algorithms can be used in procedures based on branch and bound techniques. Moreover, in practical problems often there is not enough time to find an optimal solution by an exponential algorithm, or the size of the problem is too large to use an exponential algorithm. In these cases, heuristic algorithms can be useful again. It can also occur that we do not need an optimal solution, it is sufficient to find a feasible solution the cost of which is not far from the optimal cost. One can find further details on heuristic algorithms in [9], [13]. For the PNS problem, heuristic algorithms have not been studied yet. In Section 3 we introduce two heuristic algorithms for this problem. In Section 4 tight worst-case bounds are presented for these heuristics for particular classes of PNS problems. Moreover, in Section 5 the results of an empirical analysis for the effectiveness of these heuristics are given.

2. Preliminaries

In the combinatorial approach the structure of a process can be described by the process graph (see [6]) defined as follows.

Let M be a finite nonempty set, the set of the *materials*. Moreover, let $\emptyset \neq O \subseteq \wp'(M) \times \wp'(M)$ with $M \cap O = \emptyset$, where $\wp'(M)$ denotes the set of all nonempty subsets of M. The elements of O are called *operating units*. For an operating unit $(\alpha, \beta) \in O$, α and β are called the *input-set* and *output-set* of the operating unit, respectively. Pair (M, O) is defined to be a *process graph* or P-*graph* in short. The set of vertices of this directed graph is $M \cup O$, and the set of arcs is $A = A_1 \cup A_2$ where $A_1 = \{(X, Y) : Y = (\alpha, \beta) \in O \text{ and } X \in \alpha\}$ and $A_2 = \{(Y, X) : Y = (\alpha, \beta) \in O \text{ and } X \in \beta\}$. In the rest of the paper, by path and cycle we always mean directed path and cycle.

Now, let $o \subseteq O$ be arbitrary. Let us define the following functions on the set o:

$$mat^{in}(o) = \bigcup_{(\alpha,\beta)\in o} \alpha, \qquad mat^{out}(o) = \bigcup_{(\alpha,\beta)\in o} \beta,$$

and

$$mat(o) = mat^{in}(o) \bigcup mat^{out}(o)\,.$$

Let process graphs (m, o) and (M, O) be given. (m, o) is defined to be a *subgraph* of (M, O) if $m \subseteq M$ and $o \subseteq O$.

Now, we can define the structural model of PNS for studying the problem from a structural point of view. For this purpose, let M^* be an arbitrarily fixed possibly infinite set, the set of the available materials. By a *structural model* of PNS, we mean a triplet (P, R, O) where P, R, O are finite sets, $\emptyset \neq P \subseteq M^*$ is the set of the *desired products*, $R \subseteq M^*$ is the set of the *raw materials*, and $O \subseteq \wp'(M^*) \times \wp'(M^*)$ is the set of the *available operating units*. It is assumed that $P \cap R = \emptyset$ and $M^* \cap O = \emptyset$, moreover, α and β are finite sets for every $(\alpha, \beta) = u \in O$.

Then, process graph (M, O), where $M = \cup\{\alpha \cup \beta : (\alpha, \beta) \in O\}$, presents the interconnections among the operating units of O. Moreover, every feasible process network producing the given set P of products from the given set R of raw materials using operating units from O, corresponds to a subgraph of (M, O). Examining the corresponding subgraphs of (M, O), therefore, we can determine the feasible process networks. If we do not consider further constraints such as material balance, then the subgraphs of (M, O) which can be assigned to the feasible process networks have common combinatorial properties. They are studied in [6], and their description is given by the following definition.

A subgraph (m, o) of (M, O) is called a *solution-structure* of (P, R, O) if the following conditions are satisfied:

$(A1)$ $P \subseteq m$,

$(A2)$ $\forall X \in m$, $X \in R \Leftrightarrow$ there is no arc (Y, X) in the process graph (m, o),

$(A3)$ $\forall Y_0 \in o$, $\exists$ path $[Y_0, Y_n]$ with $Y_n \in P$,

$(A4)$ $\forall X \in m$, $\exists (\alpha, \beta) \in o$ such that $X \in \alpha \bigcup \beta$.

The set of solution-structures of $\mathbf{M} = (P, R, O)$ will be denoted by $S(P, R, O)$ or $S(\mathbf{M})$.

Let us consider PNS problems in which each operating unit has a weight. We are to find a feasible process network with minimal weight where by the weight of a process network we mean the sum of the weights of the operating units belonging to the process network under consideration. Each feasible process network in such a class of PNS problems is determined uniquely from the corresponding solution-structure and vice versa. Consequently, the problem can be formalized as follows:

PNS PROBLEM WITH WEIGHTS

Let a structural model of PNS problem $\mathbf{M} = (P, R, O)$ be given. Moreover, let w be a positive real-valued function defined on O, the *weight function*. The basic problem is to find

$$\min\{\sum_{u\in o} w(u) : (m, o) \in S(P, R, O)\}. \tag{1}$$

It is known (see [2],[4], and [11]) that this problem is NP-complete. In what follows, we call the elements of $S(\mathbf{M})$ *feasible solutions* and by a PNS problem we mean a PNS problem with weights. Moreover, we extend the weight function for any subgraph of (M, O). Namely, by the weight of a subgraph we mean the sum of the weights of the operating units contained in it.

It is a basic observation that if (m, o) and (m', o') are solution-structures of $\mathbf{M}$, then $(m, o) \cup (m', o') = (m \cup m', o \cup o')$ is also a solution-structure of $\mathbf{M}$. This yields that $S(\mathbf{M})$ has the greatest element called *maximal structure*, provided that $S(\mathbf{M}) \neq \emptyset$. Indeed, the maximal structure is the union of all the solution-structures of $\mathbf{M}$. Obviously, the P-graph of an arbitrary PNS problem may contain unnecessary operating units and materials. On the basis of the maximal structure, we can disregard unnecessary operating units and materials as follows. Let $(\bar{M}, \bar{O})$ denote the P-graph of the maximal structure. Then, the P-graph of structural model $\bar{\mathbf{M}} = (P, R \cap \bar{M}, \bar{O})$ is $(\bar{M}, \bar{O})$. Since each solution-structure of $\mathbf{M}$ is a subgraph of $(\bar{M}, \bar{O})$, it is a solution-structure of $\bar{\mathbf{M}}$, and conversely. Consequently, $S(\mathbf{M}) = S(\bar{\mathbf{M}})$. On the other hand, $\bar{\mathbf{M}}$ does not contain any unnecessary operating units and materials. The structural model $\bar{\mathbf{M}}$ is called the *reduced structural model* of PNS.

To determine the reduced structural model for a PNS problem an effective procedure is presented in [7], [8]. It can decide if $S(\mathbf{M})$ is empty; if $S(\mathbf{M})$ is not empty, then the algorithm provides the corresponding maximal structure. In the rest of this work we consider reduced models only. In fact, this is not a further restriction, since one can obtain the reduced model by performing the maximal structure generation algorithm mentioned above.

3. The algorithms

The basic idea of the two algorithms is the same. Both algorithms use a cost function defined on the materials, which gives a lower bound on the producing cost of the material considered. By this function, the algorithms select operating units step by step, which form a feasible solution. First, we define the cost function, and then the algorithms are presented.

3.1. Definition of the cost function

The general definition of the cost function denoted by c was given in [10]. This definition is long and difficult. Moreover, in the worst-case bound proofs, only cycle free subclasses are investigated, and therefore, the general definition of the function c is not recalled here. On the other hand, the definition of this function is very simple for the cycle free case. Thus, we present this particular definition here for the readers convenience. We define the function c by presenting the algorithm which determines it. Before the description of the algorithm, we outline its basic idea. In each iteration step we have two sets $I^{(r)}$ and $J^{(r)}$. $I^{(r)}$ is the set of materials for which the costs have already been determined and $J^{(r)}$ is its complement. Initially, $I^{(0)}$ contains the raw materials. In the r-th step, we select a material X from set $J^{(r)}$ which is produced by operating units only, having all inputs from the set $I^{(r)}$. We determine the cost for X and move X from $J^{(r)}$ into $I^{(r)}$. At the beginning of the procedure zero is assigned as the cost of the elements of $I^{(0)}$. Now, the algorithm determining the function c is presented in details.

ALGORITHM 3.1.

- *Initialization*

  Set $I^{(0)} = R$, $J^{(0)} = M \setminus R$, $r = 0$,

  and $c(X) = 0$, for all $X \in M$.

- *Iteration (r-th iteration)*

  Terminate if $J^{(r)} = \emptyset$. Otherwise, choose a material $X \in J^{(r)}$, for which the input materials of all the operating units $u_1, \dots, u_j$ producing X directly are in $I^{(r-1)}$. For every u_t, calculate the value

  $$c_t = w(u_t) + \max\{c(V) : V \in mat^{in}(u_t)\},$$

  and let $c(X)$ be the minimal c_t value, moreover, let

  $$I^{(r+1)} = I^{(r)} \cup \{X\}, \quad J^{(r+1)} = J^{(r)} \setminus \{X\}.$$

  Set $r := r + 1$ and proceed to the next iteration.

To verify the correctness of the algorithm, we have to show that if $J^{(r)} \neq \emptyset$, then there exists an $X \in J^{(r)}$, for which the input materials of all the operating units $u_1, \ldots, u_j$ producing X directly are in $I^{(r)}$. We prove this statement by contradiction. Let us suppose that for every $X \in J^{(r)}$, an operating unit exists which produces X and has input material in $J^{(r)}$. Let us choose a material A_0 from $J^{(r)}$. Then, we have an operating unit A_1 that produces A_0 and has an input material $A_2 \in J^{(r)}$. The same statement is valid for A_2. Hence, we have an operating unit A_3 which produces A_2 and has an input material $A_4 \in J^{(r)}$. If we continue this list, we obtain a sequence of materials $A_0, A_2, \ldots$ and operating units $A_1, A_3, \ldots$ for which A_{2i} and A_{2i+2} are output and input materials of A_{2i+1}, respectively. On the other hand, the set of materials is finite, and consequently, $A_{2l} = A_{2k}$ for some $k > l$. Then, $A_{2k}, A_{2k-1}, \ldots, A_{2l+1}, A_{2l}$ is a cycle in the P-graph, which is a contradiction. Therefore, the algorithm is correct and it determines a nonnegative function c in a finite number of steps .

Now, we can present our heuristic algorithms.

3.2. Heuristic algorithms

The algorithms select one operating unit in each iteration step. The difference between the two algorithms is in the rule for selecting the operating unit. The algorithms work with two sets, the *set of the selected operating units* and the *set of the required materials.* At the beginning of the procedure the set of the selected operating units is empty and the set of the required materials is P. Later, in each iteration step, we extend the set of the selected operating units with one operating unit and delete the output materials of this operating unit from the required materials set. Moreover, every input material of the operating unit considered, which is neither raw material nor input material of any of the selected operating units is placed into the required materials set. The procedure terminates when the set of the required materials becomes empty. We obtain the feasible solution (m, o), where o is the set of the selected operating unit, and $m = mat(o)$. For completing the description of the algorithms, we have to define the rules for selecting the succeeding operating unit. We select the operating unit v for which the quotient

$$\frac{w(v) + \text{the inputs cost of } v}{\text{the number of the required outputs of } v}$$

is minimal. The difference between the two algorithms is in the calculation of the inputs cost of an operating unit. In the first algorithm, called Asum_c, this cost is estimated by $\sum_{X\in mat^{in}(v)} c(X)$. In the second algorithm, called Amax_c, this cost is estimated by $\max\{c(X) : X \in mat^{in}(v)\}$. Since this is the difference between the two algorithms, here we present one of them only, namely, the algorithm Asum_c.

ALGORITHM Asum_c

- *Initialization.* Set $N_0 = P$, $O_0 = \emptyset$ $K_0 = R$, and $i = 0$.
- *Iteration (i-th iteration)*
  - *Step 1.* Proceed to Step 3 if $N_i = \emptyset$. Otherwise, for each operating unit $u \notin O_i$ producing material from N_i, take the quotient
  $$\frac{w(u) + \sum_{X\in mat^{in}(u)} c(X)}{u(N_i)},$$
  where $u(N_i)$ denotes the number of those elements of N_i which are produced by u. Select an operating unit for which this quotient is minimal and denote it by v.
  - *Step 2.* Let $O_{i+1} = O_i \cup \{v\}$, $K_{i+1} = K_i \cup mat^{out}(v)$, and $N_{i+1} = N_i \cup mat^{in}(v) \setminus K_{i+1}$. Increase the value of i by 1, and proceed to the next iteration.
  - *Step 3.* Let $o = O_i$, and $m = mat(o)$.

First, let us consider the correctness and finiteness of the procedure. Though we consider only the algorithm Asum_c, the same proof is valid for Amax_c as well. We have to show that for every material $A \in N_i$, there exists an operating unit producing A. Let us observe that for every i, K_i contains the raw materials and the materials which are produced by some operating unit from set O_i. Therefore, there are no raw materials in N_i. On the other hand, (M, O) is the maximal structure, and hence, every material $A \notin R$ is produced by some operating unit. These observations yield the required statement. Moreover, we put one new operating unit into set O_i during each iteration, and therefore, the set K_i will contain all the materials, and the procedure terminates in the next step, provided that the procedure does not terminate earlier. Consequently, the algorithm terminates after finitely many steps and produces sets m, o. The definition of m shows that these sets determine a P-graph. We prove now that this P-graph is a feasible solution.

THEOREM 1. *Algorithms Asum_c and Amax_c result in a feasible solution.*

Proof. We prove the statement for Asum_c; the same proof is valid for Amax_c as well. We have to show that the P-graph (m, o) satisfies the conditions given in the definition of a feasible solution. First, consider condition $(A1)$. By the earlier observation on K_i, it follows that the procedure removes from the set N_i only materials which are produced by some operating units from O_i. On the other hand, $N_0 = P$ and $N_i = \emptyset$ at the end of the procedure, and therefore, for each desired product, o contains some operating unit producing it. Hence, by the definition of m, the validity of condition $(A1)$ follows. For verifying condition $(A2)$, first observe that (M, O) is the maximal structure, thus there is no operating unit producing some raw material in it. On the other hand, $o \subseteq O$ which yields that the same statement is valid for o. Moreover, in a similar way as in the case of condition $(A1)$, one can prove that all other materials from m are produced by some operating unit from o. As far as condition $(A3)$ is concerned, we show by induction on i that for each operating unit in O_i, there is a path leading from it to some desired product. The statement is obvious for $i = 1$ since O_1 contains one operating unit producing some material from $A_0 = P$. Now, suppose that the statement is valid for $i \geq 1$. We show that it is also valid for $i + 1$ provided that O_{i+1} exists. Since $O_{i+1} = O_i \cup \{v\}$, by the induction hypothesis, it is sufficient to prove that there is a path leading from v into some desired product for the operating unit v selected in the $i + 1$-th iteration step. Operating unit v has an output material B in the set N_i. This material is a desired product or it is an input material of some operating unit $u \in O_i$. In the first case there is an edge from v into a desired product and the statement is valid. In the second case there is a path from u into a desired product (by the induction hypothesis), and completing this path with (B, u) and (v, B), we obtain a path leading from v into a desired product. Therefore, condition $(A3)$ is satisfied by (m, o). The validity of condition $(A4)$ is obvious by the definition of the set m. □

4. Worst-case bounds

In the worst-case analysis of a heuristic algorithm we are to find a bound for the quotients calculated as follows. For every instance, the cost of the produced solution is divided by the cost of the optimal solution. To give the formal definition for the PNS problem we need some further notation. Let A be a heuristic PNS algorithm. For every PNS problem $\mathbf{M}$, let the weight of the solution determined by A and the weight of the optimal solution be denoted by $A(\mathbf{M})$ and $OPT(\mathbf{M})$, respectively. Then, C is called a *worst-case bound* of algorithm A if

$$A(\mathbf{M})/OPT(\mathbf{M}) \le C$$

is valid for every $\mathbf{M}$. C is called *tight* if it is the smallest worst-case bound. We can also define the worst-case bounds for some subclasses. C is called a *worst-case bound* of the algorithm for a class $\mathcal{P}$ of PNS problems if

$$A(\mathbf{M})/OPT(\mathbf{M}) \le C$$

is valid for every $\mathbf{M} \in \mathcal{P}$. C is called *tight* if it is the smallest worst-case bound on the class considered. Sometimes worst-case bounds are used in a more general sense. It is possible to use some functions depending on the problem instance instead of a general constant C in the definition. In Theorem 3 we present a result like this.

For more difficult problems, it has been proved that no heuristic algorithms exist with a constant worst-case bound under some complexity assumption (usually under the assumption $P \neq NP$). First, we prove that the PNS problem belongs to this class. To prove this statement we need some further definitions and recall some earlier results. A PNS problem is called a PNS1 problem if every material is a raw material or a desired product; moreover, every operating unit produces desired products from raw materials. In [2], it is proved that the PNS1 problem is equivalent to the set covering problem. These problems are not only equivalent from the complexity theoretical point of view, but they have the same mathematical model. We recall here this result and present the basic idea of the proof, which will also be used in the proof of Theorem 3. First, we define the set covering problem for the readers convenience. In the *set covering* problem, a finite set I and a system of its subsets $P_1, \dots, P_m$ are given where each subset has some positive cost. A set of indices $J^* \subset \{1, \dots, m\}$ is called a *cover* if $\cup\{P_j : j \in J^*\} = I$. By the cost of a cover, we mean the sum of costs of the subsets belonging to the cover. The set covering problem is to find a cover with minimal cost.

Now, we outline the equivalence proof of the PNS1 and the set covering problems. Let (P, R, O) be an arbitrary PNS1 problem with weight function w. Let $u_j = (\alpha_j, \beta_j) \in \wp'(R) \times \wp'(P)$, $j = 1, \dots, n$, denote the operating units. Then, one can prove that this problem is equivalent to the set covering problem in which the basic set is P and the system of its subsets is β_j, $j = 1, \dots, n$, and the costs are $w'(\beta_j) = w(u_j)$, $j = 1, \dots, n$, respectively.

For the other direction, consider an arbitrary set covering problem. Let P be the basic set and β_j, $j = 1, \dots, n$ the system of its subsets, with costs $w'(\beta_j)$, respectively. Let R be an arbitrary set with $R \cap P = \emptyset$.

Let us consider operating units $u_j = (R, \beta_j)$, $j = 1, \ldots, n$, and the weight function $w(u_j) = w'(\beta_j)$, $j = 1, \ldots, n$. Then, it is easy to see that the PNS1 problem (P, R, O), where $O = \{u_1, \ldots, u_n\}$ is equivalent to the set covering problem under consideration.

Finally, we have to recall Chvatal's well-known heuristic algorithm and the theorem on the worst-case bound of this algorithm.

ALGORITHM 4.1 (Chvatal [3])

- *Initialization* Let $J^* = \emptyset$.
- *Step 1.* Terminate if $P_j = \emptyset$ for all j; J^* is the cover produced by the algorithm. Otherwise, choose an index for which the quotient $|P_j|/c_j$ is maximal. Denote it by k and proceed to Step 2.
- *Step 2.* Put k into J^*, replace every P_j with $P_j \setminus P_k$ and proceed to Step 1.

PROPOSITION 1. *([3]) The tight worst-case bound of Algorithm* 4.1 *is* $\sum_{i=1}^{d} 1/i$*, where* d *is the number of elements of the largest set from the set system.*

By the above equivalence, we can prove the following theorems for the PNS problem.

THEOREM 2. *There is no polynomial time heuristic algorithm with constant worst-case bound for the PNS problem, unless P=NP.*

Proof. It is proved in [1] that there is no polynomial time heuristic algorithm with constant worst-case bound for the set covering problem unless P=NP. This result and the equivalence between the set covering and PNS1 problems validates the statement of Theorem 2. □

By Theorem 2, it is very unlikely to find a heuristic algorithm with a constant worst-case bound for the PNS problem. However, considering particular PNS classes, we can prove some tight worst-case bounds for the presented algorithms. First, we determine the worst-case bound for the PNS1 class.

THEOREM 3. *For a problem from the PNS1 class, algorithms* $Asum_c$*,* $Amax_c$ *give the same result. Moreover, they have the tight worst-case bound* $\sum_{i=1}^{m} 1/i$ *for the PNS1 class where* m *is the maximum size of the output sets.*

Proof. For PNS1 problems, each operating unit has only raw materials as input materials, and $c(X) = 0$ for every $x \in R$. This yields that the selection rules of both algorithms are reduced to the following rule. We select always the operating units for which the ratio obtained by dividing the weight of the operating unit by the number of the required output materials is minimal. If we investigate the behaviour of this algorithm by using the correspondence to the equivalent set covering problem, then it can be easily seen that the algorithm selects the operating units corresponding to the sets chosen by Algorithm 4.1. Therefore, Proposition 1 implies the validity of Theorem 3. □

Further classes of PNS problems for which we determine the worst-case bounds are the classes S_k, $k = 1, 2, \ldots$. For every fixed positive integer k, a PNS problem belongs to the class S_k if every operating unit is separator type (it has only one input material), the graph of the problem does not contain a cycle, and the number of the desired products is equal to k.

THEOREM 4. *For a problem in the class S_k, algorithms $Asum_c$, $Amax_c$ give the same result. Moreover, they have the tight worst-case bound k, for every positive integer k.*

Proof. Let k be an arbitrarily fixed positive integer. In what follows, we study the class S_k for this fixed k. Consider an arbitrary reduced S_k model. Let (M, O) be its P-graph. Since each operating unit is a separator,

$$\sum_{X \in mat^{in}(u)} c(X) = \max\{c(X) : X \in mat^{in}(u)\}$$

is valid for each operating unit u, which yields that the two algorithms are the same. In the remaining part of the proof, this algorithm is denoted by A. First, we prove a lemma on the function c.

LEMMA 1. *For every material X, the cost of every path leading from a raw material into X is at least $c(X)$.*

Proof of Lemma 1. We prove that this statement is valid for the elements of the set I_r by induction on r. Since $I_0 = R$, the statement is obviously valid for $r = 0$. Now, let $r \geq 0$ and suppose that the statement is valid for each nonnegative integer which is not greater than r. We prove it for $r+1$. During the $r+1$-th iteration, we extend the set I_r with one new material. Let us denote it by Y. Suppose that the statement is not valid. This yields that for some $Z \in R$ there is a path $[Z, Y]$ in (M, O) with cost smaller than $c(Y)$. Consider the last two vertices in this path, denote them by u and V. By the construction of function c, it follows that $V \in I_r$ and $c(Y) \leq c(V) + w(u)$. On the other hand, since $V \in I_r$, by the induction hypothesis, each path leading from a raw material to V has a cost at least $c(V)$. This yields that the cost of the path considered is at least $c(V) + w(u)$, which is a contradiction. Therefore, we proved the statement for $r + 1$, which ends the proof of Lemma 1. □

Now, we prove that algorithm A has the worst-case bound k. First, consider an optimal solution. Let P_i be a desired product for which the value of c is maximal. Since the optimal solution is a feasible solution, it must contain P_i. Furthermore, by condition $(A2)$ and by the cycle free property of the P-graph of the problem, it follows immediately that in the P-graph of the optimal solution there is a path leading from a raw material into material P_i. By Lemma 1, this yields that the cost of this path is at least $c(P_i)$. Therefore, we proved that

$$OPT(\mathbf{M}) \geq \max\{c(X) : X \in P\}.$$

Let us investigate the weight of the solution produced by algorithm A. We show the following lemma.

LEMMA 2. $A(\mathbf{M}) \leq \sum_{X \in P} c(X)$.

Proof of Lemma 2. Consider the sequence $D_i = \sum_{X \in N_i} c(X) + \sum_{u \in O_i} w(u)$, $i = 0, 1, \ldots$. We prove that this sequence is monotone decreasing as i is increasing. In the i-th step of the algorithm, we obtain N_{i+1} by choosing an operating unit u, deleting its output materials from N_i and enlarging $N_i \backslash mat^{out}(\{u\})$ by $mat^{in}(u)$ if $mat^{in}(u) \notin K_{i+1}$. Therefore, we have that

$$D_i - D_{i+1} \geq \sum_{i=1}^{l} c(X_i) - c(mat^{in}(u)) - w(u),$$

where u is the operating unit selected in the i-th iteration step and $X_1, \ldots, X_l$ are the output materials of u which are contained in N_i. On the other hand, by the selection rule of algorithm A, it follows that

$$c(mat^{in}(v)) + w(v) \geq \frac{c(mat^{in}(u)) + w(u)}{l}$$

is valid for every operating unit v producing X_i. Consequently, by the construction of function c, we obtain that

$$c(X_i) \geq \frac{c(mat^{in}(u)) + w(u)}{l}.$$

Summing up the inequalities concerning the values $c(X_i)$, we get the following inequality

$$\sum_{i=1}^{l} c(X_i) - c(mat^{in}(u)) - w(u) \geq 0 .$$

Therefore, we proved the validity of $D_i - D_{i+1} \geq 0$.

On the other hand, $D_0 = \sum_{X \in P} c(X)$ and $D_s = A(\mathbf{M})$ at the end of the procedure, and hence, by the decreasing property of sequence D_i, the statement of the lemma is valid. □

Therefore, we proved that $OPT(\mathbf{M}) \geq \max\{c(X) : X \in P\}$ and $A(\mathbf{M}) \leq \sum_{X \in P} c(X)$ are valid. These inequalities and $|P| = k$ yield that

$$\frac{A(\mathbf{M})}{OPT(\mathbf{M})} \leq k ,$$

which means that k is a worst-case bound for algorithm A.

Now, we prove that this bound is tight. We show that for every $\varepsilon > 0$, there exists a PNS problem $\mathbf{M}_\varepsilon$ having maximal structure $(M_\varepsilon, O_\varepsilon)$ from class S_k for which

$$\frac{A(\mathbf{M}_\varepsilon)}{OPT(\mathbf{M}_\varepsilon)} > k - \varepsilon .$$

First, let us observe that it is sufficient to prove the statement above for $\varepsilon < 1$. In what follows, let $0 < \varepsilon < 1$ be an arbitrarily fixed real number.

Moreover, let $0 < \delta < \frac{\varepsilon}{k-\varepsilon}$. Define the following PNS problem $\mathbf{M}$ from the class S_k. Let the set of operating units be

$$O = \{u_0, u_1, \ldots, u_k, v_1, \ldots, v_{2k}\} ,$$

where $u_0 = (\{R_0\}, \{X_1, \ldots, X_k\})$, $u_i = (\{X_i\}, \{P_i\})$, $i = 1, \ldots, k$, furthermore, $v_i = (\{R_i\}, \{Y_i\})$, and $v_{k+i} = (\{Y_i\}, \{P_i\})$, $i = 1, \ldots, k$. The P-graph of the problem is shown in *Figure 4.1.*

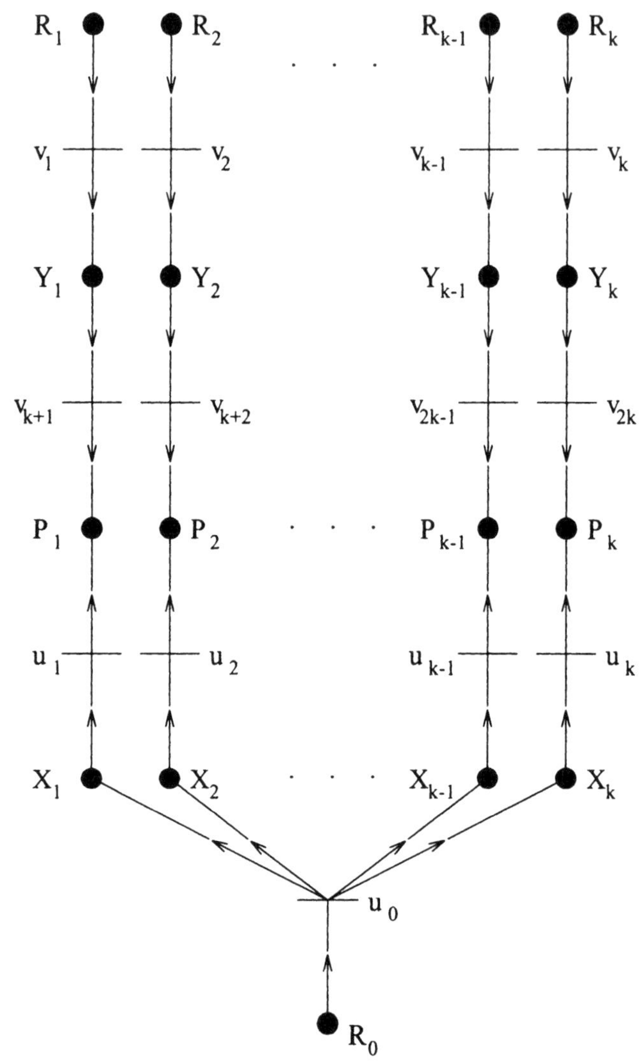

Figure 4.1. The P-graph of the defined problem

In this problem $R_0, \ldots, R_k$ are the raw materials, $P_1, \ldots, P_k$ are the desired products, moreover, $w(u_0) = 1$, $w(u_i) = \delta/k$, $i = 1, \ldots, k$, and $w(v_i) = 1/2$, $i = 1, \ldots, 2k$.

Performing algorithm A on the problem considered, we obtain the feasible solution $(\bar{m}, \bar{o})$, where $\bar{m} = \{R_1, \ldots, R_k, Y_1, \ldots, Y_k, P_1, \ldots, P_k\}$ and $\bar{o} = \{v_1, \ldots, v_{2k}\}$. The weight of this solution is equal to k, and hence, $A(\mathbf{M}) = k$. On the other hand, (m, o) is also feasible solution where $m = \{R_0, X_1, \ldots, X_k, P_1, \ldots, P_k\}$ and $o = \{u_0, u_1, \ldots, u_k\}$. The weight of this solution is $1 + \delta$, thus, $OPT(\mathbf{M}) \leq 1 + \delta$. Therefore, we obtain that

$$\frac{A(\mathbf{M})}{OPT(\mathbf{M})} \geq \frac{k}{1+\delta}$$

is valid. On the other hand,

$$\frac{k}{1+\delta} > \frac{k}{1+\varepsilon/(k-\varepsilon)} = k - \varepsilon\,.$$

This means that for the problem considered we proved that

$$\frac{A(\mathbf{M})}{OPT(\mathbf{M})} \geq k - \varepsilon\,,$$

which shows that the bound k is tight. □

By the theorem above, we obtain immediately the following corollary.

COROLLARY 1. *For class S_1, algorithm A results in an optimal solution.*

5. Empirical analysis

In this section the results of an empirical analysis are presented. In this analysis, 1000 PNS problems were generated randomly (for their generation, cf. [12]). For each of them an optimal solution, and the feasible solutions computed by algorithms Asum_c and Amax_c were determined, and the weights of the solutions were compared to each other.

The weights of the computed solutions are divided by the optimal weight. *Figure 5.1* presents the distribution of the calculated ratios.

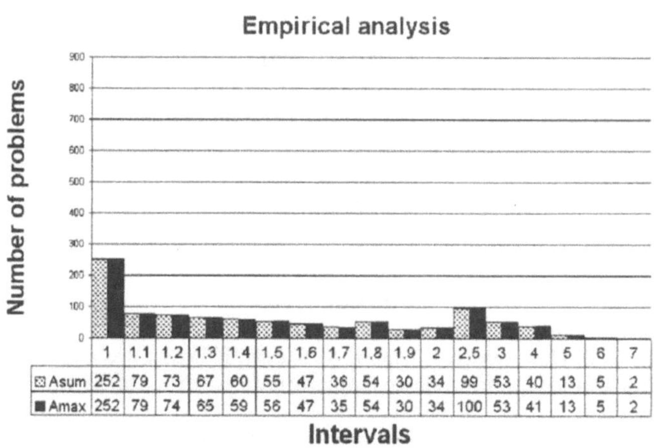

Figure 5.1. Distribution of the calculated ratios in general case

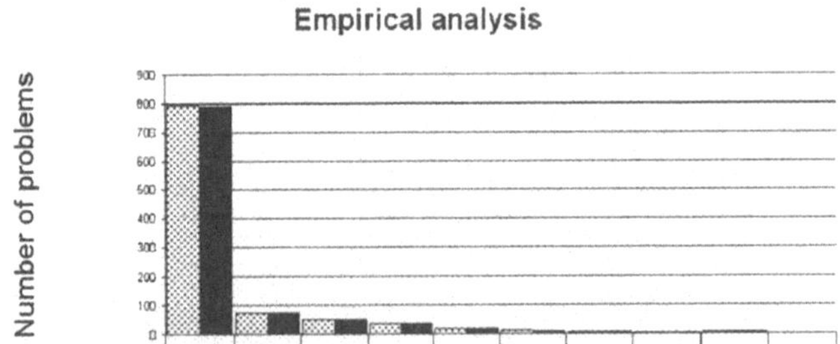

	1	1,1	1,2	1,3	1,4	1,5	1,6	1,7	1,8	1,9
Asum	792	73	52	35	18	12	8	4	5	0
Amax	791	74	52	35	20	10	8	4	5	0

Figure 5.2. Distribution of the calculated ratios in cycle free case

We also studied the cycle free case. Here, 1000 cycle free PNS problems are considered, and for each of them the optimal solution, and the feasible solutions computed by algorithms Asum_c and Amax_c are determined, and the weights of the solutions are compared to each other. *Figure 5.2* presents the distribution of the calculated ratios.

The analysis shows that the heuristic algorithms find an optimal solution for several instances (about 25 percent in the general case, and almost 80 percent in the cycle free case). We can also observe that the algorithm results in a feasible solution which is much more expensive than the optimal one, for a very few instances only. Calculating the average ratios, we obtain that in the general case it is approximately 1.621 for algorithm Amax_c, and 1.619 for Asum_c. In the cycle free case we obtained the ratios 1.042 and 1.043, respectively.

Conclusion

The results of the analysis show that although there is no general, constant worst-case bound for the heuristic algorithms considered, they work well on average on randomly generated instances.

References

1. M. Bellore, S. Goldwasser, C. Lund, A. Russel, Efficient probabilistically checkable proofs and applications to approximation, Proc. of 25-th STOC, 1993, 294-304.
2. Blázsik, Z. and B. Imreh, A note on connection between PNS and set covering problems, *Acta Cybernetica* **12**, 1996, 309-312.

3. Chvatal, V., A Greedy Heuristic for the Set-Covering Problem, *Math. Oper. Res.* **4** (1979) 233-235.
4. Fülöp, J., B. Imreh, F. Friedler, On the reformulation of some classes of PNS problems as set covering problems, *Acta Cybernetica*, **13** (1998),329-337.
5. F. Friedler, L. T. Fan, B. Imreh, Process Network Synthesis: Problem Definition, *Networks* **28** (1998), 119-124.
6. Friedler, F., K. Tarján, Y. W. Huang, and L. T. Fan, Graph-Theoretic Approach to Process Synthesis: Axioms and Theorems, *Chem. Eng. Sci.* **47**(8) (1992), 1973-1988.
7. Friedler, F., K. Tarján, Y. W. Huang, and L. T. Fan, Combinatorial Algorithms for Process Synthesis, *Computer chem. Engng.* **16** (1992), 313-320.
8. Friedler, F., K. Tarján, Y. W. Huang, and L. T. Fan, Graph-Theoretic Approach to Process Synthesis: Polynomial Algorithm for maximal structure generation, *Computer chem. Engng.* **17** (1993), 924-942.
9. Garey, M. R., D. S. Johnson, *Computers and Intractability: A Guide to the Theory of NP-completeness*, W. H. Freeman & Co, San Francisco, 1979.
10. Imreh, B., F. Friedler, L. T. Fan, An Algorithm for Improving the Bounding Procedure in Solving Process Network Synthesis by a Branch-and-Bound Method *Developments in Global Optimization*, editors: I. M. Bonze, T. Csendes, R. Horst, P. M. Pardalos, Kluwer Academic Publisher, Dordrecht, Boston, London, 1996, 301-348.
11. B. Imreh, J. Fülöp, F. Friedler, A Note on the Equivalence of the Set Covering and Process Network Synthesis Problems, *Acta Cybernetica* **14** (2000), 497-501.
12. Imreh, B., G. Magyar, Empirical Analysis of Some Procedures for Solving Process Network Synthesis Problem, *Journal of Computing and Information Technology-CIT* **6** (1998), 373-382.
13. Papadimitriou, C. H., K. Steiglitz, *Combinatorial Optimization: Algorithms and Complexity*, Prentice Hall, 1982

Heuristics for Simplified Process Network Synthesis (PNS) Problems with a Blossom-Type Algorithm for the Edge Covering Problem

Zoltán Blázsik
Research Group on Artificial Intelligence of the Hungarian Academy of Sciences, Aradi vértanúk tere 1, H-6720 Szeged, Hungary
e-mail: blazsik@sol.cc.u-szeged.hu

Kornél Keserű
Department of Informatics, József Attila University, Árpád tér 2, H-6720 Szeged, Hungary
e-mail: keseruk@sol.cc.u-szeged.hu

Zoltán Kovács
Department of Informatics, József Attila University, Árpád tér 2, H-6720 Szeged, Hungary
e-mail: kovacsz@inf.u-szeged.hu

Abstract. Every Process Network Synthesis Problem (PNS) can be transformed into a simplified form. This simplified form consist of simple operating units in which the total number of input and output materials is at most 3. This observation facilitates useful application of the Edge Covering Problem of weighted graphs, which can be used to develop a new heuristic procedure for the PNS problem.

Key words: Process Network Synthesis Problem (PNS), edge covering, heuristics

Introduction and motivation

The importance of process network synthesis (PNS) arises from the fact that networks like this are ubiquitous in the chemical and allied industries. We think that some technologies available in the world can be improved with the application of our model. Here, we can show for the first time an efficient heuristic algorithm for an approximate solution of PNS.[1]

In a manufacturing system materials of different properties are converted into desired products through various physical, chemical, or biological means. Devices in which these transformations can be carried

[1] This work has been supported by the Ministry of Culture and Education of Hungary, Grant FKFP 0008/1999, and by the Hungarian National Foundation for Scientific Research, Grant T030074.

F. Giannessi et al. (eds.), Optimization Theory, 19–31.

out are called *operating units*, and a manufacturing system can be viewed as a network of operating units called *process network*. A process design problem, in general, and flowsheeting, in particular, lead to the construction of a manufacturing system. A design problem is defined from a structural point of view by the raw materials, the desired products, and the available operating units, which all together determine the structure of the problem as a process graph containing the corresponding interconnections among the operating units. Thus the appropriate process networks can be described by some subgraph of the process graph belonging to the design problem under consideration. Naturally, the cost minimization of a process network is indeed essential. Regarding the foundations of PNS and the combinatorial model, we refer to the works [2], [4], [5], [6] and [7].

The solving of a PNS problem is difficult, in general, and it has been pointed out that the PNS problem is NP-complete (see [2], [9] and [12]). Up till now several papers have been published which solve typical PNS problem. They are based on global optimization methods (see [3] and [10]) and on a combinatorial approach (cf. [4],[5],[6], [8], and [11]). The paper is a contribution to the combinatorial approach for finding a solution.

1. Basic definition of the PNS problem

An introduction to the PNS problem and combinatorial model associated with it can be found in several basic works such as cited above. Here, we quote the necessary definitions only. Let $M \neq \emptyset$ be a finite set, the set of *materials*. Furthermore, let $\emptyset \neq O \subseteq \wp'(M) \times \wp'(M)$ with $M \cap O = \emptyset$, where $\wp'(M)$ denotes the set of all nonempty subsets of M. The elements of O are called *operating units*. For an operating unit $(\alpha, \beta) \in O$, α and β are called the *input-set* and *output-set* of the operating unit, respectively. Let us define the functions

$$mat^{in}(o) = \bigcup_{(\alpha,\beta)\in o} \alpha$$

and

$$mat^{out}(o) = \bigcup_{(\alpha,\beta)\in o} \beta.$$

The visual meaning of $mat^{in}(o)$ ($mat^{out}(o)$) is the set of all input (output) materials with respect to o. The pair (M, O) is called a *process graph* or *P-graph* for short. The set of its vertices is $M \cup O$, and the set of its arcs is $A = A_1 \cup A_2$ where $A_1 = \{(X, Y) : Y = (\alpha, \beta) \in O \ \& \ X \in \alpha\}$ and $A_2 = \{(Y, X) : Y = (\alpha, \beta) \in O \ \& \ X \in \beta\}$.

If there exist vertices X_1, X_2, ..., X_n such that (X_1, X_2), (X_2, X_3), ..., (X_{n-1}, X_n) are arcs of the process graph (M, O), then $[X_1, X_n]$ is defined to be a *path* from vertex X_1 to vertex X_n. Given process graphs (m, o) and (M, O), (m, o) is then called a *subgraph* of (M, O) if $m \subseteq M$ and $o \subseteq O$.

Now, we are ready to define the structural model of PNS. For this purpose, let M be an arbitrarily fixed finite set, the set of available materials. By the *structural model* of PNS we mean a system (P, R, O) where $\emptyset \neq P \subseteq M$ is the set of the *desired products*, $R \subseteq M$ is the set of *raw materials*, and $O \subseteq \wp'(M) \times \wp'(M)$ is the set of the available operating units. It is supposed that $P \cap R = \emptyset$ and $M \cap O = \emptyset$. Then the process graph (M, O) shows the interconnections among the operating units of O, where $M = \cup\{\alpha \cup \beta : (\alpha, \beta) \in O\}$.

Moreover, every feasible process network —which produces the given set P of products from the given set R of raw materials using operating units from O— corresponds to a subgraph of (M, O). These subgraphs, which are feasible solutions of the PNS model, can be characterized by the following combinatorial properties:

($\mathcal{A}1$) $P \subseteq m$,

($\mathcal{A}2$) $\forall X \in m,\ X \in R \Leftrightarrow$ there exists no (Y, X) arc in (m, o),

($\mathcal{A}3$) $\forall Y_0 \in o,\ \exists$ path $[Y_o, Y_n]$ with $Y_n \in P$,

($\mathcal{A}4$) $\forall X \in m,\ \exists(\alpha, \beta) \in o$ such that $X \in \alpha \cup \beta$.

We have a polynomial algorithm for the maximal structure generating problem (see [7]), so we can decide if there exists a solution to the problem. Let $S(P, R, O)$ denote the set of feasible solutions. Now we can define the following optimization problem called the PNS *problem*. Let $\mathbf{M} = (P, R, O)$ be given; moreover, let w be a positive real-valued function defined on O, the *weight function*. Then the basic problem is as follows:

$$\min\left\{\sum_{u \in o} w(u) : (m, o) \in S(P, R, O)\right\}.$$

It is easy to see that the set o determines the P-graph (m, o) uniquely for every feasible solution (m, o).

It has already been proved that the combinatorial PNS is equivalent to the Set Covering Problem in general (see [2], [9] and [12]).

2. Substitution of operating units with simpler ones

In some cases we can exploit the fact that the set of input materials and set of output materials of the operating units do not contain too many elements. The goal of the paper is to present heuristic solutions to PNS problems using only simplified operating units. In this section we propose a transformation and with the aid of this, we perform the simplification. Naturally, this modification has no effect on any feasible solutions we may find. We say that an operating unit is *simple* if the number of its input and output materials is at most 3. It is also quite obvious that every operating unit can be substituted by a network consisting of simple operating units. This observation allows us, for an arbitrary PNS problem, to construct an equivalent PNS problem consisting of simple operating units. This problem is called the *simplified* PNS *problem.* To construct the simplified problem some new operating units and materials need to be introduced, but the increase of size is not drastic. Let $\mathbf{M} = (P, R, O)$ be an arbitrary PNS problem and let $u = (\alpha, \beta) \in O$ be an operating unit with sets $\alpha = \{X_1, \ldots, X_n\}$ and $\beta = \{Y_1, \ldots, Y_m\}$. If an operating unit u is not simple, i.e., $(n = 2, m = 2)$, $(n > 2)$ or $(m > 2)$, then we define the new operating units u_1, ..., u_{n+m} as follows

$$u_i = \begin{cases} (\{X_1, X_2\}, \{Z_1\}), & \text{if } i = 1; \\ (\{X_{i+1}, Z_{i-1}\}, \{Z_i\}), & \text{if } i = 2, 3, \ldots, n-1; \\ (\{Z_{n-1}, Z_{n+m}\}, \{Z_n\}), & \text{if } i = n; \\ (\{Z_{n-1+j}\}, \{Z_{n+j}, Y_j\}), & \text{if } i = n + j \;\; j = 1, 2, \ldots, m, \end{cases}$$

where $Z_1, Z_2, \ldots, Z_{n+m}$ denote new materials and $Z_1, Z_2, \ldots, Z_{n+m} \notin M \cup O$. Next, we replace the operating unit u with the operating units $u_1, u_2, \ldots, u_{n+m}$ in the set O and let $O^* = O \backslash \{u\} \cup \{u_1, u_2, \ldots, u_{n+m}\}$ be the new set of operating units and $\{Z_1, Z_2, \ldots, Z_{n+m}\}$ be the new materials which are added to the set M. By making the substitutions of all non-simple units with simple units, we get new disjoint operating unit sets $\{u_1, u_2, \ldots, u_{n+m}\}$ and disjoint new material sets $\{Z_1, Z_2, \ldots, Z_{n+m}\}$ for any two different units. Let $\mathbf{M}^* = (P^*, R^*, O^*)$ be the derived new PNS problem where the set of product $P^* = P$ and the set of raw materials $R^* = R$. For any operating unit $u \in O$ let $\pi(u) = \{u_1, u_2, \ldots, u_{n+m}\}$ if u is non-simple and $\pi(u) = \{u\}$ if u is simple. In addition, for any subset of operating units $o \subset O$, let $\Pi(o) = \cup\{\pi(u) : u \in o\}$.

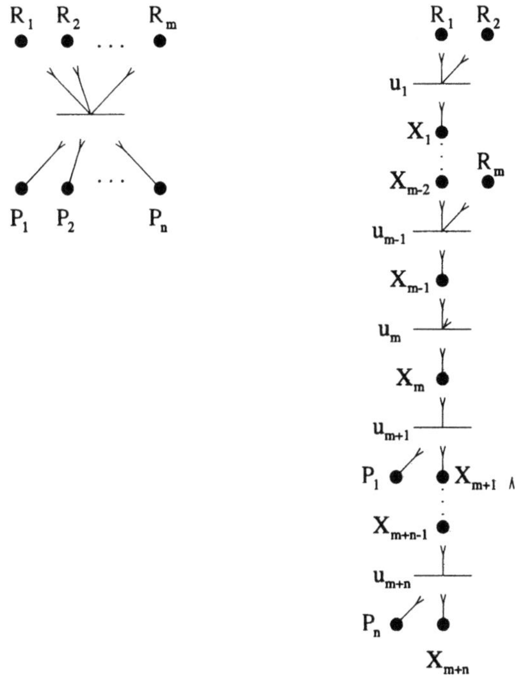

Figure 1. Substitution of an operating unit

After the substitution, we have the following statement.

THEOREM 1. *There exists a bijective mapping of* $S(\mathbf{M})$ *onto* $S(\mathbf{M}^*)$.

Sketch of the Proof. Firstly, we construct a mapping φ of $S(\mathbf{M})$ onto $S(\mathbf{M}^*)$ and prove that this mapping is injective and surjective. For every feasible solution $(m, o) \in S(\mathbf{M})$ let $\varphi((m, o)) = (m^*, o^*)$, so that $o^* = \Pi(o)$ and $m^* = mat^{in}(o^*) \cup mat^{out}(o^*)$.

Let $(m, o) \in S(\mathbf{M})$ be an arbitrary feasible solution. It can then be shown that $\varphi((m, o))$ is a feasible solution of $\mathbf{M}^*$. Obviously, $\varphi((m, o))$ is a P-graph which is a subgraph of (M^*, O^*). Consequently, it is sufficient to check that $\varphi((m, o))$ satisfies conditions $(A1)$ - $(A4)$. Condition $(A1)$ is clearly valid, since $P^* = P \subseteq m \subseteq m^*$. To validate condition $(A2)$ let us observe that $mat^{out}(o) \cap m = mat^{out}(\Pi(o)) \cap m$ and $mat^{out}(\Pi(o)) \backslash m \cap M = \emptyset$. For condition $(A3)$, let u^* be an arbitrary operating unit in the set of operating units o^* and let Y_0 be an operating unit in the set o such that $u^* \in \pi(Y_0)$. Moreover, if $[Y_0, Y_n]$ is an arbitrary path in (m, o) with $Y_n \in P$, then there exists a path $[u^*, Y_n]$ in (m^*, o^*) as every subpath $[Y_{i-1}, Y_i, Y_{i+1}]$ of path $[Y_0, Y_n]$ can be replaced with path $[Y_{i-1}, u_{i_1}, X_{i_1}, u_{i_2}, X_{i_2}, ..., u_{i_r}, Y_{i+1}]$, where $\{u_{i_1}, u_{i_2}, ..., u_{i_r}\} \subset \pi(Y_i)$ and $X_{i_j} \in mat^{out}(u_{i_j}) \backslash m$. Hence, we obtain a path $[u^*, Y_n]$ in (m^*, o^*) where $Y_n \in P$, which implies the validity of $(A3)$. Finally, it is obvious that $(A4)$ holds as $m^* = mat^{in}(o^*) \cup mat^{out}(o^*)$.

Now, we prove that φ is an injective mapping. To this end, let $(m,o) \neq (m',o') \in S(\mathbf{M})$ be two different feasible solutions. Since the sets of operating units are different $(o \neq o')$, so the sets of simple operating units are also different, i.e. $\Pi(o) \neq \Pi(o')$. Hence, the images of feasible solution under φ are also different, more precisely, $(\varphi((m,o)) = (mat^{in}(\Pi(o)) \cup mat^{out}(\Pi(o)), \Pi(o)) \neq (mat^{in}(\Pi(o')) \cup mat^{out}(\Pi(o')), \Pi(o')) = \varphi((m',o'))$.

Now, we show that φ is an surjective mapping. For this purpose, let $(m^*,o^*) \in S(\mathbf{M}^*)$ be an arbitrary feasible solution and let us construct a feasible solution $(m,o) \in S(\mathbf{M})$, such that $\varphi((m,o)) = (m^*,o^*)$. Let $o = \{v : v \in O \text{ and } \exists u \in o^*, \text{ where } u \in \pi(v)\}$ and let $m = M \cap m^*$. We can demonstrate the validity of our four conditions again. Firstly, we show that (m,o) is a feasible solution of $\mathbf{M}$. Condition $(A1)$ is clearly valid, since $P \subseteq M$ and $P \subseteq m^*$. To prove condition $(A2)$ by contradiction, let us suppose that $\exists X \in m$ such that $X \notin mat^{out}(o)$ and $X \notin R$. Then $X \in m^*$ and $X \notin mat^{out}(o^*)$ which is a contradiction. Moreover, $R \cap mat^{out}(o) = \emptyset$ holds since $R \cap mat^{out}(o^*) = \emptyset$. Consequently, condition $(A2)$ holds.

Let $u \in o$ and $Y_0 \in \pi(u)$ be arbitrary operating units. Furthermore, let $[Y_0, Y_n]$ be an arbitrary path in (m^*,o^*) with $Y_n \in P$. From path $[Y_0, Y_n]$ let us eliminate all materials and operating units which are in (m^*,o^*). However, if there are materials Y_i, Y_{i+j} side by side in the modified path, then write $u \in o$ between them Y_i which is $Y_i \in mat^{in}(\{u\})$ and $Y_{i+j} \in mat^{out}(\{u\})$ then, insert u at the beginning of the path. We can always make this change because of property $(A2)$ and the nature of the substitutions. The modified path is a path in (m,o) from u to $Y_n \in P$ and this means that property $(A3)$ is satisfied. To prove condition $(A4)$ let us observe that for every $X \in M \cap m^*$, there exists an operating unit $u \in o^*$ and, therefore, that $X \in mat^{in}(\{u\}) \cup mat^{out}(\{u\})$, thus $X \in mat^{in}(\{v\}) \cup mat^{out}(\{v\})$, where $u \in \pi(v)$. Finally, we can show that $\varphi((m,o)) = (m^*,o^*)$, where it is sufficient to prove equality $\Pi(o) = o^*$. It is clear that $\Pi(o) \supseteq o^*$. Otherwise, if an operating unit $u \in o$, then $\pi(u) \subseteq o^*$ by property $(A2)$ and the nature of substitutions, so $\Pi(o) \subseteq o^*$.

3. Application of Algorithm for Minimum-cost Edge Covering

A heuristic algorithm is developed for the simplified PNS problem. In every iteration step the procedure has a set of materials called the set of *required materials*. (In the first step this set is equal to P.) The procedure constructs a graph G for this set in the following way. Two required materials are connected with an edge if there exists an operating unit, the so-called *parent operating unit*, which has these two materials as outputs.

Recall that we have several operating units with exactly two materials as outputs. The cost of this edge is the weight of the minimal weight parent operating unit. Excluding the isolated vertices and applying this algorithm, an edge cover is counted. Then the succeeding set of the required materials contains the materials which are input materials for the current collection of operating units or materials excluded as singular points in this step, but they are not output materials for any operating unit which has already been selected in some iteration step. The procedure is terminated when the current set of required materials is empty. Then the operating units which have been selected in some iteration step constitute a feasible solution.

What is the idea behind of our heuristic algorithm? There is no way of finding an efficient algorithm for the solution of a PNS problem because it is an NP-hard (see [2]) problem. In some well-selected particular cases we can be find useful results [1].

In general, heuristics seem to be suitable tools. In what follows we search for an optimal set of operating units which produce the required material set at every step. But we prefer an operating unit if its two outputs are the required material set. Our main tool for the exact optimization at every step is the Algorithm for the Minimum Cost Edge Covering (see [13] and [14]). But for this algorithm we can find very few applications in graph theory.

4. Blossom Algorithm for the Minimum Cost Edge Covering problem

We will consider only undirected networks here. Let $G = (N, A, c = (c_{ij}))$ be a given undirected network where c is the vector of edge weights or edge cost coefficients, $|N| = n$, and $|A| = m$. Our network may contain parallel edges, but we assume that there are no self loops or isolated nodes in G. Each c_{ij} could be real, but we use the case $c_{ij} \geq 0$ only. A subset of edges $E \subseteq A$ is said *to cover* the nodes in G if each node in N is incident with at least one edge in E, and such a set being called an *edge covering* or a *dominating edge set* in G. A *matching* is an edge covering iff it is a *perfect matching*.

Since it is assumed that there are no isolated nodes in G, A itself is an edge cover in G. We have an efficient Blossom-type algorithm for finding a minimum cost edge covering (see [13]).

This is fairly similar to the famous Edmonds' matching algorithm. It is also a well-known fact that every solution edge set of the minimum-cost-edge covering problem is a star forest ($c_{ij} > 0$). (see [14])

Defining an edge covering vector $x = (x_{ij})$ in G, this problem can be expressed by the following integer program:

$$\text{Minimize } z(x) = \sum (c_{ij} x_{ij} : (i,j) \in A)$$

$$\text{subject to } x(i) \geq 1 \text{ for all } i \in N$$

$$x(i) := \sum (x_{ij} : \text{ over } j \text{ such that } (i,j) \in A)$$

$$0 \leq x_{ij} \leq 1, \text{ for all } (i,j) \in A$$

$$x_{ij} \text{ integer for all } (i,j) \in A \qquad (*)$$

Let $Y \subseteq N$ with $|Y| \geq 3$ and odd. Define

$$Y^+(x) = \sum (x_{ij} : (i,j) \in A \text{ with at least one of } i \text{ or } j \text{ or both in } Y).$$

If we change constrain (*) in the integer program above to

$$Y^+(x) \geq (|Y| + 1)/2 \text{ for every } Y \qquad (**)$$

we have an LP. $(**)$ is called the *covering blossom inequality*, which is similar to the *matching blossom inequality* of Edmonds'. The optimal feasible solution of this LP is an integer vector and a minimum cost edge covering vector in G. There is an efficient algorithm (MCEC) for the solution of the dual of our LP problem.

5. Heuristics for simplified PNS problems

Let (P, R, O) be the structural model of a simplified-type PNS problem. R: raw materials, P: desired products, $P \cap R = \emptyset$, $M \supseteq P \cup R$ the set of materials. The type of every operating unit $u \in O$ is $(1,2)$ or $(2,1)$.

$$\text{For every } m_q \in M \setminus R \text{ there exists } (\alpha, \beta) \in O \text{ such that } m_q \in \beta \qquad (***)$$

ALGORITHM 1.

Notation: RM_i : Required Materials after Step i.

PM_i : Produced Materials after Step i.

$G_i(N_i, A_i)$ $(c(e) : e \in A_i)$.

We connect two material points $m_q, m_r \in RM_{i-1}$, with an edge e iff there exists $(\alpha, \beta) \in O$ and $m_q, m_r \in \beta$ $(\beta = \{m_q, m_r\})$.

$(\alpha, \beta)_{q,r}$:= one of the minimal weight operating units which produces m_q and m_r, such an operating unit exists.

$c(e) = w(\alpha, \beta)_{q,r}$ if $e = (m_q, m_r)$

$A_i = \{e :: e = (m_q, m_r)\ m_q, m_r \in RM_{i-1}\ m_q \neq m_r : and : \exists(\alpha; m_q, m_r) \in O\}$

If $A_i = \emptyset$ then G_i doesn't exist.

$N_i = \{$nodes incident with edges of $A_i\}$ $(\subseteq RM_{i-1})$

Step 1: *Initialization* $O_0 := \emptyset$, $RM_0 := P$, $PM_0 := R$, $i := 0$

Step 2: *Construction of* G_i

If there exists a G_i, (apply Algorithm MCEC to G_i), $\longrightarrow$ Step 3.

If G_i doesn't exist $\longrightarrow$ Step 4

Step 3: Apply Algorithm MCEC to G_i, and let E_i denote the solution set of edges. Let

$O_i := O_{i-1} \cup \{(\alpha, \beta)_{q,r} : (m_q, m_r) \in E_i\}$,

$RM_i := (P \cup mat^{in} O_i) \setminus mat^{out} O_i$,

$PM_i := R \cup mat^{out} O_i$.

If $RM_i = \emptyset$, then $O^1 = O_i$, the solution set of operating units of Algorithm 1.

If $RM_i \neq \emptyset$, then $i := i + 1 \longrightarrow$ Step 2.

Step 4: Let u^i one of minimal weight operating units which produce one of materials of RM_{i-1}. Moreover, let

$O_i := O_{i-1} \cup u^i$,

$RM_i := (P \cup mat^{in} O_i) \setminus mat^{out} O_i$,

$PM_i := R \cup mat^{out} O_i$,

If $RM_i = \emptyset$, then $O^1 = O_i$, the solution set of operating units of Algorithm 1.

If $RM_i \neq \emptyset$, then $i := i + 1 \longrightarrow$ Step 2.

ALGORITHM 2.

From among the minimum cost operating units we choose an operating unit such that the input material was already produced or is a raw material. When there is no such operating unit, we select an input material which can be produced in the cheapest way.

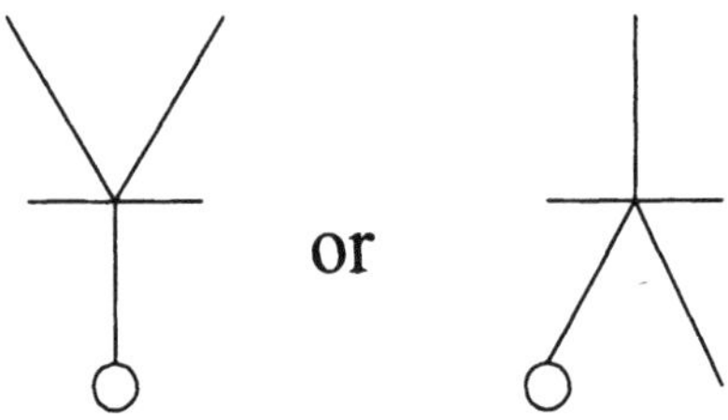

Figure 2. Algorithm 2

ALGORITHM 3.

As can be seen, only the edge costs of G_i are different from Algorithm 1. We constructed G_i as we did in Step 2 of Algorithm 1 but we used another edge cost function. Utilising the edge cost function of Algorithm 1, $c((p_k; p_l))$, we define the new edge cost function to get:

$$c'((p_k; p_l)) = \frac{\sum_j c(p_k; p_j) + \sum_j c(p_j; p_l)}{d(p_k) + d(p_l)}$$

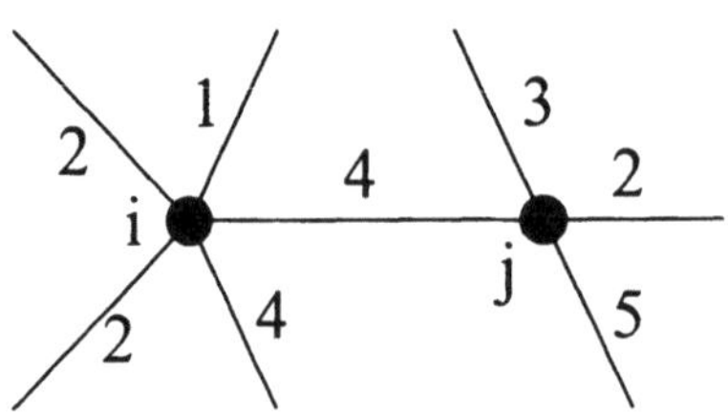

Figure 3. An example.

$c(i;j) = 4, \ c'(i;j) = \frac{(1+2+2+4+4)+(3+2+5+4)}{5+4} = 3$

6. Empirical analysis of the algorithms

This section contains some results of our computational experiments. We compared the solutions of different PNS problem classes consisting of 100-100 randomly generated problems. These problems are generated in accordance with the following rules:

- The number of materials, raw materials, desired products and operating units are all given.
- The operating unit weights are generated randomly within the range of 1 and a given integer number.
- The ratio of the (1,2) and (2,1) operating units is generated randomly.
- Raw materials are not produced in the system.
- In some cases the generated problem does not satisfy the condition (***) in Section 5. It means that some problems may be unsolvable.

In the comparison study a 450 MHz Celeron PC was used. The results were expressed in 0.01 units. The tables show the number of solvable problems with the algorithm, the average value of the operating units in the solutions, the average cost of the solutions, the average operating unit costs and the average difference between the costs of the optimal solution.

We did not compare the running times of the algorithms because solving 100 problems heuristically required just several seconds while finding the optimal solution took several minutes. In some cases we were able to find an optimal solution by using the heuristics.

In the analysis $|M|$ denotes the number of materials, $|R|$ the number of raw materials, $|P|$ the number of desired products, and $|O|$ the number of operating units. Let $w(O)$ also denote the operating unit weights. Tables show the solutions of 100 problems of the following type:

	B&B	Alg 1	Alg 2	Alg 3
solvable	93	87	87	87
$\lvert O\rvert$	4.27	6.31	6.31	6.31
$w(O)$	33.85	53.41	53.77	53.77
$\frac{w(O)}{\lvert O\rvert}$	7.93	8.46	8.52	8.52
$\frac{w(O)}{w(O^{opt})}$	1.00	1.58	1.59	1.59

Table 1. $|M| = 18, |R| = 2, |P| = 3, |O| = 40, 1 \leq w(O) \leq 25$.

	B&B	Alg 1	Alg 2	Alg 3
solvable	93	82	82	82
$\|O\|$	5.04	5.76	5.72	5.72
$w(O)$	74.52	98.63	97.74	97.74
$\frac{w(O)}{\|O\|}$	14.79	17.12	17.09	17.09
$\frac{w(O)}{w(O^{opt})}$	1.00	1.32	1.31	1.31

Table 2. $|M| = 17, |R| = 2, |P| = 4, |O| = 38, 1 \leq w(O) \leq 50.$

	B&B	Alg 1	Alg 2	Alg 3
solvable	87	75	75	75
$\|O\|$	6.30	7.80	7.81	7.81
$w(O)$	99.25	132.35	131.27	131.27
$\frac{w(O)}{\|O\|}$	15.75	16.97	16.81	16.81
$\frac{w(O)}{w(O^{opt})}$	1.00	1.33	1.32	1.32

Table 3. $|M| = 25, |R| = 4, |P| = 5, |O| = 45, 1 \leq w(O) \leq 50.$

Conclusion

The results of the analysis show that the heuristics work well in most cases, by using comparatively short CPU time and memory. Using the heuristics large PNS problems consisting of thousands of materials and operating units can be solved quite quickly.

References

1. Blázsik, Z., Cs. Holló, B. Imreh, On Decision-Mappings Related to Process Network Synthesis Problem, *Acta Cybernetica* **13** (1998), 319-328.
2. Blázsik, Z., B. Imreh, A Note on Connection between PNS and Set Covering Problems, *Acta Cybernetica* **12** (1996), 309-312.
3. Floudas, C. A., I. E. Grossmann, Algorithmic Approaches to Process Synthesis: Logic and Global Optimization, *AiChE Symposium Series* No. 304, **91** (Eds: L. T. Biegler and M. F. Doherly), (1995), 198-221.
4. Friedler, F., L. T. Fan, B. Imreh, Process Network Synthesis: Problem Definition, *Networks* **28** (1998), 119-124.
5. Friedler, F., K. Tarján, Y. W. Huang, L. T. Fan, Combinatorial Structure of Process Network Synthesis, Sixth SIAM Conference on Discrete Mathematics, Vancouver, Canada, 1992.

6. Friedler, F., K. Tarján, Y. W. Huang, L. T. Fan, Graph-Theoretic Approach to Process Synthesis: Axioms and Theorems, *Chem. Eng. Sci.* **47**(8) (1992), 1973-1988.
7. Friedler, F., K. Tarjan, Y. W. Huang, L. T. Fan, Graph-Theoretic Approach to Process Synthesis: Polynomial Algorithm for Maximal Structure Generation, *Computers chem. Engng.* **17**(9) (1993), 929-942.
8. Friedler, F., J. B. Varga, E. Fehér, L. T. Fan, Combinatorially Accelerated Branch-and-Bound Method for Solving the MIP Model of Process Network Synthesis, presented at the International Conference on State of the Art in Global Optimization: Computational Methods and Applications, Princeton University, Princeton, NJ, U.S.A., April 28-30, 1995; also to be published in *Nonconvex Optimization and its Applications*, Kluwer Academic Publishers, Norwell, MA, U.S.A. (in press).
9. Fülöp, J., F. Friedler, B. Imreh, On the reformulation of some classes of PNS-problems as set covering problems, *Acta Cybernetica* **13** (1998), 329-337.
10. Grossmann, I. E., V. T. Voudouris, O. Ghattas, Mixed-Integer Linear Programming Reformulations for Some Nonlinear Discrete Design Optimization Problems, *Recent Advances in Global Optimization* (ed.: C. A. Floudas and P. M. Pardalos) Princeton University Press, New Jersey, 1992.
11. Imreh, B., F. Friedler, L. T. Fan, An Algorithm for Improving the Bounding Procedure in Solving Process Network Synthesis by a Branch-and-Bound Method, *Developments in Global Optimization* (eds.: I. M. Bonze, T. Csendes, R. Horst, P. M. Pardalos), Kluwer Academic Publishers, 1997, 315-348.
12. Imreh, B., J. Fülöp, F. Friedler, A note on the Equivalence of the Set Covering and Process Network Synthesis Problems, *Acta Cybernetica*, to appear
13. Murty, Katta G., Clovis Perin, A 1-Matching Blossom-Type Algorithm for Edge Covering Problems, *Networks* **12** (1982), 379-391.
14. Murty, Katta G., *Network Programming*, Prentice Hall, 1992.

Lower and Upper Bounds on the Probability of the Union of Some Events with Applications

József Bukszár
Institute of Mathematics, University of Miskolc
H-3515 Miskolc-Egyetemváros

Abstract. Lower and upper bounds on the probability of the union of some events are presented. The bounds are based on special hypergraphs called (h, m)-hypermultitrees. A new probability bounding technique based on the bounds is described and it is applied to estimate the value of multivariate normal distribution functions.

1. Introduction

Let $A_1, ..., A_n$ be arbitrary probability events. We present lower and upper bounds on the probability $P(A_1 \cup ... \cup A_n)$ based on some of the terms $P(A_{k_1} \cap ... \cap A_{k_i})$ $(1 \leq k_1 < ... < k_i \leq n,\ i = 1, ..., d)$, where $d \geq 1$ is an integer. (In applications d is much less than n.) For a pair of integers h, m with $h + m + 1 = d$, $h \geq 0$ and $m \geq 1$, we introduce a class of hypergraphs on n vertices, called (h, m)-hypermultitrees, and assign a lower (upper) bound to each hypergraph of the class for even (odd) h.

The problem of finding such upper bounds was first treated by G. Boole (see [1], [2]). C.E. Bonferroni (see [3]) generalized Boole's results by presenting lower and upper bounds based on binomial moments $S_1, S_2, \ldots, S_d$, where

$$S_k = \sum_{1 \leq i_1 < ... < i_k \leq n} P(A_{i_1} \cap \cdots \cap A_{i_k}), \quad k = 1, \ldots, n.$$

Then several Bonferroni-type bounds were published (see [17], [10]), and finally, A. Prékopa (see [14]) proved that some previously known bounds based on S_1, S_2 and S_1, S_2, S_3 are sharp in the sense that no better bound can be given on the basis of the knowledge of the first some binomial moments. Furthermore, A. Prékopa (see [15]) presented a method to obtain the sharp lower and upper bounds based on $S_1, S_2, \ldots, S_d$. D. Hunter (see [9]) and K.J. Worsley (see [22]) gave an upper bound based on S_1 and on the individual probabilities $P(A_i \cap A_j)$ $(1 \leq i < j \leq n)$ rather than on S_2. I. Tomescu (see [21]) generalized the Hunter-Worsley bound by using $S_1, S_2, \ldots, S_{d-1}$ and the individual probabilities $P(A_{k_1} \cap ... \cap A_{k_d})$ $(1 \leq k_1 < ... < k_d \leq n)$. Our bounds are generalization of Tomescu's bounds, they can be obtained by setting $m = 1$.

F. Giannessi et al. (eds.), *Optimization Theory*, 33–43.

Our bounds have the advantage over the previously known ones that only $\mathcal{O}(n)$ ($\mathcal{O}(n^2)$) $P(A_{i_1} \cap ... \cap A_{i_k})$, $1 \leq k \leq d$ probabilities are required to compute a new upper (lower) bound. That is obtained by using $h = 0$ ($h = 1$). The computation of Tomescu bounds requires the knowledge of $\mathcal{O}(n^{d-1})$ similar probabilities.

This property of the new bounds is advantageous in applications where the probabilities $P(A_{i_1} \cap ... \cap A_{i_k})$ required to compute the bounds have to be evaluated. It is the case, for instance, if we want to present lower and upper bounds on network reliability or on the values of multivariate distribution functions. Here, the computation time of the bounds is practically determined by the time required for evaluating the terms $P(A_{i_1} \cap ... \cap A_{i_k})$.

In the third section recursive algorithms on m are presented to find $(0, m)$-hypermultitrees and $(1, m)$-hypermultitrees the bounds —that are based on— are improved whenever m is increased.

In the forth section we demonstrate probability bounding by hypermultitrees through computing bounds on the values of multivariate normal distribution functions. These bounds serve as a basis of simulation schemes on the values of multivariate normal distribution functions (see [19]).

2. (h, m)-hypermultitrees

To introduce the concept of (h, m)-hypermultitree we need two definitions.

DEFINITION 1. *Let m be a positive integer. An* m-multicherry *is a hypergraph of the form $(V, \mathcal{E}_2, ..., \mathcal{E}_{m+1})$, where $V = \{v_1, ..., v_{m+1}\}$ is the set of vertices and the family of hyperedges $\mathcal{E}_i$ is the set of all subsets of $\{v_1, ..., v_{m+1}\}$ containing i vertices with v_{m+1} included, i.e. $\mathcal{E}_i = \{H \mid v_{m+1} \in H \subset \{v_1, ..., v_{m+1}\}, |H| = i\}$. The vertex v_{m+1} is called the* dominating vertex *of the m-multicherry.*

The m-multicherry with dominating vertex v_{m+1} and with non-dominating vertices $v_1, ..., v_m$ is denoted by $(\{v_1, ..., v_m\}, v_{m+1})$.

The 3-multicherry $(\{1, 2, 3\}, 4)$ is illustrated in *Figure 1* by its edges and vertices. Its hyperedges containing more than two vertices $\{1, 2, 4\}$, $\{1, 3, 4\}$, $\{2, 3, 4\}$ and $\{1, 2, 3, 4\}$ are not marked in the figure because of the perspicuity.

Note that a 1-multicherry is a single edge together with its incident vertices.

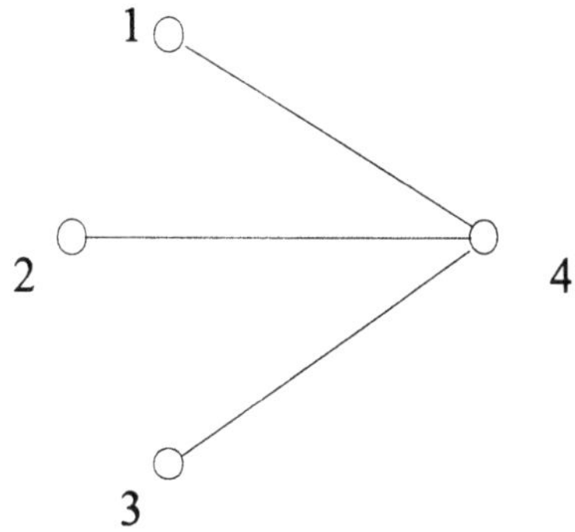

Figure 1

DEFINITION 2. *Let m be a positive integer. An m-multitree is a hypergraph of the form $(V, \mathcal{E}_2, ..., \mathcal{E}_{m+1})$, where V is the set of vertices and $\mathcal{E}_i$'s are sets of hyperedges containing i vertices. An m-multitree is recursively defined by the following two rules.*

i) The smallest m-multitree $\Delta = (V, \mathcal{E}_2, ..., \mathcal{E}_{m+1})$ has m vertices and $\mathcal{E}_i$ is the family of all subsets of V containing i vertices (here $\mathcal{E}_{m+1} = \emptyset$.)

ii) From an m-multitree $\Delta = (V, \mathcal{E}_2, ..., \mathcal{E}_{m+1})$ we can obtain a new m-multitree $\Delta' = (V', \mathcal{E}'_2, ..., \mathcal{E}'_{m+1})$ by adjoining an m-multicherry $(\{v_1, ..., v_m\}, v_{m+1})$, where $v_1, ..., v_m \in V$ and v_{m+1} is a new vertex (i.e. $v_{m+1} \notin V$). More precisely

$$V' = V \cup \{v_{m+1}\}, \quad \mathcal{E}'_i = \mathcal{E}_i \cup \{H \mid v_{m+1} \in H \subset \{v_1, ..., v_{m+1}\}, |H| = i\}.$$

A 3-multitree $\Delta = (V, \mathcal{E}_2, \mathcal{E}_3, \mathcal{E}_4)$ is illustrated in *Figure 2*. Given 1, 2, 3 and the hyperedges $\{1,2\}$, $\{1,3\}$,$\{2,3\}$, $\{1,2,3\}$, we subsequently adjoin the 3-multicherries $(\{1,2,3\},4)$, $(\{1,3,4\},5)$ and $(\{2,3,5\},6)$ as shown in the figure. The edges of a 3-multicherry are drawn in the same line-style. The vertices and hyperedges of Δ are $V = \{1,2,3,4,5,6\}$, $\mathcal{E}_2 = \{\{1,2\}, \{1,3\}, \{1,4\}, \{1,5\}, \{2,3\}, \{2,4\}, \{2,6\}, \{3,4\}, \{3,5\}, \{3,6\}, \{4,5\}, \{5,6\}\}$, $\mathcal{E}_3 = \{\{1,2,3\}, \{1,2,4\}, \{1,3,4\}, \{2,3,4\}, \{1,3,5\}, \{1,4,5\}, \{3,4,5\}, \{2,3,6\}, \{2,5,6\}, \{3,5,6\}\}$ and $\mathcal{E}_4 = \{\{1,2,3,4\}, \{1,3,4,5\}, \{2,3,5,6\}\}$.

Note that a 1-multitree is a usual tree. It can be proved that an m-multitree is determined by its vertices and edges.

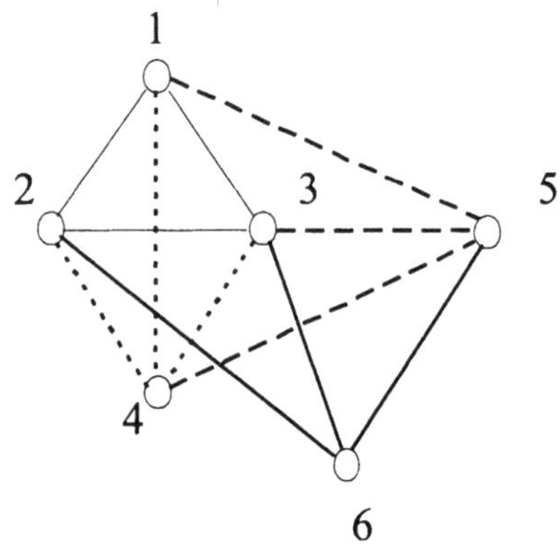

Figure 2

Remark 1. If $\Delta = (V, \mathcal{E}_2, ..., \mathcal{E}_{m+1})$ is an m-multitree with $|V| = n$, then $|\mathcal{E}_i| = \binom{m}{i} + \binom{m}{i-1}(n-m)$ for all $i = 2, .., m+1$.

DEFINITION 3. *Let $h \geq 0$ and $m \geq 1$ be arbitrary integers. An (h,m)-hypermultitree is a hypergraph of the form $(V, {}_h\mathcal{E}_2, ..., {}_h\mathcal{E}_{m+1})$, where V is the set of vertices and ${}_h\mathcal{E}_i$'s are sets of hyperedges containing $h+i$ vertices. An (h,m)-hypermultitree is defined recursively by the following rules.*

(i) The $(0,m)$-hypermultitrees are the same as the m-multitrees.

(ii) The smallest (h,m)-hypermultitree $\Delta = (V, {}_h\mathcal{E}_2, ..., {}_h\mathcal{E}_{m+1})$ has $h+m$ vertices and ${}_h\mathcal{E}_i$ consists of all subsets of V containing $h+i$ vertices (here ${}_h\mathcal{E}_{m+1} = \emptyset$).

(iii) From an (h,m)-hypermultitree $\Delta = (V, {}_h\mathcal{E}_2, ..., {}_h\mathcal{E}_{m+1})$ we can obtain a new (h,m)-hypermultitree in the following manner. Let $\Gamma = (V, {}_{h-1}\mathcal{E}^*_2, ..., {}_{h-1}\mathcal{E}^*_{m+1})$ be an arbitrary $(h-1,m)$-hypermultitree with the same set of vertices as in Δ. By adjoining a new vertex v to Δ and the hyperedges of Γ extended by v, we obtain the new (h,m)-hypermultitree $\Delta' = (V', {}_h\mathcal{E}'_2, ..., {}_h\mathcal{E}'_{m+1})$, i.e.

$$V' = V \cup \{v\} \qquad {}_h\mathcal{E}'_i = {}_h\mathcal{E}_i \cup \{E \cup \{v\} \mid E \in {}_{h-1}\mathcal{E}^*_i\}\,.$$

The (h,m)-hypermultitrees are generalizations of Tomescu's hypertrees, which are the $(h,1)$-hypermultitrees in our definition.

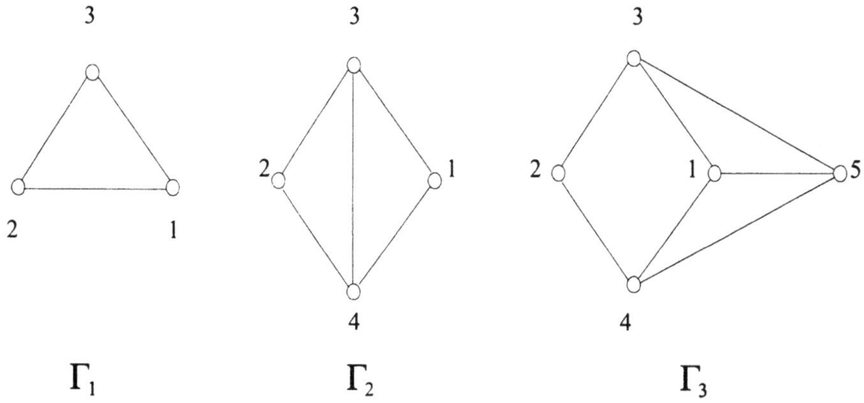

Figure 3

EXAMPLES. $\Delta = (V,\ {}_1\mathcal{E}_2,\ {}_1\mathcal{E}_3)$ with ${}_1\mathcal{E}_2 = \{\{1,2,3\}\}$ and ${}_1\mathcal{E}_3 = \emptyset$ is an $(1,2)$-hypermultitree by (ii). From Δ we can obtain the $(1,2)$-hypermultitree $\Delta' = (V',\ {}_1\mathcal{E}'_2,\ {}_2\mathcal{E}'_3)$ on the basis of $(0,2)$-hypermultitree $\Gamma_1 = (\{1,2,3\},\ \{\{1,2\},\{1,3\},\ \{2,3\}\},\ \{\{1,2,3\}\})$ shown in *Figure 3* by adjoining vertex 4 by (iii), where $V = \{1,2,3,4\}$, ${}_1\mathcal{E}'_2 = \{\{1,2,3\},\ \{1,2,4\},\ \{1,3,4\},\ \{2,3,4\}\}$, ${}_1\mathcal{E}'_3 = \{\{1,2,3,4\}\}$. From Δ' we obtain $\Delta'' = (V'',\ {}_1\mathcal{E}''_2,\ {}_1\mathcal{E}''_3)$ on the basis of Γ_2, where $V'' = \{1,2,3,4,5\}$, ${}_1\mathcal{E}''_2 = {}_1\mathcal{E}'_2 \cup \{\{1,3,5\},\ \{1,4,5\},\ \{2,3,5\},\ \{2,4,5\},\ \{3,4,5\}\}$, ${}_1\mathcal{E}''_3 = {}_1\mathcal{E}'_3 \cup \{\{1,3,4,5\},\ \{2,3,4,5\}\}$. Finally, we obtain $\Delta''' = (V''',\ {}_1\mathcal{E}'''_2,\ {}_1\mathcal{E}'''_3)$ from Δ'' on the basis of Γ_3, where $V''' = \{1,2,3,4,5,6\}$, ${}_1\mathcal{E}'''_2 = {}_1\mathcal{E}''_2 \cup \{\{1,3,6\},\ \{1,4,6\},\ \{1,5,6\},\ \{2,3,6\},\ \{2,4,6\},\ \{3,5,6\},\ \{4,5,6\}\}$, ${}_1\mathcal{E}'''_3 = {}_1\mathcal{E}''_3 \cup \{\{1,3,5,6\},\ \{1,4,5,6\},\ \{2,3,4,6\}\}$.

LEMMA 1. *Let $\Delta = (V,\ {}_h\mathcal{E}_2, \ldots,\ {}_h\mathcal{E}_{m+1})$ be an arbitrary (h,m)-hypermultitree with n vertices. Then*

$$|{}_h\mathcal{E}_i| = \sum_{k=0}^{h+1} \binom{m}{i-1+k} \binom{n-m}{h+1-k}.$$

For the proof see [5].

DEFINITION 4. *Let $A_1, \ldots, A_n$ be arbitrary events. Then the* weight *of (h,m)-hypermultitree $\Delta = (V,\ {}_h\mathcal{E}_2, \ldots,\ {}_h\mathcal{E}_{m+1})$ is*

$$w(\Delta) = \sum_{\{i_1,\ldots,i_{h+2}\}\in {}_h\mathcal{E}_2} P(A_{i_1}\cap\ldots\cap A_{i_{h+2}}) - \sum_{\{i_1,\ldots,i_{h+3}\}\in {}_h\mathcal{E}_3} P(A_{i_1}\cap\ldots\cap A_{i_{h+3}}) + \ldots$$

$$\ldots + (-1)^{m+1} \sum_{\{i_1,\ldots,i_{h+m+1}\}\in {}_h\mathcal{E}_{m+1}} P(A_{i_1}\cap\ldots\cap A_{i_{h+m+1}})$$

THEOREM 1. *Let $A_1, \dots, A_n$ be arbitrary events and let $\Delta = (V, {}_h\mathcal{E}_2, \dots, {}_h\mathcal{E}_{m+1})$ be an arbitrary (h, m)-hypermultitree. The following inequalities hold. (i) If h is even, then*

$$P\left(\bigcup_{i=1}^{n} A_i\right) \leq \sum_{k=1}^{h+1} (-1)^{k-1} S_k - w(\Delta).$$

(ii) If h is odd, then

$$P\left(\bigcup_{i=1}^{n} A_i\right) \geq \sum_{k=1}^{h+1} (-1)^{k-1} S_k + w(\Delta),$$

where $S_k = \sum_{1 \leq i_1 < \dots < i_k \leq n} P(A_{i_1} \cap \dots \cap A_{i_k})$.

For the proof see [5]. The theorem has been proved for the special case of $m = 1$ by I. Tomescu and in the more special case of $m = 1$ and $h = 0$ by D. Hunter and K. J. Worsley. In the case of $h = 0$, $m = 2$ some other results are detailed in [6]. The above lemma implies that replacing h by $h+1$ in the bounds requires the knowledge of much more additional probabilities than replacing m by $m+1$. Consequently, from practical point of view, it is worth increasing m rather than h to improve on the bounds. The algorithms described in the next section are partially based on this principle.

3. Algorithms

In some applications, e.g. in computing bounds on the values of multivariate distribution functions or on the reliability of a network etc., the probabilities $P(A_{k_1} \cap \dots \cap A_{k_i})$ required to compute the bounds have to be evaluated. The evaluation is usually quite expensive, and the time it requires is practically the time needed to compute the bounds. Consequently, it is a crucial question whether "good" bounds can be presented by evaluating only a "few" terms $P(A_{k_1} \cap \dots \cap A_{k_i})$. In the view of the above theorem, the greater the weight of the used hypermultitree is, or simply spoken, the heavier the used hypermultitree is the better the bound will be. To compute an upper bound on $P(A_1 \cup \dots \cup A_n)$ we present an algorithm that finds a heavy m-multitree (i.e. $(0, m)$-hypermultitree). Obtaining this bound requires the evaluation of $P(A_{k_1})$ $(1 \leq k_1 \leq n)$, $P(A_{k_1} \cap A_{k_2})$ $(1 \leq k_1 < k_2 \leq n)$ and $\mathcal{O}(n)$ terms of $P(A_{k_1} \cap \dots \cap A_{k_i})$ $(1 \leq k_1 < \dots < k_i \leq n,\ i = 3, \dots, m+1)$ out of all the $\mathcal{O}(n^{m+1})$. To compute a lower bound on $P(A_1 \cup \dots \cup A_n)$ we present an algorithm that finds a heavy $(1, m)$-hypermultitree. Obtaining the bound requires the evaluation of $P(A_{k_1} \cap \dots \cap A_{k_i})$ $(1 \leq$

$k_1 < ... < k_i \leq n$, $i = 1,2,3$) and $\mathcal{O}(n^2)$ terms of $P(A_{k_1} \cap ... \cap A_{k_i})$ ($1 \leq k_1 < ... < k_i \leq n$, $i = 4, ..., m+2$) out of all the $\mathcal{O}(n^{m+2})$. The algorithms are such recursions on m, the m-multitree ($(1,m)$-hypermultitree) bound they provide are improved whenever m is increased.

Our algorithm that finds a heavy m-multitree is as follows. The probabilities $P(A_i \cap A_j)$ ($1 \leq i < j \leq n$) are evaluated first. Then the weight function on the edges of the complete graph Γ with vertex set $\{1, ..., n\}$ is introduced: the edge incident to the vertices i and j has the weight $P(A_i \cap A_j)$. The maximum weight (1-multi)tree on Γ is found by Prim's algorithm and the Hunter-Worsley bound is obtained with the maximum weight tree (see the above theorem). Then, the maximum weight tree is extended to a 2-multitree by the following recursion. In the first step the maximum weight edge is taken together with the vertices incident to it. In all other steps a 2-multicherry is adjoined where an edge is in the tree and the other one is the heaviest possible. The r-multitree is extended to an $(r+1)$-multitree ($r = 2, ..., m-1$) as follows. In the first step the $(r+1)$-multitree is taken the $r+1$ vertices of which are those selected in the first two steps of the recursion producing the r-multitree. (The $(r+1)$-multitree with $r+1$ vertices has a unique structure.) In all other steps an $(r+1)$-multicherry is adjoined which is an extension of an r-multicherry of the r-multitree with the heaviest possible edge.

To compute the weight of the m-multitree, additionally, we need to evaluate as many probabilities $P(A_{k_1} \cap ... \cap A_{k_i})$ ($1 \leq k_1 < ... < k_i \leq n$, $i = 3, ..., m+1$) as the m-multitree has the hyperedges containing i vertices, namely $\binom{m}{i} + \binom{m}{i-1}(n-m)$, which is $\mathcal{O}(n)$ (see the above remark).

Our algorithm that finds a heavy $(1,m)$-hypermultitree is as follows. The probabilities $P(A_i \cap A_j \cap A_k)$ ($1 \leq i < j < k \leq n$) are evaluated first and a $(1,1)$-hypermultitree is constructed by the following greedy algorithm. In the first step a hyperedge $\{h, l, m\}$ is taken for which $P(A_h \cap A_l \cap A_m) = \max\{P(A_i \cap A_j \cap A_k) \mid (1 \leq i < j < k \leq n)\}$ and the $(1,1)$-hypermultitree $\{\{h,l,m\}, \{\{h,l,m\}\}, \emptyset\}$ is constructed. In the other steps a vertex and some hyperedges are adjoined by (iii) in the definition as follows. Assume that a $(1,1)$-hypermultitree has already been constructed the vertex set of which is denoted by V. Let k be a vertex in $\{1, ..., n\} \setminus V$. A tree is assigned to vertex k as follows. A weight function is introduced on the edges of the complete graph with vertex set V: the edge incident to the vertices i and j has the weight $P(A_i \cap A_j \cap A_k)$. The maximum weight tree is found on the complete graph by Prim's algorithm and this tree is assigned to vertex k. The vertex is adjoined from $\{1, ..., n\} \setminus V$ which the heaviest tree is assigned to and we adjoin the hyperedges obtained by extending the edges of the tree (assigned to the adjoined vertex) extended by the adjoined vertex. To obtain a $(1, m)$-hypermultitree from the $(1, 1)$-hypermultitree constructed above we extend the (1-multi)trees which were assigned to the adjoined vertices to m-multitrees as described previously.

To compute the weight of the $(1, m)$-hypermultitree obtained by the above algorithm, additionally, we need to evaluate as many probabilities $P(A_{k_1} \cap ... \cap A_{k_i})$ $(1 \leq k_1 < ... < k_i \leq n,\ i = 4, ..., m+2)$ as the $(1, m)$-hypermultitree has the hyperedges containing i vertices, namely $\sum_{k=0}^{2} \binom{m}{i-2+k}\binom{n-m}{2-k}$, which is $\mathcal{O}(n^2)$ (see the above lemma). This property of the new bounds allows using them in practice either by estimating the value of multivariate normal distribution function. Some tests results are presented in the next section.

4. Application to estimate the value of the multivariate normal distribution function

Let us introduce the notation $A_i = P(\xi_i < x_i)$ $(i = 1, ..., n)$ and designate the n-variate standard normal distribution function by $F(x_1, ..., x_n)$. Hence

$$F(x_1, ..., x_n) =$$
$$P(\xi_1 < x_1, ..., \xi_n < x_n) = P(A_1 \cap ... \cap A_n) = 1 - P(\overline{A}_1 \cup ... \cup \overline{A}_n),$$

thus, our upper (lower) bounds on $P(\overline{A}_1 \cup ... \cup \overline{A}_n)$ are lower (upper) bounds on the distribution function values. The bounds were computed by the algorithm described in the previous section by means of a Fortran code. The required probabilities $P(\overline{A}_k)$ $(1 \leq k \leq n)$, $P(\overline{A}_{k_1} \cap \overline{A}_{k_2})$ $(1 \leq k_1 < k_2 \leq n)$ were computed by the IMSL subroutines mdnor and mdbnor. The probabilities $P(\overline{A}_{k_1} \cap \overline{A}_{k_2} \cap \overline{A}_{k_3})$ $(1 \leq k_1 < k_2 < k_3 \leq n)$ were computed by Schervish's Fortran code, NORINT7 (see [18]) with accuracy 10^{-6}, where the fact is used that $P(\overline{A}_{i_1} \cap ... \cap \overline{A}_{i_k}) = P(\xi_{i_1} \geq x_{i_1}, ..., \xi_{i_k} \geq x_{i_k}) = P(\xi_{i_1} < -x_{i_1}, ..., \xi_{i_k} < -x_{i_k})$ $(1 \leq i_1 < ... < i_k \leq n)$. Finally, the probabilities $P(\overline{A}_{k_1} \cap ... \cap \overline{A}_{k_i})$ $(1 \leq k_1 < ... < k_i \leq n,\ i \geq 4)$ were computed by Genz's Fortran code, SADMVN (see [7]) with accuracy 10^{-6} whenever $i = 4, 5$ and 10^{-5} whenever $i = 6, 7, 8$.

In each test, in addition to the bounds, an estimated value of the normal probability distribution function value was computed by Szántai's method (see [19]). The sample size in each case is 100,000. In the table the average differences between the bounds and the estimated value are indicated. All computations have been carried out by a Pentium

200MHz computer. We generate $n = 40$ dimensional examples in both tests.

Test 1. For each individual test 40 random x_i's were generated uniformly from $[2,3]$ and a random correlation matrix was generated by using the method described by Marsaglia and Olkin (see [11]). With this method, a lower triangular matrix $\widehat{C}$ is generated first, with elements uniformly random from $[-1,1]$. The rows of $\widehat{C}$ are then scaled so that they have unit 2-norms. The result is a lower triangular matrix C that is used for producing a random correlation matrix $\Sigma = CC^t$. For each test run, fifty of these random correlation matrices and $\mathbf{x}$ vectors were generated.

	Average difference between bound and estimated value	Average time (seconds)
Hunter-Worsley lower	0.037702	0.06
2-multitree lower	0.025482	0.09
3-multitree lower	0.017672	0.18
4-multitree lower	0.012275	0.47
5-multitree lower	0.008310	1.23
6-multitree lower	0.005429	3.27
7-multitree lower	0.003156	8.88
Tomescu upper	0.016810	8.48
$(1,2)$-hypermultitree upper	0.011606	9.83
$(1,3)$-hypermultitree upper	0.009149	14.61
Szántai's method	–	32.65

Table 1

The average standard deviation of Szántai's method is 0.0002202.

Test 2. For each individual test 40 random x_i's were generated uniformly from $[-2,2]$. The correlations between x_i and x_j are $\sqrt{\frac{i}{j}}$ whenever $i \neq j$.

	Average difference between bound and estimated value	Average time (seconds)
Hunter-Worsley lower	0.508777	0.06
2-multitree lower	0.014971	0.11
3-multitree lower	0.000534	0.27
4-multitree lower	0.000096	0.66
5-multitree lower	0.000067	1.43
6-multitree lower	0.000048	3.13
7-multitree lower	0.000027	8.24
Tomescu upper	5.238051	7.42
$(1,2)$-hypermultitree upper	0.145856	8.57
$(1,3)$-hypermultitree upper	0.003992	12.75
Szántai's method	–	35.92

Table 2

The average standard deviation of Szántai's method is 0.0000010.

5. Concluding remarks

We executed the algorithms described in the third section to obtain the data in the tables above. The algorithms are recursions on m, the m-multitree ($(1,m)$-hypermultitree) bound they provide are improved whenever m is increased. The algorithm can be executed by increasing m step by step till the lower and upper bounds are close enough to each other. The property of the bounds according to which they are based on only few probabilities of the intersections of at most d events ($d = m+1$ by m-multitree and $d = m+2$ $(1,m)$-hypermultitree) enables us to reach a great m without spending too much time.

On the other hand, the bounds may serve as a basis of a simulation scheme estimating the values of multivariate distribution functions.

References

1. Boole, G. *Laws of Thought*, American reprint of 1854 edition, Dover, New York, 1854.
2. Boole, G. *Of Propositions Numerically Definite*, Transactions of the Cambridge Philosophical Society, Part II, XI, 1868.

3. Bonferroni, C. E. *Teoria Statistica Delle Classi e Calcolo Delle Probabilitá* Volume in onore di Riccardo Dalla Volta, Universitá di Firenze, **1-62**, 1937.
4. J. Bukszár, *Probability Bounds with Multirees*, RUTCOR Research Report, 05-99, 1999.
5. J. Bukszár, *Hypermultitrees and Bonferroni Inequalities*, Journal of Combinatorial Theory (submitted).
6. Bukszár, J. and Prékopa, A. *Probability Bounds with Cherry Trees*, RUTCOR Research Report, 04-99, 1999.
7. Genz, A. *Numerical Computation of the Multivariate Normal Probabilities*, J. Comput. Graph. Stat., **1**, pp. 141-150, 1992.
8. Hailperin, T. *Best Possible Inequalities for the Probability of a Logical Function of Events*, The American Monthly, **72**, pp. 343-359.
9. Hunter, D. *An Upper Bound for the Probability of a Union*, J. Appl. Prob., Vol. **13** pp. 597-603, 1976.
10. Kounias, S. and Marin, J. *Best Linear Bonferroni Bounds*, SIAM J. Appl. Math., Vol. **30**, No. 2 307-323, 1976.
11. Marsaglia, G. and Olkin, I. *Generating Correlation Matrices*, SIAM Journal of Scientific and Statistical Computing **5**, pp. 470-475, 1984.
12. Prékopa, A. *Stochastic Programming*, Kluwer Academic Publishers, Dordrecht, 1995.
13. Prékopa, A. Vizvári, B. and Regős, G. *Lower and Upper Bounds on Probabilities of Boolean Functions of Events*, RUTCOR Research Report, pp. 36-95.
14. Prékopa, A. *Boole-Bonferroni Inequalities and Linear Programming*, Operations Research, Vol. **36**, pp. 145-162, 1988.
15. Prékopa, A. *Sharp Bounds on Probabilities Using Linear Programming*, Operations Research, Vol. **38**, pp. 227-239, 1990.
16. Prékopa, A. *The Discrete Moment Problem and Linear Programming*, Discrete Applied Mathematics, Vol. **27**, pp. 235-254, 1990.
17. Sathe, Y. S., Pradhan, M., Shah, S. P. *Inequalities for the Probability of the Occurrence of at least m out of n Events*, Journal of Applied Probability **17**, pp. 1127-1132, 1980.
18. Schervish, M. *Multivariate Normal Probabilities with Error Bound*, Applied Statistics **33**, pp. 81-87, 1984.
19. Szántai, T. *Improved Bounds and Simulation Procedures on the Value of the Multivariate Normal Probability Distribution Function*, Proceedings of the VII. International Conference on Stochastic Programming, Vancouver, Canada Aug. 8-16, 1998 (submitted).
20. Szántai, T. *Evaluation of a Special Multivariate Gamma Distribution*, Mathematical Programming Study, **27**, pp. 1-16, 1986.
21. Tomescu, I. *Hypertrees and Bonferroni Inequalities*, Journal of Combinatorial Theory, Series B **41**, pp. 209-217, 1986.
22. Worsley, K. J. *An Improved Bonferroni Inequality and Applications*, Biometrika, Vol. **69**, pp. 297-302, 1982.

The Linear Complementarity Problem and the Class of Generalized Positive Subdefinite Matrices

Jean-Pierre Crouzeix
Université Blaise Pascal
Clermont-Ferrand, France

Sándor Komlósi
University of Pécs
Pécs, Hungary

Abstract. It is well known that the study of the Linear Complementarity Problem can be linked with the study of an appropriate quadratic programming problem. The solution set to a given LCP always contains the KKT-set of the corresponding quadratic programming problem. The reverse implication holds only for some very special classes of matrices as positive semidefinite, P-, row adequate, row sufficient and positive subdefinite ($PSBD$) matrices. The paper extends the concept of $PSBD$ matrices and proves that the above coincidence still holds for this larger class of matrices.

Key words: Linear Complementarity Problem, Generalized Positive Subdefinite Matrices.

1. Introduction

The link between Complementarity Problems and Nonlinear Programming Problems is well known. For the sake of completeness let us recall very briefly this connection. Consider the following simple optimization problem:

$$\text{Min } [f(x) : x \geq 0], \tag{NLP}$$

where f is differentiable on the nonnegative orthant of R^n. Then, the well-known Karush-Kuhn-Tucker optimality condition for (NLP) reads as follows: if $a \geq 0$ is an optimal solution of the above optimization problem, then there exists a Lagrange multiplier vector $\lambda \geq 0$ such that

$$\nabla f(a) - \lambda = 0,$$
$$\langle \lambda, a \rangle = 0.$$

Moreover, the condition is sufficient when f is pseudoconvex.

If we replace the gradient map $\nabla f(x)$ by $F(x)$, the KKT-condition leads to find $a \in R^n$ such that

$$\begin{aligned} \langle F(a), a \rangle &= 0, \\ F(a) \geq 0,\ a &\geq 0. \end{aligned} \tag{CP}$$

F. Giannessi et al. (eds.), Optimization Theory, 45–63.

The problem of finding the solutions of (CP) with arbitrary mapping $F(x)$ is called a *Complementarity Problem*. The special case, where $F(x)$ is an affine map, is called a *Linear Complementarity Problem*.

2. The Linear Complementarity Problem

In the Linear Complementarity Problem, $F(x)$ has the following form:

$$F(x) = Mx + q,$$

where M is an arbitrary quadratic matrix of order n, and q is an arbitrary n-dimensional vector. The problem itself consist of finding a vector $a \in R^n$ such that

$$\begin{aligned} \langle Ma + q, a\rangle &= 0, \\ Ma + q \geq 0,\ a &\geq 0. \end{aligned} \qquad LCP(M,q)$$

$LCP(M,q)$ is feasible if there is $x \in R^n$ so that $x \geq 0$ and $Mx + q \geq 0$. Applying the Farkas-Lemma, the feasibility is equivalent to the following condition:

$$u \geq 0,\ M^T u \leq 0 \quad \Rightarrow \quad \langle q, u\rangle \geq 0. \qquad (F)$$

Let us consider the polyhedral convex cone

$$C_M = \left\{u \geq 0 : M^T u \leq 0\right\},$$

and its polar cone

$$C_M^0 = \{u^* : \langle u, u^*\rangle \leq 0 \ \ \forall u \in C_M\}.$$

Then, the feasibility condition (F) can be expressed as follows.

PROPOSITION 1. *$LCP(M,q)$ is feasible if and only if $-q \in C_M^0$.*

2.1. Solvability

Let $S(M,q)$ denote the solution set of $LCP(M,q)$. One of the main objectives of the theory is to find conditions on M and q ensuring the nonemptiness of the solution set $S(M,q)$. A frequently used approach is the investigation of the following quadratic programming problem:

$$\text{Min}\ [f(x) = \langle Mx + q, x\rangle : x \geq 0, Mx + q \geq 0]\ . \qquad QP(M,q)$$

We denote by $S'(M,q)$ the set of optimal solutions of $QP(M,q)$. It is clear that $a \in S(M,q)$ if and only if $f(a) = 0$ and a is feasible. Hence, $a \in S(M,q)$ if and only if $f(a) = 0$ and $a \in S'(M,q)$. It is worth noticing that $QP(M,q)$ has some very remarkable properties, namely, the feasible set is polyhedral and the objective function is bounded from below. By the well-known Frank-Wolfe theorem [14], it follows that if $QP(M,q)$ is feasible, then it is solvable, as well, i.e.

$$S'(M,q) \neq \emptyset \quad \text{if and only if} \quad -q \in C_M^0.$$

Assume that $a \in S'(M,q)$. Then, in view of KKT-condition for optimality there exist $u, v \in R^n$ such that

$$(M + M^T)a + q - M^T u - v = 0\,, \qquad (KKTa)$$

$$a, u, v, Ma + q \geq 0\,, \qquad (KKTb)$$

$$\langle a, v\rangle = \langle u, Ma + q\rangle = 0\,. \qquad (KKTc)$$

We denote by $S''(M,q)$ the set of points a for which such u and v exist. $S''(M,q)$ is called the set of *KKT-stationary points*. It is obvious that

$$S(M,q) \subseteq S'(M,q) \subseteq S''(M,q),$$

where strict inclusions can occur. We are interested in conditions implying that

$$S(M,q) = S''(M,q).$$

Assume that $a \in S''(M,q)$ and (a,u,v) fulfils conditions $(KKTa)$, $(KKTb)$ and $(KKTc)$. Then, we have:

$$0 \leq f(a) = \langle Ma + q, a\rangle = \left\langle M^T(u-a), a\right\rangle = \frac{1}{2}\langle q, a - u\rangle\,, \qquad (2.1)$$

$$0 = f(a) + \langle u, v\rangle + \left\langle M^T(u-a), u-a\right\rangle. \qquad (2.2)$$

More precisely, for all i, we have [3]:

$$0 = (Ma+q)_i a_i + u_i v_i + (M^T(u-a))_i (u-a)_i\,. \qquad (CPV)$$

Notice that the quantities $f(a)$, $\langle u, v\rangle$ in (2.2) and $(Ma+q)_i a_i$ and $u_i v_i$ in (CPV) are nonnegative. This observation is the starting point of the study of different classes of matrices or maps we consider now.

2.1.1. *Positive semidefinite matrices*

Assume that M is positive semidefinite (in short: PSD) and conditions $(KKTa)$, $(KKTb)$ and $(KKTc)$ hold. Then, $\left\langle M^T(u-a), u-a\right\rangle \geq 0$. Hence, by (2.2), we have

$$0 = f(a) = \langle u, v\rangle = \left\langle M^T(u-a), u-a\right\rangle.$$

It follows that for PSD matrices $S(M,q) = S''(M,q)$. It is worth noticing that in this case the map $F(x) = Mx + q$ is monotone and the function $f(x) = \langle Mx + q, x\rangle$ is convex on R^n.

2.1.2. *P-matrices*
A matrix M is said to be a P-matrix [13] if each of its principal minors admits positive determinant. It is known that M is a P-matrix if and only if

$$x_i(M^T x)_i \leq 0 \quad \text{for all } i \quad \Rightarrow \quad x = 0\,. \qquad (P)$$

For such a matrix assume that conditions $(KKTa)$, $(KKTb)$, $(KKTc)$ hold. Then, it follows from (CPV) that $u = a$. Hence, (2.1) implies that $f(a) = 0$. Thus, $S''(M,q) = S(M,q)$.

2.1.3. *Row adequate matrices*
Recall [15] that a matrix M is said to be *row adequate* if

$$x_i(M^T x)_i \leq 0 \quad \text{for all } i \quad \Rightarrow \quad M^T x = 0\,. \qquad (RA)$$

For such a matrix assume that $(KKTa)$ to $(KKTc)$ hold. Then, (CPV) implies $M^T(u - a) = 0$. Hence, (2.2) implies $0 = f(a) = \langle u, v\rangle$. Here, again, $S(M,q) = S''(M,q)$.

2.1.4. *Row sufficient matrices*
A matrix M is said to be *row sufficient* [3] if

$$x_i(M^T x)_i \leq 0 \quad \text{for all } i \quad \Rightarrow \quad \left\langle M^T x, x\right\rangle = 0\,. \qquad (RS)$$

Here, again, in view of (2.2) and (CPV) it is straightforwardly deduced that for such matrices we have

$$S(M,q) = S''(M,q)\,.$$

It is easy to check that $(P) \Rightarrow (RA) \Rightarrow (RS)$, and thus, the class of row sufficient matrices contains the class of the row adequate matrices and P-matrices, as well.

2.1.5. *Pseudomonotone maps*
In the three subsections above only the matrix M was involved, and not q. Assume that M is not PSD since this case is treated in Subsection 2.1.1 and exclude the trivial case $q = 0$.

Let us consider the condition

$$\langle Mx, x\rangle < 0 \text{ and } \langle q, x\rangle \leq 0 \Rightarrow M^T x \leq 0. \qquad (PM)$$

Assume that conditions $(KKTa)$ to $(KKTc)$ hold in addition to (PM). Then, it follows from (2.1) and (2.2) that $f(a) = 0$. Hence, here again, $S(M,q) = S''(M,q)$.

In fact, condition (PM) characterizes pseudomonotone maps on R^n_+ [10]. It is clear that (PM) implies

$$\langle Mx, x\rangle < 0 \Rightarrow \text{ either } M^Tx \leq 0 \text{ or } M^Tx \geq 0. \qquad (PSBD)$$

Such matrices, called *positive subdefinite* $(PSBD)$ after MARTOS [17] who introduced these matrices in the symmetric case, have been studied by CROUZEIX at al. [10] in the non-symmetric case. $PSBD$ matrices which are not PSD matrices are called *merely positive subdefinite* $(MPSBD)$. We briefly describe the properties of such matrices in the next section.

3. Positive subdefinite matrices and matrices with exactly one negative eigenvalue

In the sequel we use the following notations:

Given an $n \times n$ matrix A, we denote by $A^\dagger$ its *pseudoinverse* and by A^s the symmetric matrix

$$A^s = A^T(A + A^T)^\dagger A.$$

It is known [10] that $A^s = (A^T)^s$ when

$$A(R^n) = A^T(R^n) = (A + A^T)(R^n).$$

Given an $n \times n$ symmetric matrix H, its *inertia* is the triple

$$\text{Iner}\,(H) = (\nu_-(H), \nu_0(H), \nu_+(H)),$$

where $\nu_-(H), \nu_0(H)$ and $\nu_+(H)$ are the number of negative, zero and positive eigenvalues of H, respectively. We have

$$\nu_-(H) + \nu_0(H) + \nu_+(H) = n.$$

For positive subdefinite matrices we have the following characterization.

PROPOSITION 2. [10] *The matrix M is positive subdefinite if one and only one of the following conditions holds.*

(i) $M + M^T$ *is positive semidefinite,*

(ii) $M = ab^T \neq 0$ *with* $a, b \in R^n, a \neq tb$ *for all* $t > 0$ *and either* $b \geq 0$ *or* $b \leq 0$,

(iii) $\text{rank}(M) \geq 2$, $M(R^n) = M^T(R^n) = (M + M^T)(R^n)$, $\nu_-(M + M^T) = 1$ *and* $M^s \leq 0$.

The above theorem demonstrates that positive subdefiniteness of M can be checked in a finite number of steps, since computing the inertia of $M + M^T$ is a finite procedure. Checking that M is a P-matrix requires also a finite number of steps. This is not the case with row adequate and row sufficient matrices. The following theorem shows that testing pseudomonotonicity of the affine map $F(x) = Mx + q$ on the nonnegative orthant is still a finite procedure.

PROPOSITION 3. [10] *Assume that* M *is not* PSD, $\text{rank}(M) \geq 2$ *and* $q \neq 0$. *Then,* $F(x) = Mx + q$ *is pseudomonotone on the nonnegative orthant of* R^n *if and only if the following condition holds.*

$$q \in M(R^n) = M^T(R^n) = [M + M^T](R^n),$$

$\nu_-(M + M^T) = 1$ *and* $M^s \leq 0$. *If* $\hat{x}$ *is such that* $q = M\hat{x}$, *then* $\langle M^s\hat{x}, \hat{x}\rangle \leq 0$ *and* $M^s\hat{x} \leq 0$.

It is obvious from Propositions 2 and 3 that symmetric matrices having one and only one negative eigenvalue play an important role. For such matrices the following result is of great importance.

PROPOSITION 4. [10] *Assume that* H *is symmetric and* $\nu_-(H) = 1$. *Then there exists a closed convex cone* T *such that*

$$\{h : \langle Hh, h\rangle \leq 0\} = T \cup (-T). \qquad (CD)$$

T *and its polar cone*

$$T^0 = \{h^* : \langle h, h^*\rangle \leq 0 \text{ for all } h \in T\}$$

can be represented with the help of any $q \in T^0$ *in the following way:*

$$T = \{h : \langle Hh, h\rangle \leq 0 \quad \text{and} \quad \langle q, h\rangle \leq 0\}, \qquad CR(q)$$

$$T^0 = \left\{h^* : \left\langle h^*, H^\dagger h^*\right\rangle \leq 0 \text{ and } \left\langle H^\dagger q, h^*\right\rangle \leq 0\right\} \cap H(R^n). \quad PCR(q)$$

Moreover, in this case, the interior of T *is given by*

$$\text{int}(T) = \{h : \langle Hh, h\rangle < 0 \quad \text{and} \quad \langle q, h\rangle < 0\}.$$

The quadratic form $h(x) = \langle Hx, x\rangle$ *is quasiconvex on the two sets* T *and* $-T$, *whereas* $f(x) = -\sqrt{-\langle Hx, x\rangle}$ *is convex on* T *and* $-T$.

Note that $H^{\dagger}q \in T$ and thus $\left\langle H^{\dagger}q, q\right\rangle \leq 0$. Decomposition (CD) is called the *canonical decomposition* of the "non-positive part" of the quadratic form $\langle Hh, h\rangle$, whereas representations $CR(q)$ and $PCR(q)$ are called *canonical representations* of T and T^0, respectively. We close this section with recalling one more important result.

PROPOSITION 5. [5, 6, 12] *Assume that H is symmetric, $\nu_-(H) \geq 1$ and $q \neq 0$. Then, the three following conditions are equivalent:*

(i) $\langle q, h\rangle = 0 \;\Rightarrow\; \langle Hh, h\rangle \geq 0,$

(ii) $\nu_-(H) = 1,\ q \in H(R^n)$ *and* $\left\langle q, H^{\dagger}q\right\rangle \leq 0,$

(iii) $\nu_- \begin{pmatrix} H & q \\ q^T & 0 \end{pmatrix} = 1\,.$

Other conditions equivalent to (i)-(iii) can be found in [16].

4. Comparison of the different classes of matrices

Since we want to compare the class of P-matrices, row adequate and row sufficient matrices with the class of merely positive subdefinite matrices, the following simple statement will be useful in the sequel.

PROPOSITION 6. *Consider the following matrix:*

$$C = \begin{bmatrix} A & 0 \\ 0 & B \end{bmatrix},$$

where A and B are square matrices of arbitrary order. Then, the following statements hold:

(i) C is a P-matrix whenever A and B are P-matrices.

(ii) C is a row adequate matrix whenever A and B are row adequate matrices.

(iii) C is a row sufficient matrix whenever A and B are row sufficient matrices.

EXAMPLE 1. Consider the following matrix:

$$C = \begin{bmatrix} A & 0 \\ 0 & B \end{bmatrix},$$

where

$$A = B = \begin{bmatrix} 1 & -1 \\ 4 & 1 \end{bmatrix}.$$

It can be shown that this matrix is a P-matrix and thus C is also a P-matrix. By computing the inertia of $C + C^T$ we will get that this matrix has exactly 2 negative eigenvalues. It proves that a P-matrix need not be a $PSBD$ matrix.

EXAMPLE 2. Consider matrix C from Proposition 6 with

$$A = B = \begin{bmatrix} 0 & -2 \\ 1 & 0 \end{bmatrix}.$$

It can be verified easily that C is a row sufficient matrix (and fails to be a P-matrix). Since $\text{Iner}(C + C^T) = (2, 0, 2)$, therefore C is not $PSBD$.

EXAMPLE 3. Consider the following matrix

$$C = \begin{bmatrix} 0 & -1 \\ -1 & 0 \end{bmatrix}.$$

By applying Proposition 2, it can be verified that C is $PSBD$, and then, an easy computation shows that it fails to be row sufficient.

These examples demonstrate that the matrix classes above are really very different.

5. New Condition Ensuring $S(M, q) = S''(M, q)$

It is the key point in the Cottle-Pang-Venkatesvaran approach [3] that condition

$$x_i(M^T x)_i \leq 0 \quad \text{for all } i \quad \Longrightarrow \quad \left\langle x, M^T x \right\rangle = 0,$$

is sufficient for the inclusion

$$S''(M, q) \subseteq S(M, q). \tag{A}$$

We recall that the reverse inclusion always holds.

In this section, we present a more sophisticated analysis of the KKT-conditions and provide a different approach to give sufficient condition for (A) to be hold. This new approach is based on the observation that condition $\left\langle w, M^T w \right\rangle = 0$, with $w = u - a$, is sufficient but not necessary for having inclusion (A). We may have $f(a) = 0$ and $\left\langle w, M^T w \right\rangle < 0$. It follows that the "negative part" of the quadratic form $\left\langle w, M^T w \right\rangle$ may deserve special attention. The concept of positive subdefiniteness has been proved suitable in investigating condition (A). This approach has been elaborated by CROUZEIX at al. in [10]. The main result of the subsequent sections is the extension of the notion of positive subdefiniteness in such a way that for this broader class of matrices, called *generalized positive subdefinite* matrices, inclusion (A) is still valid.

For further purposes we introduce the following notations: given $t \in R$, we set $t^+ = \max(t, 0)$ and $t^- = \max(0, -t)$. We extend this notations for vectors, too: for any $x \in R^n$ we set $x^+ = (x_1^+, x_2^+, \dots, x_n^+)$ and $x^- = (x_1^-, x_2^-, \dots, x_n^-)$. The result below is the starting point of further investigations [11].

LEMMA 1. *Let a be a KKT-stationary point of problem $QP(M, q)$ with Lagrange multiplier vectors u, $v \geq 0$. Set $w = u - a$. Then,*

$$f(a) = \langle a, Ma + q \rangle = \left\langle (M^T w)^+, w^- \right\rangle, \tag{C1}$$

$$\langle u, v \rangle = \left\langle (M^T w)^-, w^+ \right\rangle, \tag{C2}$$

$$0 = \left\langle (M^T w)^+, w^+ \right\rangle = \left\langle (M^T w)^-, w^- \right\rangle. \tag{C3}$$

Proof. The componentwise form of condition $(KKTa)$ reads as follows:

$$v_i + (M^T w)_i = (Ma + q)_i, \quad \text{for all } i, \tag{5.1}$$

whereas simple rearrangement of (CPV) gives

$$-w_i (M^T w)_i = a_i (Ma + q)_i + u_i v_i \geq 0, \quad \text{for all } i. \tag{5.2}$$

It is easy to see that (5.2) implies $(C3)$.

Assume that $(M^T w)_i > 0$. Then, in virtue of (5.1) and (5.2), we have $w_i \leq 0$ and $(Ma + q)_i > 0$. The complementarity condition $(KKTc)$ yields that the dual variable $u_i = 0$. Next (5.2) implies

$$a_i (Ma + q)_i = -w_i (M^T w)_i.$$

Finally, we have

$$a_i (Ma + q)_i = (M^T w)_i^+ w_i^- \tag{5.3}$$

$$u_i v_i = (M^T w)_i^- w_i^+ . \tag{5.4}$$

Assume that $(M^T w)_i < 0$. Then, in virtue of (5.1) and (5.2), we have $w_i \geq 0$ and $v_i > 0$. The complementarity condition $(KKTc)$ yields that the primal variable $a_i = 0$. In this case, (5.3) and (5.4) hold, too.

Finally, assume that $(M^T w)_i = 0$. Then, from (5.2), it follows that $a_i(Ma+q)_i = u_i v_i = 0$, and thus, (5.3) and (5.4) trivially hold and the thesis follows. □

This lemma leads to introduce the following definition:

DEFINITION 1. *We say that the pair (M,q) is* appropriate, *if*

$$\langle Mw, w\rangle < 0 \quad and \quad \langle q, w\rangle \leq 0 \quad \Longrightarrow \quad \left\langle (M^T w)^+, w^- \right\rangle = 0. \tag{APP}$$

The following statements are simple consequences of the definition:

1) (M,q) is appropriate, if the matrix $M+M^T$ is positive semidefinite,

2) (M,q) is appropriate, if M is $MPSBD$ with the additional property that
$$M^T w \geq 0 \quad \Longrightarrow \quad \langle q, w\rangle \geq 0,$$
which, by Farkas-lemma, is equivalent to
$$q \in M(R^n_+).$$

3) (M,q) is appropriate, if the map $F(x) = Mx + q$ is pseudomonotone on the nonnegative orthant. This is true when F is monotone because in this case M is positive semidefinite. When F is merely pseudomonotone it follows from the characterization of such maps [10] that
$$\langle Mw, w\rangle < 0 \ \text{ and } \ \langle q, w\rangle \leq 0 \quad \Longrightarrow \quad M^T w \leq 0.$$

The following theorem provides some important properties of appropriate pairs.

THEOREM 1. *Assume that (M,q) is appropriate. Then,*

(i) $\langle q, w\rangle = 0 \quad \Longrightarrow \quad \left\langle M^T w, w \right\rangle \geq 0,$ *hence* $\nu_-(M + M^T) \leq 1.$

If, in addition, we have $\nu_-(M + M^T) = 1$, then,

(ii) $\langle Mw, w\rangle \leq 0 \quad and \quad \langle q, w\rangle \leq 0 \quad \Longrightarrow \quad \left\langle (M^T w)^+, w^- \right\rangle = 0,$

(iii) $\langle Mw, w\rangle = \langle q, w\rangle = 0 \quad \Longrightarrow \quad (M^T w)_i w_i = 0$ *for all* i.

Proof. First, we prove (i). Assume for contradiction, that $\langle Mw, w\rangle < 0$ and $\langle q, w\rangle = 0$. Then, by (APP), we have

$$\left\langle (M^T w)^+, w^- \right\rangle = 0 .$$

Obviously, we may replace w with $-w$, and thus, we obtain

$$\left\langle (M^T w)^-, w^+ \right\rangle = 0 .$$

As a simple consequence, we get

$$\left\langle M^T w, w \right\rangle = \left\langle (M^T w)^+, w^+ \right\rangle + \left\langle (M^T w)^-, w^- \right\rangle \geq 0 ,$$

contradicting to the hypothesis that $\langle Mw, w\rangle < 0$. Then, Proposition 5 implies that $(M + M^T)$ has at most one negative eigenvalue and it proves (i).

It is easy to see that (ii) can be obtained by continuity arguments from Proposition 4 and 5. For proving (iii) assume that $\langle Mw, w\rangle = \langle q, w\rangle = 0$, then (ii) applied to w and $-w$ implies that

$$\left\langle (M^T w)^+, w^- \right\rangle = \left\langle (M^T w)^-, w^+ \right\rangle = 0 .$$

Hence,

$$0 = \langle Mw, w\rangle = \left\langle M^T w, w \right\rangle = \left\langle (M^T w)^+, w^+ \right\rangle + \left\langle (M^T w)^-, w^- \right\rangle .$$

It follows that

$$0 = \left\langle (M^T w)^+, w^+ \right\rangle = \left\langle (M^T w)^-, w^- \right\rangle ,$$

and the thesis follows. □

Since, the case where $(M + M^T)$ is positive semidefinite is well studied, we focus now on the case where $\nu_-(M + M^T) = 1$ and $q \neq 0$. In the sequel of this section, we use the notation of Proposition 4. Next, we give a characterization of appropriacy.

THEOREM 2. *Assume that* $\nu_-(M + M^T) = 1$ *and* $q \neq 0$. *Then,* (M, q) *is appropriate if and only if* $q \in T^0$ *and for all* $i = 1, 2, \ldots, n$, *there exist* $s_i, t_i \geq 0$ *such that* $s_i + t_i = 1$ *and*

$$-s_i e_i + t_i m_i \in T^0 , \qquad (CA)$$

where the vectors e_i, m_i *denote the* i*-th columns of the identity matrix and* M, *respectively.*

Proof. *Necessity:* Assume that (M, q) is appropriate. Then, by the previous theorem we have that

$$\langle q, w\rangle = 0 \Longrightarrow \langle Mw, w\rangle \geq 0,$$

or equivalently,

$$\langle Mw, w\rangle < 0 \Longrightarrow \langle q, w\rangle \neq 0.$$

Taking into account the canonical decomposition (CD), a simple consequence of the above implication is that we have either $q \in T^0$ or $q \in -T^0$. Let T denote that part of the decomposition (CD) for which we have $q \in T^0$. It follows that T enjoys the canonical representation $CR(q)$.

Set for $i = 1, 2, \ldots, n$,

$$W_i = \{w : w_i < 0 \text{ and } (M^T w)_i > 0\}.$$

It is easy to see that (M, q) is appropriate if and only if $T \cap W_i = \emptyset$ for all i. It is not difficult to prove that $W_i = \emptyset$ if and only if there exist $s_i, t_i \geq 0$ such that $s_i + t_i = 1$ and $-s_i e_i + t_i m_i = 0$. Assume now that $W_i \neq \emptyset$. Then, by a separation theorem, this is equivalent to the existence of $d_i \in R^n$, $d_i \neq 0$ so that

$$\sup\{\langle d_i, w\rangle : w \in T\} \leq 0 \leq \inf\{\langle d_i, w\rangle : w \in W_i\}.$$

It follows from the one hand that

$$d_i \in T^0, \tag{5.5}$$

and from the other hand, that

$$0 \leq \inf\{\langle d_i, w\rangle : w_i \leq 0 \text{ and } (M^T w)_i \geq 0\}.$$

By the Farkas-Lemma, it follows the existence of $s_i, t_i \geq 0$ such that $s_i + t_i > 0$ and $-s_i e_i + t_i m_i = d_i$. Since T^0 is a cone you can normalize the coefficients s_i and t_i satisfying the sum-condition $s_i + t_i = 1$ and it is just the thesis.

Sufficiency: Assume that the assumptions of the present theorem together with (CA) hold and we have

$$\left\langle M^T w, w\right\rangle < 0 \text{ and } \langle q, w\rangle \leq 0.$$

Taking into account the definition of T it follows that $w \in \text{int}(T)$. Since

$$-s_i e_i + t_i m_i \in T^0$$

it follows that

$$\langle -s_i e_i + t_i m_i, w\rangle = -s_i w_i + t_i (M^T w)_i \le 0.$$

Assume first that $s_i \neq 0$. In this case we have

$$w_i \ge \frac{t_i}{s_i}(M^T w)_i.$$

If $t_i \neq 0$, then

$$(M^T w)_i \le \frac{s_i}{t_i} w_i.$$

In both cases we have that for all $i = 1, 2, \ldots, n$

$$(M^T w)_i^+ (w^-)_i = 0,$$

which gives the desired result $\left\langle (M^T w)^+, w^- \right\rangle = 0.$ □

According to Theorem 2, we introduce the following classes of maps (M, q).

$$(APP1): \quad \begin{cases} \nu_-(M + M^T) = 1, \ (M + M^T) \ \text{ is non-singular,} \\ \bar{x} = -(M + M^T)^{-1} q, \\ \left[(M + M^T)^{-1}\right]_{ii} \le 0 \ \text{ and } \ \bar{x}_i \le 0, \quad \text{for all } i\,. \end{cases}$$

$$(APP2): \quad \begin{cases} \nu_-(M + M^T) = 1, \ M(R^n) \subseteq (M + M^T)(R^n), \\ \exists \bar{x} \ \text{ such that } \ q = -(M + M^T)\bar{x}, \\ (M^s)_{ii} \le 0 \ \text{ and } \ m_i \bar{x}_i \le 0, \quad \text{for all } i\,. \end{cases}$$

PROPOSITION 7. *If (M, q) satisfies $(APP1)$ or $(APP2)$, then (M, q) is appropriate.*

Proof. $(APP1)$ corresponds to $t_i = 0$ for all i, whereas $(APP2)$ corresponds to $s_i = 0$. □

Pseudomonotone maps on the nonnegative orthant of R^n are a particular case of $(APP2)$ as seen from Proposition 3. The main result of this section is the following.

THEOREM 3. *If (M, q) is appropriate, then*

$$S(M, q) = S''(M, q)\,.$$

Proof. The conclusion holds if M is PSD. Assume that $\nu_-(M + M^T) \geq 1$. It is clear, by continuity that

$$\langle Mw, w\rangle \leq 0 \text{ and } \langle q, w\rangle \geq 0 \implies \left\langle (M^T w)^+, w^- \right\rangle = 0.$$

Let $a \in S''(M, q)$. Let $u, v \geq 0$ the associate multipliers and set $w = u - a$. Then, (2.1) and (2.2) together imply

$$\langle Mw, w\rangle \leq 0 \quad \text{and} \quad \langle q, w\rangle \geq 0.$$

Then, $\left\langle (M^T w)^+, w^- \right\rangle = 0$, hence $(C1)$ implies $f(a) = 0$ and $a \in S(M, q)$. □

6. The structure of $S(M, q)$, when (M, q) is appropriate

It is known that $S(M, q)$ is a polyhedral convex set when M is PSD, a $P-$matrix, a row-adequate matrix or column sufficient matrix [3]. This is also the case when $F(x) = Mx + q$ is pseudomonotone on R^n_+ [10]. The next result is concerned with the structure of the solution set $S(M, q)$ when (M, q) is appropriate.

THEOREM 4. *Assume that (M, q) is appropriate, $q \neq 0$ and $\nu_-(M + M^T) = 1$. Then, there are two polyhedral convex sets (possibly empty) S_+ and S_- such that*

$$S(M, q) = S_- \cup S_+.$$

Proof. Set $\bar{x} = -(M + M^T)^\dagger q$ so that $(M + M^T)\bar{x} = -q$. Then,

$$f(x) = \langle Mx + q, x\rangle = \langle M(x - \bar{x}), x - \bar{x}\rangle - \langle M\bar{x}, \bar{x}\rangle.$$

We recall that $\left\langle (M + M^T)^\dagger q, q\right\rangle \leq 0$. Hence $\langle M\bar{x}, \bar{x}\rangle \leq 0$. Set $u = x - \bar{x}$. In virtue of Theorem 1, we have the following coincidences:

$$S(M, q) = S'(M, q) = S''(M, q),$$

and thus, a belongs to $S(M, q)$ if and only if it is an optimal solution of the problem

$$\operatorname{Min}\left[f(x) : Mx + q \geq 0,\ x \geq 0\right].$$

It follows that $u + \bar{x} \in S(M, q)$ if and only if u is an optimal solution of the problem

$$\operatorname{Min}\left[g(u) = \langle Mu, u\rangle : Mu \geq M^T\bar{x},\ u \geq -\bar{x}\right].$$

Let $\hat{S}$ denote the optimal solution set for this later problem. Since the optimal value in the first case is 0, for any $u \in \hat{S}$ we have

$$g(u) = \langle Mu, u \rangle = \langle M\bar{x}, \bar{x} \rangle \leq 0 ,$$

and thus, $\hat{S} \subseteq T \cup (-T)$. It follows that $u + \bar{x} \in S(M,q)$ if and only if u is an optimal solution of one of the problems:

$$\text{Min} \left[g(u) = \langle Mu, u \rangle : Mu \geq M^T\bar{x},\ u \geq -\bar{x},\ u \in T \right] ,$$

or

$$\text{Min} \left[g(u) = \langle Mu, u \rangle : Mu \geq M^T\bar{x},\ u \geq -\bar{x},\ u \in -T \right] .$$

It is known that g is quasiconvex on each of the sets T and $-T$. Hence, the two optimal sets are convex. Call these sets $\hat{S}_+$ and $\hat{S}_-$. Since

$$T = \{v : \langle Mv, v \rangle \leq 0 \ \text{ and } \ \langle q, v \rangle \leq 0\} ,$$

it follows that $\hat{S}_+$ and $\hat{S}_-$ coincide with the optimal solution sets of the following problems, respectively.

$$\text{Min} \left[g(u) = \langle Mu, u \rangle : Mu \geq M^T\bar{x},\ u \geq -\bar{x},\ \langle q, u \rangle \leq 0 \right] ,$$

$$\text{Min} \left[g(u) = \langle Mu, u \rangle : Mu \geq M^T\bar{x},\ u \geq -\bar{x},\ \langle q, u \rangle \geq 0 \right] .$$

Since the optimal set of a problem of minimizing a quadratic form over a polyhedral set is polyhedral, then $\hat{S}_+$ and $\hat{S}_-$ are polyhedral. It is obvious that $\hat{S} = \hat{S}_- \cup \hat{S}_-$. If $u \in \hat{S}_+$ or $\hat{S}_-$ then KKT conditions hold for the problem in x. Since $S''(M,q) = S(M,q)$, $u + \bar{x} \in S(M,q)$. Take $S_+ = \bar{x} + \hat{S}_+$ and $S_- = \bar{x} + \hat{S}_-$. □

7. Generalized Positive Subdefinite Matrices

Careful examinations of the conditions figuring in Theorem 2 lead to introduce the following concept.

DEFINITION 2. *A matrix M is called* generalized positive subdefinite *($GPSBD$) if there exist nonnegative multipliers s_i, t_i, $s_i + t_i = 1$, $i = 1, 2, \ldots, n$ such that*

$$\langle Mw, w \rangle < 0 \quad \Rightarrow \begin{cases} \textit{either} & -s_i w_i + t_i (M^T w)_i \leq 0 \textit{ for all } i , \\ \textit{or} & -s_i w_i + t_i (M^T w)_i \geq 0 \textit{ for all } i , \end{cases} \qquad (GPSBD)$$

where w_i denotes the i-th component of w. M is called non-degenerate *if for all w satisfying $\langle Mw, w \rangle < 0$ at least one of the inequalities in ($GPSBD$) holds in a strict way.*

Remark 1. It is easy to verify that matrix M is $PSBD$ if and only if M is non-degenerate $GPSBD$ with multipliers $s_i = 0$ and $t_i = 1$ for all i.

For $GPSBD$ matrices we have the following *decomposition-property*.

COROLLARY 1. *If a matrix M is $GPSBD$ with $\{w : \langle Mw, w\rangle < 0\} \neq \emptyset$, then the following decomposition holds:*

$$\{w : \left\langle M^T w, w\right\rangle \leq 0\} = Z \cup (-Z), \tag{7.1}$$

where Z is a closed cone given by

$$Z = \{w : \langle Mw, w\rangle \leq 0 \ \text{ and } \ \langle -s_i e_i + t_i m_i, w\rangle \leq 0, \ \text{ for all } i\} .$$

Non-degenerate $GPSBD$ matrices enjoy the following *eigenvalue-property*.

THEOREM 5. *Let M be a non-degenerate $GPSBD$ matrix. Then we have*

$$\nu_-(M + M^T) \leq 1.$$

Proof. Assume for contradiction that there exist w_1, w_2, two orthogonal eigenvectors of H corresponding to negative eigenvalues h_1 and h_2, respectively: $Hw_i = h_i w_i$, $i = 1, 2$. It follows that for all convex combination of the two eigenvectors $w(\lambda) = \lambda w_1 + (1-\lambda)w_2$, $\lambda \in [0, 1]$ we have that

$$\langle Mw(\lambda), w(\lambda)\rangle < 0.$$

Taking into account ($GPSBD$) it follows that for any $\lambda \in [0, 1]$ we have either

$$\langle -s_i e_i + t_i m_i, w(\lambda)\rangle \geq 0, \ \text{ for all } i\,,$$

or

$$\langle -s_i e_i + t_i m_i, w(\lambda)\rangle \leq 0, \ \text{ for all } i\,.$$

Without loss of the generality we may assume that

$$\langle -s_i e_i + t_i m_i, w_1\rangle \geq 0, \ \text{ for all } i\,,$$

and

$$\langle -s_i e_i + t_i m_i, w_2\rangle \leq 0, \ \text{ for all } i\,.$$

It follows that there exists $\lambda_0 \in (0, 1)$ such that

$$\langle -s_i e_i + t_i m_i, w(\lambda_0)\rangle = 0, \ \text{ for all } i\,,$$

which contradicts to the non-degeneracy assumption. □

COROLLARY 2. *Let M be a matrix such that $\nu_-(M+M^T) = 1$. Then, M is GPSBD if and only if there exist nonnegative multipliers s_i, t_i, $s_i + t_i = 1$, $i = 1, 2, \ldots, n$ such that either*

$$-s_i e_i + t_i m_i \in T^0, \qquad \textit{for all } i\,,$$

or

$$-s_i e_i + t_i m_i \in -T^0, \qquad \textit{for all } i\,,$$

M is non-degenerate GPSBD if and only if there exist at least one index j such that

$$-s_j e_j + t_j m_j \neq 0.$$

Proof. Since in this case decomposition (7.1) coincides with the canonical decomposition (CD) it follows that we have either $Z = T$ or $Z = -T$ and it proves the thesis.

An immediate consequence of Theorem 2 and Corollary 2 is the next assertion.

COROLLARY 3. *Let M be a matrix such that $\nu_-(M+M^T) = 1$. Then, (M, q) is appropriate if and only if M is GPSBD and*

$$q \in T^0\,.$$

The following example shows that the class of $GPSBD$ matrices is larger then the class of $PSBD$ matrices.

EXAMPLE 4. Consider the following matrix:

$$M = \begin{bmatrix} 0 & 2 & 0 \\ -1 & 0 & -1 \\ 0 & 1 & 0 \end{bmatrix}.$$

The inertia can be computed very easily by the Cottle-algorithm [2], which gives that $\text{Iner}(M + M^T) = (1, 1, 1)$. Moreover we have that $\left\langle M^T w, w \right\rangle = w_1 w_2$ and

$$M^T w = \begin{bmatrix} -w_2 \\ 2w_1 + w_3 \\ -w_2 \end{bmatrix}.$$

It is easy to check that M is not $PSBD$, but with the choice $s_1 = t_2 = s_3 = 0$ and $t_1 = s_2 = t_3 = 1$ condition $GPSBD$ holds.

This example also demonstrates that the 'range condition', $M(R^n) = M^T(R^n) = (M + M^T)(R^n)$, which is characteristic for $MPSBD$ matrices (cf. Proposition 2) does not hold in general for merely $GPSBD$ matrices.

In virtue of Proposition 1, Theorems 3 and 5, and Corollary 3 the following statement is true.

THEOREM 6. *Let M be a non-degenerate $GPSBD$ matrix which is not PSD. Then, for any $q \in -C_M^0 \cap T^0$ we have*

$$S''(M,q) = S(M,q) \neq \emptyset.$$

8. Acknowledgment

The paper presents results from the joint work started in May of 1998, during the stay of the second author at the Blaise Pascal University, Clermont Ferrand and continued during the stay of the first author in March of 1999 in the Computer and Automation Institute of the Hungarian Academy of Sciences, Budapest. Financial supports are greatly acknowledged provided by the two institutions and the Hungarian National Research Fund (OTKA T 025442).

References

1. Y. Chabrillac and J.-P. Crouzeix, Definiteness and semidefiniteness of quadratic forms revisited, *Linear Algebra and its Applications*, **63** (1984) 283–292.
2. W. Cottle, Manifestations of the Schur complement, *Linear Algebra and its Applications*, **8** (1974) 189–211.
3. W. Cottle, J. S. Pang and V. Venkateswaran, Sufficient matrices and the linear complementarity problem, *Linear Algebra and its Applications*, **114/115** (1989) 231–249.
4. W. Cottle, J. S. Pang and R. E. Stone, *The linear complementarity problem*, Academic Press, New York, 1992.
5. J.-P. Crouzeix, Contribution à l'étude des fonctions quasi-convexes, *Thèse d'Etat, U.E.R des Sciences Exactes et Naturelles*, Université de Clermont-Ferrand II, 1977.
6. J.-P. Crouzeix and J. A. Ferland, Criteria for quasiconvexity and pseudoconvexity: relationships and comparisons, *Mathematical Programming*, **23** (1982) 193–205.
7. J.-P. Crouzeix and S. Schaible, Generalized monotone affine maps, *SIAM J. Matrix Anal. Appl.*, **17** (1996) 992–997.
8. J.-P. Crouzeix, P. Marcotte and D. Zhu, Conditions ensuring the applicability of cutting-plane methods for solving variational inequalities, *Mathematical Programming*, **88** (2000) 521–539.

9. J.-P. Crouzeix, Characterizations of generalized convexity and generalized monotonicity, in: *Generalized Convexity, Generalized Monotonicity: Recent Results* (eds. J.-P. Crouzeix, J.-E. Martínez Legaz and M. Volle), Kluwer Academic Publishers, Dordrecht, 1998, pp. 237–256.
10. J.-P. Crouzeix, A. Hassouni, A. Lahlou and S. Schaible, Positive Sub-Definite Matrices, Generalized Monotonicity and Linear Complementarity Problems, To appear in *SIAM J. Matrix Anal. Appl.*
11. J.-P. Crouzeix, *Manuscript*, September 1998.
12. J.-A. Ferland, Matrix-theoretic criteria for the quasiconvexity of twice continuously differentiable functions, *Linear Algebra and its Applications*, **38** (1981) 51-63.
13. M. Fiedler and V. Pták, On matrices with non-positive off-diagonal elements and positive principal minors, *Czechoslovak Math. J.*, **12** (1962) 382–400.
14. M. Frank and P. Wolfe, An algorithm for quadratic programming, *Naval Res. Logistic Quart.*, **3** (1956), 992–997.
15. A. W. Ingleton, The linear complementarity problem, *J. London Math. Soc.*, **2** (1970) 330–336.
16. S. Komlósi, On pseudoconvex functions, *Acta Sci. Math. (Szeged)* **57** (1993) 569–586.
17. B. Martos, Sub-definite matrices and quadratic forms, *SIAM J. Appl. Math.*, **17** (1969) 1215–1223.
18. H. Samelson, R. M. Thrall and O. Wesler, A partition theorem for Euclidean n-space, *Proc. Amer. Math. Soc.* **9** (1958) 805–807.

Computer Experiences with Successive Regression Approximations for Solving Equations

István Deák
Operations Research Group,
Department of Differential Equations,
Technical University of Budapest,
H-1111 Budapest, XI. Muegyetem rkp. 3.,
e-mail: deak@math.bme.hu

Abstract. In a recent paper the successive regression approximations method was proposed for solving equations for one-dimensional nonlinear functions with deterministic function values. Here we apply this method to noisy function values. A set of Fortran subroutines, performing the actual computations has been developed, among them a subroutine, capable of computing the gradient of the multidimensional normal distribution. A short description of the method, its convergence, the subroutines and computer experiences are given in this paper. Numerical examples and results of computer runs are presented up to fifteen dimensions. According to the numerical results the error of the approximate root depends inversely on the square root of the number of points used in the approximation.

1. Introduction.

One of the basic problems in numerical analysis is solving an equation. Solving equations frequently appears as a subproblem in optimization problems, too and accordingly there are several well known numerical techniques for finding the root of an equation

$$f(x) = 0, \tag{1}$$

where $f : R^1 \to R^1$. Generally it is assumed, that analytical inversion is not possible, only function values $f(x)$ can be computed (for corresponding methods see [17] for example). Difficulties arise, however, when the values of $f(x)$ can not be exactly evaluated. This is the case when applying STABIL-type stochastic programming models ([20], [21]), where during actual numerical optimization one of the troublespots is the problem of finding the root of the equation

$$f(x) = \Phi(\mathbf{x}_k + x(\mathbf{d}_k - \mathbf{x}_k)) - p = 0 \tag{2}$$

for the unknown x, where $\mathbf{x}_k, \mathbf{d}_k$ are given vectors, p is a reliability level, usually near to 1, and Φ is the distribution function of the n-dimensional normal probability distribution. Solving (2) corresponds to finding the intersection of a line with the surface of the set of feasible solutions, which is given as $\{\mathbf{x}|p = \Phi(\mathbf{x})\}$. There exist some Monte

F. Giannessi et al. (eds.), *Optimization Theory*, 65–80.

Carlo methods for computing estimators η of $\Phi(\cdot)$, where η is a random variable with $E(\eta) = p, D^2(\eta) = \sigma^2$, with σ depending on the sample size (see [2], [3], [4], [6], [14], [23], for recent developments [10], [8]).

Least squares approximations (or in a probabilistic wording regression estimators) are well known tools in numerical computations (see [1], [16]), but were not used previously in a successive way for solving nonlinear equations. There seems to be some ambiguity in using the terms of least squares and regression function – here the first term will be applied in the case of deterministic function values, the second one in the case of function values with a random error, but mainly the term regression is used. A new heuristic approach, using regression estimators was suggested for solving (2), or equivalently (1) in [5], and it proved to be rather competitive to existing techniques according to extensive numerical tests (see [7], [9]). A simplified version was given in [13], where its convergence was proved, too. In this paper, the computational results and experiences are reported, mainly based on those, appearing in [12]. In addition another algorithm is implemented and tested for computing the gradient of the n-dimensional normal distribution function, making use of the analytical derivative of the regression function.

Function values with an additive random error are obtained during Monte Carlo computations, whose error is $O(N^{-1/2})$, basically inversely proportional to the amount of sampling work. Using Monte Carlo computations for evaluating $f(x)$, and employing some kind of heuristic line search techniques for solving (1) gives no such assurance with regard to the error. It seems to be, that according to numerical experiments the new equation solving algorithm gives the same – probabilistic – error bound: the accuracy of the result is inversely proportional to the square root of the number of points, where the function value is evaluated.

First of all in the next section it is described, how a least squares approximation is computed for a given set of points and function values. In Section 3. an iterative algorithm, called successive regression approximation (or SRA) is suggested for solving an equation of the form $f(x) = 0$, then in the subsequent section results concerning the convergence of the procedure are recalled. In the next section the subroutines and their parameters are described, then results of computer tests, including numerical examples are presented.

2. Computing a least squares–regression approximation

We make the following assumption: $f(x)$ is a continuous, bounded, strictly monotone increasing function, which has a unique root Θ, that is

$$f(\Theta) = 0.$$

Assume, that all points x_i used to produce a least squares approximation are different from each other, and the function value $f(x)$ can be evaluated exactly for any x.

The function $f(x)$ is approximated by a linear least squares function $g(x)$. For a given set S_k of k points x_i and function values $f_i = f(x_i)$ consider the L_2 minimum norm approximation of the form $g(x) = \alpha x + \beta$, that is for $S_k = \{x_i, f_i\}_{i=0}^{k-1}$ let $g(x)$ be the solution of the problem

$$\min_{\alpha,\beta} \sum_{i=0}^{k-1} [f_i - (\alpha x_i + \beta)]^2 . \tag{3}$$

In order to solve this problem, we consider the first order necessary conditions of optimality, that can be obtained by differentiating the sum in (3) with respect to α, β, and setting the resulting derivatives to zero. This means solving the system of equations

$$\sum_{i=0}^{k-1} x_i [f_i - (\alpha x_i + \beta)] = 0, \tag{4}$$

$$\sum_{i=0}^{k-1} [f_i - (\alpha x_i + \beta)] = 0, \tag{5}$$

for the unknowns α, β. Introduce the following notations:

$$m_0 = \frac{1}{k}\sum_{i=0}^{k-1} f_i , \qquad m_1 = \frac{1}{k}\sum_{i=0}^{k-1} x_i f_i ,$$
$$M_0 = \frac{1}{k}\sum_{i=0}^{k-1} 1 = 1 , \quad M_1 = \frac{1}{k}\sum_{i=0}^{k-1} x_i , \qquad M_2 = \frac{1}{k}\sum_{i=0}^{k-1} x_i^2 , \tag{6}$$
$$\mathbf{M} = \begin{pmatrix} M_2 & M_1 \\ M_1 & M_0 \end{pmatrix} ,$$

by the aid of which the system of equations can be rewritten as

$$\begin{aligned} m_1 &= \alpha M_2 + \beta M_1, \\ m_0 &= \alpha M_1 + \beta M_0, \end{aligned} \tag{7}$$

or equivalently one can have the same equations in matrix form

$$(m_1, m_0)' = \mathbf{M}(\alpha, \beta)'. \tag{8}$$

Using these notations the solution of the system can be written as

$$(\alpha, \beta)' = \mathbf{M}^{-1}(m_1, m_0)'.$$

Finally, using the determinant $|\mathbf{M}| = M_2 - M_1^2$ the values of the coefficients in $g(x)$ can be given as

$$\alpha = \frac{-m_0 M_1 + m_1}{M_2 - M_1^2}, \tag{9}$$

$$\beta = \frac{m_0 M_2 - m_1 M_1}{M_2 - M_1^2}. \tag{10}$$

Using the approximating linear function $g_k(x) = \alpha x + \beta$ the original equation $f(x) = 0$ can be replaced by an approximate equation $g_k(x) = 0$, whose solution is

$$x_k = -\frac{\beta}{\alpha} = -\frac{m_0 M_2 - m_1 M_1}{-m_0 M_1 + m_1} \tag{11}$$

Similarly to the linear approximating function the formulas for the quadratic approximations can easily be developed. The function $f(x)$ is now approximated by a regression function of the form $q(x) = \alpha x^2 + \beta x + \gamma$. Note here, that both $g(x)$ and $q(x)$ are linear regressions in their parameters α, β, γ, but their form is different, one of them is a linear function, the other is quadratic, so when using $q(x)$ we will talk about a quadratic approximation, though it is linear regression. The function $q(x)$ is the best approximation in L_2, if the coefficients α, β, γ are determined as the solution of the problem

$$\min_{\alpha,\beta,\gamma} \sum_{i=0}^{k-1} \left[f_i - (\alpha x_i^2 + \beta x_i + \gamma) \right]^2, \tag{12}$$

from which the first order necessary conditions of optimality are

$$\sum_{i=0}^{k-1} x_i^2 \left[f_i - (\alpha x_i^2 + \beta x_i + \gamma) \right] = 0,$$

$$\sum_{i=0}^{k-1} x_i \left[f_i - (\alpha x_i^2 + \beta x_i + \gamma) \right] = 0,$$

$$\sum_{i=0}^{k-1} \left[f_i - (\alpha x_i^2 + \beta x_i + \gamma) \right] = 0.$$

These equations are to be solved for the unknowns α, β, γ. Now the following notations are introduced, in addition to the previously given ones:

$$m_2 = \frac{1}{k}\sum_{i=0}^{k-1} x_i^2 f_i, \quad M_3 = \frac{1}{k}\sum_{i=0}^{k-1} x_i^3, \quad M_4 = \frac{1}{k}\sum_{i=0}^{k-1} x_i^4,$$

$$\mathbf{M} = \begin{pmatrix} M_4 & M_3 & M_2 \\ M_3 & M_2 & M_1 \\ M_2 & M_1 & M_0 \end{pmatrix}$$

by the aid of which the system of equations can be rewritten as

$$\begin{aligned} m_2 &= \alpha M_4 + \beta M_3 + \gamma M_2 \\ m_1 &= \alpha M_3 + \beta M_2 + \gamma M_1 \\ m_0 &= \alpha M_2 + \beta M_1 + \gamma M_0, \end{aligned} \tag{13}$$

or equivalently one can have the same equations in matrix form

$$(m_2, m_1, m_0)' = \mathbf{M}(\alpha, \beta, \gamma)', \tag{14}$$

Using these notations the solution of the system can be written as

$$(\alpha, \beta, \gamma)' = \mathbf{M}^{-1}(m_2, m_1, m_0)'.$$

Finally, using the determinant

$$|\mathbf{M}| = M_4(M_2 - M_1^2) - M_3(M_3 - M_1 M_2) + M_2(M_1 M_3 - M_2^2)$$

the values of the coefficients in $q(x)$ can be given as

$$\begin{aligned} \alpha &= \frac{m_2(M_2 - M_1^2) - m_1(M_3 - M_1 M_2) + m_0(M_1 M_3 - M_2^2)}{|\mathbf{M}|}, \\ \beta &= \frac{-m_2(M_3 - M_1 M_2) + m_1(M_4 - M_2^2) - m_0(M_1 M_4 - M_2 M_3)}{|\mathbf{M}|}, \\ \gamma &= \frac{m_2(M_1 M_3 - M_2^2) - m_1(M_1 M_4 - M_2 M_3) + m_0(M_2 M_4 - M_3^2)}{|\mathbf{M}|}. \end{aligned} \tag{15}$$

Generally the approximate quadratic equation $q(x) = 0$ has two roots, so a truncation is performed, so that only the increasing part of the function is used as an approximation (for the detailed description of the truncation see [9]).

3. Successive regression approximation algorithm for solving an equation

In what follows an algorithm for determining a solution Θ of the equation $f(x) = 0$ is suggested. The basic point of the suggested algorithm is, that we take a "global" kind of approximation, since least square approximations are influenced by all previously computed points. This is in contrast to some conventional methods (e.g. Newton's), where a "use and forget" approach is employed; once a point, and the function value at that point (or the first and/or the second derivative) is used to build an approximation, and subsequently a new approximation point is evaluated, thereafter previous points and values are disregarded.

This method based on regression is somewhat similar to the stochastic approximation's basic Robbins–Monroe method (see [22], [15]), but it is basically different, – the updating function is not just a weighted sum of previous points, but a more complicated function, given in (11).

The basic features of the proposed algorithm consist of starting with an arbitrary initial set of points and function values $S_k = \{x_i, f_i\}_{i=0}^{k-1}$, a linear least square function $g_k(x) = \alpha x + \beta$ is computed, the approximate equation $g_k(x) = 0$ is solved yielding x_k, then this point with its function value is added to the set S_k, and finally the approximating least squares function is updated. Convergence proof is outlined in the next section.

Assume, that we are given an initial set of k points and corresponding function values $S_k = \{x_i, f_i\}_{i=0}^{k-1}$, where $k \geq 1$, from which a linear least squares approximation can be determined.

ALGORITHM SRA (successive regression approximation)

0. Initialization. Set the iteration counter to the number k of points given.
1. Compute the coefficients of $g_k(x) = \alpha x + \beta$ from S_k, using (9) and (10).
2. Determine an approximate root x_k from $g_k(x) = 0$ from (11).
3. If x_k is good enough, then STOP, otherwise attach the new point x_k to the set S_k, that is let $S_{k+1} = S_k \cup \{x_k, f_k\}$, increase the iteration counter k and go back to Step 1.

The same algorithm, when $g(x)$ is replaced by the quadratic function $q(x)$ is quite similar, the only difference is, that the truncation is to be taken into consideration.

4. Convergence of the successive regression approximations in case of linear approximations

In this section results from [13] are recalled, that show the convergence of the previously given algorithm SRA for the case of linear regression approximations. They show, that the points $x_i, i = 0, 1, \ldots$ produced by the previously given algorithm SRA, converge to the root in the case of linear regression approximation for deterministic function values.

The assumptions that we need to prove the convergence results are simple

A1 $f(x), x \in R^1$ is a continuous, strictly monotone increasing function, which has a unique root Θ, that is

$$f(\Theta) = 0.$$

A2 Furthermore, $f(x)$ is differentiable almost everywhere, $0 < \delta_L < f'(x) < \delta_U < \infty$ is required in all points of differentiability, and in points y where the function is not differentiable, the same bounds apply to the left hand side $f'_-(y)$ and right hand side $f'_+(y)$ derivatives $\delta_L < f'_-(y) < \delta_U, \delta_L < f'_+(y) < \delta_U$.

A3 (assumption on the set S). Let the set $S_k = \{x_i, f_i\}_{i=0}^{k-1}$ be given, where $f_i = f(x_i), i = 0, 1, \ldots, k-1$, with $x_0 < x_1 < \ldots < x_{k-1}$, thus $f_i < f_{i+1}, i = 0, 1, \ldots, k-2$, and x_k is computed from S_k by SRA. Furthermore, it is assumed, that the points $x_i, i = 0, 1, \ldots, k-1$ are different. (Ordering the values x_i and assuming all points to be different is simply for convenience.)

THEOREM 1. *Let $S_k = \{x_i, f_i\}_{i=0}^{k-1}$ a set of k points and the corresponding function values be given, x_k is computed from S_k by SRA, x_{k+1} is computed from $S_{k+1} = S_k \cup \{x_k, f_k\}$ by SRA, then*

$$x_{k+1} = x_k - f_k \sum_{i=0}^{k-1} \frac{(x_k - x_i)^2}{\sum_{i=0}^{k-1} \sum_{j=i+1}^{k} (x_j - x_i)(f_j - f_i)}. \tag{16}$$

This theorem implies the following corollary, that sheds some light on the behaviour of the point sequence, produced by SRA.

COROLLARY 1. *If $x_k < \Theta$, then $x_k < x_{k+1}$ (or alternatively, if $x_k > \Theta$, then $x_k > x_{k+1}$).*

It can be shown, that the set of points computed by SRA is bounded and then based on this result we have:

THEOREM 2. *Assume, that the function $f(x)$ is subject to the assumptions A1 and A2 and the function values $f(x)$ can be evaluated exactly. The point sequence $\{x_n\}_{n=k}^{\infty}$ generated by SRA from an initial set S_k, converges to the root Θ.*

5. Description of the subroutines

The subroutines developed in FORTRAN for equation solving and gradient computation are described here.

ROOTLIN - computes an approximate root using the linear approximation $g(x)$.

ROOTSQU - computes an approximate root, using the quadratic approximation $q(x)$.

INLIZE - is a subroutine, which by increasing and decreasing the steplength in a heuristic way tries to find an interval containing the root.

NORGRAD - is a subroutine, that for a given point computes the gradient of the n-dimensional normal distribution function, by applying the quadratic approximations along each coordinate axis.

The parameters are described as they appear in the subroutine ROOTLIN, but mostly they have the same meaning for all subroutines. (The exception is the parameter IFLAG, see there.) The first parameters describe the distribution function Φ:

ND - number of dimensions of the distribution function.

XV,YV - are the two vectors, that span the line XV(.) + XLAM*(YV(.) - XV(.)), along which the root is looked for,

TMAT - is a lower triangular matrix, that is the Cholesky factor of the correlation matrix COR of the distribution function.

The other parameters are governing the successive regression approximation.

IDIR - can be 1 or 2, if IDIR=1, the procedure executes a fixed number (KITER) of iterations, then stops (in this case the parameter EPS is meaningless). If IDIR=2, then ROOTLIN evaluates the data and determines the number of necessary iterations to compute an approximate root XROOT, at which the error (the absolute value of the function value f(XROOT)) is smaller, than EPS (in the case of the choice IDIR=2 the parameter value KITER becomes meaningless).

IPRINT - may be 0,1,2,3,4, it regulates the amount of printing. For IPRINT=4 all available data are printed (though generally only up to iteration 40), for IPRINT=1 no output is produced.

IFLAG - is an important flag, if ROOTLIN (or ROOTSQU) returns IFLAG=-1 or IFLAG=-2, then something went wrong, no approximate root has been returned. ROOTLIN returns the value IFLAG=1, if it judges, that the returned XROOT is really an approximate root of the original equation. In the subroutine INLIZE a value is given to IFLAG, if it is equal to 1, then the subroutine found two values, that bracket the root, if it is equal to -1, then INLIZE signifies, that something might be wrong (either there are two roots, or no root at all). If INLIZE returns a value IFLAG=-1, the regression can still be computed (either ROOTLIN or ROOTSQU can be called in), and frequently it still finds a root.

PRELAB - is a prescribed reliability level, it must be between the bounds 0.10, 0.99.

KITER - is the number of iterations to be performed, it should be greater than 10 and smaller than 100000.

KINIT - is the number of initially given points, that are given prior to ROOTLIN and ROOTSQU (the corresponding $x, f(x)$ and error values should be given in the arrays AXVAL, AFVAL, ERRAR though). ROOTLIN or ROOTSQU can be used without assigning a number to this parameter (no prior knowledge or points are needed except XV and YV), in this case INLIZE will compute some points from XV and YV (and this is the recommended use). However, if the user has some points on the line (and the corresponding function values), he can store them in AXVAL, AFVAL, ERRAR, give the number of points in KINIT, and "comment out" (delete) the call to INLIZE in ROOTLIN or ROOTSQU at the very beginning of these subroutines, and then call immediately ROOTLIN or ROOTSQU (this is not recommended).

EPS - is the required accuracy of the final result (it is meaningful only for the parameter choice IDIR=2).

AXVAL - is an array of x values (the sequence of the approximate roots), computed by ROOTLIN, ROOTSQU or INLIZE.

AFVAL - is an array of function values, computed at the points AXVAL (the sequence of the function values at the approximate roots), by ROOTLIN, ROOTSQU or INLIZE.

ERRAR - is an array of the errors of the computed function values (it is equal to three times the sample standard deviation of the computed function value).

XROOT - is the approximate root.

Now we give the declarations of the four subroutines, together with some comments:

SUBROUTINE ROOTLIN(ND, XV, YV, TMAT, IDIR, IPRINT, IFLAG, PRELAB, KITER, KINIT, EPS, AXVAL, AFVAL, ERRAR, XROOT)

This subroutine computes the approximate root of the function $f(x)$, as it is defined in equation (2), by using linear approximating functions.

SUBROUTINE ROOTSQU(ND, XV, YV, TMAT, IDIR, IPRINT, IFLAG, PRELAB, KITER, KINIT, EPS, AXVAL, AFVAL, ERRAR, XROOT)

This subroutine computes the approximate root of the function $f(x)$, as it is defined in equation (2), by using quadratic approximating functions.

SUBROUTINE INLIZE(ND, XV, YV, TMAT, IPRINT, PRELAB, KINIT, IFLAG, AXVAL, AFVAL, ERRAR)

This subroutine initializes the set of initially given points, prior to calling ROOTLIN or ROOTSQU. First, by stepping up and down tries to find two values, that "bracket" the root, that is for one of them the function value is smaller than 0, and for the other one is bigger than 0. Second, it examines and changes if necessary, these values, until the bracketing interval is "big enough", that is the negative value is smaller than 3 times the error of the function evaluation, and the positive value is bigger than 3*ERR. Third, the bracketing interval is decreased, so that the negative function value becomes bigger than BLOW, the positive function value smaller than BUPP (function values —and corresponding x values— outside this interval are deleted). These bounds BLOW and BUPP are preset in the program, and are dependent on the reliability level, but can be changed; for PRELAB=0.9 the values BLOW=-0.1 and BUPP=0.044 are used.

The IFLAG parameter gives a warning (it returns a value -1), if it could not find a bracketing interval, but ROOTLIN or ROOTSQU still could be called in, and it could be still successful at finding the root.

SUBROUTINE NORGRAD(ND, XXV, TMAT, IDIR, IPRINT, KITER, EPS, FACT, PRELAB, GRADV)

This subroutine computes an approximate gradient to the normal distribution at the point XXV. The method used in the evaluation is the following. Along the coordinate axis i a set of points are determined, that converge to the point, where the gradient is required. The function values are evaluated then in each of these points, and a quadratic approximation of the type $q(x)$ is computed from the points and function values, the derivative of $q(x)$ is computed and this is offered as the i-th component of the gradient.

The structure is very similar to that of the root finding subroutines: first a crude, heuristic search is done to find a "meaningful" interval, where the function values are about the same size, as the error in the function evaluation, and then KITER points are generated.

The description of the parameters of the subroutine NORGRAD is given next, since some of them have meanings different from the above given ones; ND, TMAT, IPRINT and PRELAB are the same ones.

XXV is the point, where the gradient is computed.

IDIR is a directing parameter, presently it can have the value 1 only. (In a later version it is planned, that for a different value of IDIR the required accuracy of the gradient can be specified.)

KITER is the number of points, where the function value is to be evaluated in one direction (so altogether 2*ND*KITER points will be computed), presently it can have a value between 10 and 100. Actually the number of function evaluation is bigger, since the initialization procedures, that is executed for each coordinate axis, also determines some points.

EPS is presently a dummy variable, no accuracy can be given.

FACT is a multiplicative fbe actor, presently automatically computed by NORGRAD, so no value has to given here either.

GRADV is the result, that is the subroutine NORGRAD returns the gradient of the ND-dimensional normal distribution function at the point XXV in the vector GRADV.

The subroutines used for computing the multidimensional normal distribution function values are those coming from NORSET (subroutines NORDIS, and NORDIS1, NORDIS2, NORDIS3, NORDIS4). The general recommendation is to run these subroutines using the minimum sample size (see [12] for reasons).

6. Numerical experiences

Numerical results and experience with randomly generated problems of type (2) are presented and discussed in this section.

6.1. Remark on the performance of the subroutines

Before the actual results the different types of the general situation is discussed. The set $X = \{\mathbf{x} \mid \Phi(\mathbf{x}) \geq p = PRELAB\}$ is a convex set (see [20]), so for an arbitrary line in the n-dimensional space there are three different cases. In the first case (i) there are two intersections of the line with the bounding surface $\{\mathbf{x} \mid \Phi(\mathbf{x}) - p = 0\}$ (two roots), in the second one (ii) there is only one intersection, and finally (iii) there are cases, when no intersection exists.

The root finding procedure (especially the initialization subroutine) heavily depends on the nature of the two points given initially (XV and YV, spanning the line). If one of the points is inside X and the other one is outside (one has a probability greater than p, and at the other point the function value is smaller than p), then no difficulties arise, the root (or one of the roots) is found always. In the case, when e.g. in both points probabilities are smaller than p, then - if the points are not "far away" from the set X and we have case (i) or (ii), then one of the roots is found with high probability (in about 80% of the cases). However in the case, when the boundary $B = \{\mathbf{x} \mid \Phi(\mathbf{x}) - p = 0\}$ does not separate the two points, then the procedure may not find the root, even in case (i) or (ii). Note further, that in case (i) ROOTLIN frequently brakes down, while ROOTSQU finds the root. There are two points, where the progress of the iterative procedure is checked upon, in the 10-th iteration and at the 100-th iteration, where in case of unsatisfactory behaviour (not small enough function values at the most recent approximate root) the procedure is aborted.

It must be emphasized, that during actual optimization a bracketing point pair is almost always is available, so the above described unsatisfactory events do not practically limit the usefulness of the subroutine package.

6.2. Numerical results

Table 1 shows results for the root finding procedure, where a two-dimensional normal distribution was investigated, with the correlation coefficient equal to 0.99. Table 2 presents data on the initial point-pairs for the examples given in Table 1..Table 3 contains root-evaluations for a 10 dimensional distribution, in this case all correlations were equal to 0.9, in Table 4 a 15-dimensional problem was computed, the correlation coefficients were 0.2 up to dimension 11, and two additional two-dimensional problems were linked. (Examples were taken from [6].)

Entries in Table 1 are as follows: in the first column the number of the example is given, then the type of the estimator (linear or quadratic) appears, k denotes the number of points, where the function value was evaluated, next x_r is the final approximate root, $f(x_r)$ is the function value at this approximate root, σ_f denotes the standard deviation of this function value $f(x_r)$, and σ_1 is the standard deviation of one function evaluation. Finally, in the last column the time required to compute the approximate root x_r is given.

As an interesting numerical result the gradient vector of the multidimensional normal distribution for the above given $ND = 15$ dimensional example, at the point x_r given in Table 4 is also reported, it is: (0.0, 0.201, -0.016, 0.0, -0.009, 0.0, 0.0, -0.016, -0.022, -0.009, 0.0, 0.0, 0.0, 0.0, 0.0), where 0.0 means a value, with an absolute value smaller than 0.001. This gradient was computed using 20 points on each coordinate axis (10 in each direction), and the necessary time to compute this fifteen dimensional gradient was 23.5 sec. The situation shown here is typical, almost all of the components are very small, except one or two, which are at least an order or two greater, than the remaining components. So it may be a good idea to make a small preliminary sample, and then disregard all those components, where the gradient is significantly smaller than at the other components.

However, we must remark here, that the gradient can not be considered stable, since for repeated runs variations of up to 40% occur according to computer experiences, though an increasing direction can be easily found in almost all cases.

exmpl.	est.	k	x_r	$f(x_r)$	σ_f	σ_1	time
1	LIN	31015	-0.028815	0.000000	0.000014	0.0015	1.85 sec
	SQU	34506	-0.028795	-0.000025	0.000014	0.0015	2.07 sec
2	LIN	13749	0.330353	-0.000004	0.000014	0.0018	0.82 sec
	SQU	35651	0.330361	0.000001	0.000014	0.0017	2.14 sec
3	LIN	16625	0.020190	0.000010	0.000012	0.0014	1.00 sec
	SQU	19885	0.020231	-0.000031	0.000012	0.0012	1.20 sec
4	LIN	20097	1.79617	0.000032	0.000014	0.0016	1.21 sec
5	LIN	31356	-1.19462	-0.000005	0.000014	0.0015	1.84 sec
	SQU	29938	-1.19459	-0.000015	0.000014	0.0017	1.77 sec
6	LIN	13417	2.74267	-0.000001	0.000014	0.0016	0.80 sec
	SQU	18115	2.74262	0.000015	0.000014	0.0013	1.09 sec

Table 1. $ND = 2$ dimensional examples

exmpl.	XV	YV
1	1.1808 7.3912	-2.3190 -1.2651
2	-0.6728 4.1696	5.2430 4.5170
3	1.4183 1.3369	1.9670 -0.8405
4	-6.3409 4.1002	-0.5936 2.5311
5	-2.7596 5.6660	-6.1423 5.8367
6	5.3247 3.6140	3.8505 6.6916

Table 2. Points determining a line

est.	k	x_r	$f(x_r)$	σ_f	σ_1	time
LIN	10	15.010	0.000347	0.00011	0.0012	0.23 sec
SQU	10	14.998	-0.000555	0.00010	0.0015	0.23 sec

Table 3. $ND = 10$ dimensional examples

est.	k	x_r	$f(x_r)$	σ_f	σ_1	time
LIN	10	33.082	-0.00029	0.000092	0.0009	0.63 sec
SQU	10	33.099	0.00022	0.000102	0.0015	0.63 sec

Table 4. $ND = 15$ dimensional examples

7. Conclusions

Numerical results illustrate, that roots with four accurate digits (measured by accuracy in function value of the root, that is the standard deviation of the function value at the root is less than 0.0001) can be computed up to 10 dimensions in less than 1 sec (the real concern here is the time required to compute the normal distribution function, not the time necessary for SRA). Results in Table 1 substantiate the claim, made in the Introduction, that the convergence is of the order $1/\sqrt{k}$.

The root finding procedure is very reliable, it can compute the root with prescribed error tolerance, and it works properly in all dimension, if a bracketing pair is given, but this is generally the case in stochastic programming optimization examples.

As far as gradient evaluation is concerned, a satisfactory direction can be computed in about 1 sec up to 10 dimensions and in 10 sec up to 15 dimensions. This is not an outstanding speed, but it corresponds to previous computational experiences with a more sophisticated algorithm [9], [7]. Here, with more carefully chosen parameter it might be possible to achieve an improvement of at least one order of magnitude.

Acknowledgment

The computer program was implemented and tested at the Institut fur Operations Research, University of Zurich, where the author enjoyed a nine-month visit courtesy of Prof. P. Kall.

References

1. Bjorck, A.: Numerical methods for least squares problems, SIAM, 1996, pp.408.
2. Deák, I.: Computation of multiple normal probabilities, in: Symp. on Stochastic Programming, Oberwolfach, 1979, ed. P. Kall, Springer Lecture Notes in Economics and Math. Systems, Springer, 107-120.
3. Deák, I.: Three digit accurate multiple normal probabilities, Numerische Math. 35 (1980) 369-380.
4. Deák, I.: Computing probabilities of rectangles in case of multidimensional normal distribution, J. Stat. Comp. and Simul. 26 (1986), 101-114.
5. Deák, I.: Procedures to solve STABIL on a parallel computer, Working Paper, 1989, University of Wisconsin, Department of Industrial Engineering, 89-10, pp. 26., lecture at the Vth International Conference on Stochastic Programming, Ann Arbor, MI, 1989.
6. Deák, I.: Random number generators and simulation, in: Mathematical Methods of OR (series editor A. Prékopa), Akadémiai Kiadó (Publishing House of the Hungarian Academy of Sciences), Budapest, 1990. pp. 342.
7. Deák, I.: Regression estimators for multinormal distributions: computer experiences in root finding, in: Stochastic programming methods and technical applications, (eds. K. Marti, P. Kall), Lecture Notes in Economics and Mathematical Systems, Springer, V. 458, (1998) pp. 279-293.
8. Deák, I.: Normal probabilities – computer experiences and program description, University of Zurich, Institut of Operations Research, Manuskripte, 1997, pp. 38.
9. Deák, I.: Linear regression estimators for multinormal distributions in optimization of stochastic programming models, European J. Operational Research, 111 (1998) 555-568.
10. Deák, I.: Probabilities of simple n-dimensional sets in case of normal distribution, IIE Trans. on Operations Engineering, 1998, submitted.
11. Deák, I.: Subroutines for computing normal probabilities of sets – computer experiences, Proc. 7th Int. Conference on Stochastic Programming, 1998, Vancouver B. C., submitted.
12. Deák, I.: Evaluating roots: Computer experiences and program description, University of Zurich, Institut of Operations Research, Manuskripte, 1998, pp. 21.
13. Deák, I.: Successive regression approximations for solving equations, Submitted to JOTA, 1999.
14. Gassmann, H.: Conditional probability and conditional expectation of a random vector, in: Numerical techniques for stochastic optimization (eds. Y. Ermoliev, R. Wets), Springer series in computational mathematics, Springer Verlag, 1988, 237-254.
15. Kushner, H. J., Clarck, D. S.: Stochastic approximation methods for constrained and unconstrained systems, Springer Verlag, in: Applied Mathematical Sciences 26, 1978, p. 261.
16. Lawson, C. L., Hanson, R. J.: Solving least squares problems, SIAM Classics in Applied Mathematics, 1995, pp. 337.
17. Luenberger, D. G.: Linear and nonlinear programming, Reading MA, Addison-Wesley, 1984.
18. Mayer, J.: Computational techniques for probabilistic constrained optimization problems, in: Lecture Notes on Economics and Mathematical Systems, Springer, V379. (1992), 141-164.
19. Monahan, J., Genz, A.: Spherical-radial integration rules for Bayesian computation, J. American Stat. Ass. 92 (1997) 664-674.
20. Prékopa, A.: Stochastic Programming, in: Mathematics and its Applications 324, Kluwer, 1995.

21. Prékopa, A., Ganczer, S. Deák, I., Patyi, K.: The STABIL stochastic programming model and its experimental application to the electrical energy sector of the Hungarian economy, in: Proc. of the International Symp. on Stochastic Programming, ed. M. Dempster (1980) Academic Press, 369-385.
22. Robbins, H., Monro, S.: A stochastic approximation method, Ann. Math. Stat. 22 (1951) 400-407.
23. Szántai, T.: Evaluation of a special multivariate gamma distribution function, Math. Programming Study 27 (1986) 1-16.
24. Tong, Y. L.: The multivariate normal distribution, Springer Series in Statistics, 1990, pp. 271.
25. Wasan, M. T.: Stochastic approximation, University Press, Cambridge, 1969.

A Bundle of Smooth Lagrangians in Mathematical Programming

V. F. Demyanov
Applied Mathematics Dept.
St.Petersburg State University
Staryi Peterhof, St.Petersburg
198904 RUSSIA

Abstract. A constrained minimization problem in a finite-dimensional space is considered. A general scheme is proposed to remove some or all constraints. A duality problem is stated and sufficient conditions for the zero duality gap are formulated. In the case of smooth mathematical programming problems this general approach leads to Exact Nonsmooth Penalty functions. It is shown that a smooth Lagrange function exists in the case where the so-called 'abstract' constraint is absent or convex. If the abstract constraint is not convex, then, generally speaking, it is impossible to find a smooth Lagrangian such that the minimizer of the constrained problem is an inf-stationary point of the Lagrangian. However, a bundle of smooth Lagrangians exists such that the minimizer of the constrained problem is an inf-stationary point of each of these Lagrangians. This new phenomenon is promising and may be used to construct numerical algorithms.

Key words: Constrained minimization problem, Lagrange multipliers, exact penalization, dual problems, bundle of Lagrangians, calmness.

AMS Classification: 49K99, 90C99.

1. Introduction

One of the most important contributions of Mechanics to Mathematics is the Lagrange multiplier rule. This rule is extremely popular in Mathematical Programming. Let us consider the problem of minimizing a function $f : X \to \mathbb{R}$ on the set

$$\Omega = \{x \in X \mid h_i(x) \leq 0 \quad \forall i \in I\}$$

where $X \subset \mathbb{R}^n$ is a closed set, $h_i : X \to \mathbb{R}$, $I = 1 : N$.

Assume that $x^* \in \Omega$ is a minimizer of the function f on the set Ω. If f and h_i's are smooth, $X = \mathbb{R}^n$ then under some additional conditions (called *constraint qualifications*) coefficients $\lambda_1, ..., \lambda_N$ exist such that the minimizer x^* is a stationary point of the function $L(x, \lambda) = f(x) + \sum_{i \in I} \lambda_i h_i(x)$ on $\mathbb{R}^n$. Thus, the constrained optimization problem is reduced to an unconstrained one. The function

[1] The research was supported by the Russian Foundation for Fundamental Studies (grant RFFI No. 97-01-00499).

F. Giannessi et al. (eds.), Optimization Theory, 81–98.

$L(x, \lambda)$ is called a *Lagrange function* or just a *Lagrangian.* It turns out that, in general, every minimizer generates its own Lagrangian (in some cases, e.g., if f and h_i's are convex, Lagrangians for all minimizers coincide).

The same result holds if X is a convex set: under some conditions there exist coefficients $\lambda_1, ..., \lambda_N$ such that the minimizer x^* is a stationary point of $L(x, \lambda)$ on X (to be more precise, x^* is an inf-*stationary* point, i.e. a necessary condition for a minimum of L on X is satisfied). The case of smooth functions f and h_i's and a convex set X is discussed in detail by R. T. Rockafellar in [1] where nonsmooth and nonlinear Lagrangians are also described. A. M. Rubinov, B. M. Glover and X.-Q. Yang [2] considered the case where f and h_i's are not necessarily smooth, X is not necessarily convex and $L(x, \lambda)$ is some composite (not necessarily smooth) function

$$L(x, \lambda) = H(f(x), h_1(x), ..., h_N(x), \lambda_1, ..., \lambda_N).$$

Conditions on H are stated under which the duality gap is zero:

$$\inf_{x \in X} \sup_{\lambda \in \Lambda} L(x, \lambda) = \sup_{\lambda \in \Lambda} \inf_{x \in X} L(x, \lambda).$$

Here

$$\Lambda = \{\lambda = (\lambda_1, ..., \lambda_N) \mid \lambda_i \geq 0 \quad \forall i \in I\}.$$

In the present paper only the case of smooth functions f and h_i's but nonconvex X is discussed. Let $x^* \in \Omega$ be a minimizer of f on Ω. We will show that, in general, there exist no coefficients $\lambda_1, ..., \lambda_N$ such that the point x^* is an inf-stationary point of the function $L(x, \lambda) = f(x) + \sum_{i \in I} \lambda_i h_i(x)$ on X. It is also shown that there exists no smooth composition $H(x) = f(x) + H_1(h_1, ..., h_N)$ having this property. However, it is demonstrated in the sequel that it is possible to construct several smooth Lagrangians $L_k(x) = f(x) + \sum_{i \in I} \lambda_i^{(k)} h_i(x)$ and several convex cones A_k $(k \in K)$ such that for every $k \in K$ the point x^* is an inf-stationary point of $L_k(x)$ on the set $x^* + A_k$. Thus, if M_k is the set of all inf-stationary points of $L_k(x)$ on the set $x^* + A_k$ then $x^* \in \bigcap_{k \in K} M_k$.

Thus, there exist a bundle of smooth Lagrangians and a family of convex cones such that the problem of minimizing the function f on the set Ω is reduced to that of finding the sets of inf-stationary points of several smooth functions on respective convex cones. To the author's best knowledge this is a new phenomenon to be appreciated and used for constructing applicable algorithms aimed at finding a common minimizer of several "simple" problems.

2. Exact Penalty Functions

Let $X \subset \mathbb{R}^n$ be a given set, a function $d(x)$ be defined and nonnegative on X,

$$\Omega = \{x \in X \mid d(x) = 0\}. \tag{2.1}$$

Assume that $\Omega \neq \emptyset$. Note that every closed set $\Omega \subset \mathbb{R}^n$ can be described in the form (2.1) (e.g., if one takes $X = \mathbb{R}^n$, $d(x) = \min_{y \in \Omega} ||x - y||$).

Let a function f be defined and real-valued on X. Consider the following problem:

Find

$$\inf_{x \in \Omega} f(x) = f^*. \tag{P}$$

For $A \geq 0$ let us construct the functions

$$F(x, A) = f(x) + Ad(x), \tag{2.2}$$

$$\psi(A) = \inf_{x \in X} F(x, A) \tag{2.3}$$

and

$$\phi(x) = \sup_{A \geq 0} F(x, A). \tag{2.4}$$

Note that

$$F(x, A) = f(x) \qquad \forall\, x \in \Omega.$$

Since

$$\psi(A) = \inf_{x \in X} F(x, A) \leq \inf_{x \in \Omega} F(x, A) = \inf_{x \in \Omega} f(x) = f^* \tag{2.5}$$

then

$$\psi^* = \sup_{A \geq 0} \psi(A) \leq f^*. \tag{2.6}$$

As

$$\phi(x) = \sup_{A \geq 0} F(x, A) = \begin{cases} f(x) & \forall x \in \Omega, \\ +\infty & \forall x \in X \setminus \Omega, \end{cases}$$

then

$$\inf_{x \in X} \phi(x) = \inf_{x \in \Omega} f(x) = f^*. \tag{2.7}$$

Relations (2.6) and (2.7) imply

$$f^* = \inf_{x \in X} \sup_{A \geq 0} F(x, A) \geq \sup_{A \geq 0} \inf_{x \in X} F(x, A) = \psi^*. \tag{2.8}$$

The following properties are trivial:

1. The function $\psi(A)$ is nondecreasing as A increases:

$$if \quad A' \geq A \quad then \quad \psi(A') \geq \psi(A).$$

2. As a corollary,

$$\sup_{A \geq 0} \psi(A) = \lim_{A \to \infty} \psi(A) = \psi^*. \tag{2.9}$$

3. Let $\{A_k\}$, $\{x_k\}$ be sequences such that

$$A_k \underset{k \to \infty}{\longrightarrow} \infty, \qquad x_k \in X,$$

$$\psi(A_k) = F(x_k, A_k) - \varepsilon_k = f(x_k) + A_k d(x_k) - \varepsilon_k,$$

$$\varepsilon_k \geq 0, \quad \varepsilon_k \underset{k \to \infty}{\longrightarrow} 0. \tag{2.10}$$

If f is bounded from below on X:

$$f(x) \geq -B > -\infty \quad \forall\, x \in X \tag{2.11}$$

then

$$d(x_k) \to 0. \tag{2.12}$$

The problem of finding $\psi^* = \sup_{A \geq 0} \psi(A)$ (see (2.6)) is called *dual* to the problem (P).

LEMMA 2.1 (Zero duality gap). *If $X \subset \mathbb{R}^n$ is a closed set, the functions $f(x)$ and $d(x)$ are continuous on X, f is bounded from below on X and (in the case of the unboundedness of X)*

$$f(x) \underset{||x|| \to \infty, x \in X}{\longrightarrow} \infty \tag{2.13}$$

then

$$f^* = \inf_{x \in X} \sup_{A \geq 0} F(x, A) = \sup_{A \geq 0} \inf_{x \in X} F(x, A) = \psi^*. \tag{2.14}$$

Proof. Take any sequences $\{A_k\}$ and $\{x_k\}$ satisfying (2.10). By virtue of (2.9)

$$\psi(A_k) = f(x_k) + A_k d(x_k) - \varepsilon_k \to \psi^* \leq f^*. \tag{2.15}$$

(2.11) implies $f^* > -\infty$, then (2.15) and (2.8) yield

$$f(x_k) \leq \psi^* + \varepsilon_k \leq f^* + \varepsilon_k. \tag{2.16}$$

Hence, $f(x_k)$ is bounded, therefore (via (2.13)), the sequence $\{x_k\}$ is also bounded. Without loss of generality we can assume that $x_k \to x^* \in X$. By the continuity of $d(x)$ we conclude that $d(x^*) = 0$, i.e. $x^* \in \Omega$. It follows from (2.15) that $\psi(A_k) \geq f(x_k) - \varepsilon_k$ which implies

$$\psi^* \geq f(x^*). \tag{2.17}$$

Since $x^* \in \Omega$ then $f(x^*) \geq \inf_{x\in\Omega} f(x) = f^*$. Therefore (2.17) means that

$$\psi^* \geq f^*. \tag{2.18}$$

Inequalities (2.6) and (2.18) yield (2.14) thus completing the proof.

Let

$$\Omega_\varepsilon = \{x \in X \mid d(x) \leq \varepsilon\},\ \varepsilon \geq 0,\ f_\varepsilon = \inf_{x\in\Omega_\varepsilon} f(x). \tag{2.19}$$

THEOREM 2.1 *Let all the hypothesis of Lemma 2.1 hold. If*

$$\limsup_{\varepsilon\downarrow 0} \frac{f^* - f_\varepsilon}{\varepsilon} \leq C < \infty \tag{2.20}$$

then there exist $A^* < \infty$ and $x^* \in \Omega$ such that

$$\psi(A) = \inf_{x\in X} F(x, A) = F(x^*, A) = f(x^*) = f^* \quad \forall A \geq A^*. \tag{2.21}$$

Proof. We shall argue by contradiction. Assume the contrary. Then (2.5) implies that there exists a sequence $\{A_k\}$ such that

$$A_k \to \infty,\ \psi(A_k) = \inf_{x\in X} F(x, A_k) < f^* \quad \forall\, k.$$

Therefore one can find a sequence $\{x_k\}$ such that

$$x_k \in X,\ F(x_k, A_k) = f(x_k) + A_k d(x_k) < f^* \quad \forall\, k,$$

$$\psi(A_k) = F(x_k, A_k) - \varepsilon_k,\ \varepsilon_k \to 0.$$

Due to (2.12),

$$d(x_k) \to 0. \tag{2.22}$$

We have

$$x_k \in \Omega_{d(x_k)}, \quad f(x_k) \geq \inf_{x \in \Omega_{d(x_k)}} f(x) = f_{d(x_k)},$$

$$f_{d(x_k)} + A_k d(x_k) \leq f(x_k) + A_k d(x_k) < f^*.$$

Hence,

$$\frac{f^* - f_{d(x_k)}}{d(x_k)} \geq A_k.$$

Since $A_k \to \infty$, (2.22) implies

$$\limsup_{\varepsilon \downarrow 0} \frac{f^* - f_\varepsilon}{\varepsilon} \geq \lim_{k \to \infty} \frac{f^* - f_{d(x_k)}}{d(x_k)} = \infty .$$

This relation contradicts (2.20). The existence of $x^* \in \Omega$ such that

$$\psi(A) = F(x^*, A) = f(x^*) = f^* > -\infty$$

follows from the closedness of Ω and relations (2.11) and (2.13).

Remark 2.1 If condition (2.20) is satisfied, the problem (P) is called *calm*. The function $F(x, A)$ is called a *penalty function.* If $A^* < \infty$ exists such that (2.21) holds then $F(x, A)$ is called an *exact penalty function.* The constraint $d(x) = 0$ is called *functional*, while the constraint $x \in X$ is called *nonfunctional* or *abstract.* Theorem 2.1 means that under the calmness condition (2.20) the constrained optimization problem (P) can be replaced by the following problem:

Find

$$\inf_{x \in X} F(x, A) \tag{$P1$}$$

(where $A \geq A^*$), i.e. the functional constraint $d(x) = 0$ is removed.

In [3],[4] conditions different from (2.20) are stated under which there exists an exact penalty function such that the set of local and global minimizers of problem (P) coincides with the set of local and global minimizers of problem (P1).

3. Mathematical Programming Problems

Now we consider a special case of problem (P). Let Ω be defined by

$$\Omega = \{x \in X \mid h_i(x) \leq 0 \quad \forall i \in I\}, \tag{3.1}$$

where $X \subset \mathbb{R}^n$ is a closed set, $h_i : X \to \mathbb{R}$, $I = 1 : N$. The problem of minimizing a function $f : X \to \mathbb{R}$ on a set Ω described by (3.1) is called a *Mathematical Programming Problem.* Equality-type constraints can also be included in (3.1) (each equality being replaced by two inequalities). There are many ways to describe the set Ω (given by (3.1)) in the form (2.1). For example, take

$$d(x) = d_1(x) = \sum_{i\in I} h_i^+(x) \tag{3.2}$$

where

$$h_i^+(x) = \max\{0, h(x)\}.$$

One can also take

$$d(x) = d_2(x) = \max_{i\in I} h_i^+(x). \tag{3.3}$$

Some properties of functions (3.2), (3.3) (as well as some others) are discussed in [5]. Note that

$$d_1(x) \leq Nd_2(x), \quad d_2(x) \leq d_1(x). \tag{3.4}$$

Take $\varepsilon \geq 0$ and put

$$\Omega_{1\varepsilon} = \{x \in X \mid d_1(x) \leq \varepsilon\}, \quad f_{1\varepsilon}(x) = \inf_{x\in\Omega_{1\varepsilon}} f(x), \tag{3.5}$$

$$\Omega_{2\varepsilon} = \{x \in X \mid d_2(x) \leq \varepsilon\}, \quad f_{2\varepsilon}(x) = \inf_{x\in\Omega_{2\varepsilon}} f(x). \tag{3.6}$$

Observe that

$$\Omega_{2\varepsilon} = \{x \in X \mid h_i(x) \leq \varepsilon \quad \forall i \in I\}.$$

Due to (3.4)

$$\Omega_{2\varepsilon} \subset \Omega_{1\varepsilon}, \quad \Omega_{1\varepsilon} \subset \Omega_{2,N\varepsilon}.$$

Therefore if the calmness condition (2.20) holds for $\Omega_{1\varepsilon}$ described by (3.5), it also holds for $\Omega_{2\varepsilon}$ described by (3.6) (may be with a different constant C).

Let us construct the function

$$\Phi(x, \lambda) = f(x) + \sum_{i\in I} \lambda_i^{(k)} h_i^+(x). \tag{3.7}$$

LEMMA 3.1 (Zero duality gap.) *Let X be a closed set, the functions $f(x)$ and $h_i(x)$ $(i \in I)$ be continuous on X, f be bounded from below on X and, in the case of the unboundedness of X, (2.13) hold. Then*

$$f^* = \inf_{x \in X} \sup_{\lambda \in \Lambda} \Phi(x, \lambda) = \sup_{\lambda \in \Lambda} \inf_{x \in X} \Phi(x, \lambda), \tag{3.8}$$

where

$$\Lambda = \{\lambda = (\lambda_1, ..., \lambda_N) \mid \lambda_i \geq 0 \quad \forall i \in I\}.$$

Proof. Take

$$d(x) = d_1(x) = \sum_{i \in I} h_i^+(x).$$

Put

$$\psi_1(\lambda) = \inf_{x \in X} \Phi(x, \lambda),$$

$$F(x, A) = f(x) + A \sum_{i \in I} h_i^+(x),$$

$$\psi(A) = \inf_{x \in X} F(x, A).$$

By Lemma 2.1

$$f^* = \inf_{x \in X} \sup_{A \geq 0} F(x, A) = \sup_{A \geq 0} \inf_{x \in X} F(x, A) = \psi^*. \tag{3.9}$$

Since

$$F(x, A) = \Phi(x, \lambda_A)$$

where $\lambda_A = (A, ..., A) \in \mathbb{R}^N$ then

$$\psi_1(\lambda_A) = \psi(A),$$

therefore (3.9) yields

$$f^* = \sup_{A \geq 0} \psi(A) \leq \sup_{\lambda \in \Lambda} \psi_1(\lambda) = \psi_1^*. \tag{3.10}$$

(Note that $\lambda_A \in \Lambda$ if $A \geq 0$.) On the other hand,

$$\inf_{x \in X} \sup_{\lambda \in \Lambda} \Phi(x, \lambda) \geq \sup_{\lambda \in \Lambda} \inf_{x \in X} \Phi(x, \lambda) = \psi_1^*. \tag{3.11}$$

As

$$\varphi_1(x) = \sup_{\lambda \in \Lambda} \Phi(x, \lambda) = \begin{cases} f(x) & \forall x \in \Omega, \\ +\infty & \forall x \in X \setminus \Omega, \end{cases}$$

then

$$\inf_{x\in X}\sup_{\lambda\in\Lambda}\Phi(x,\lambda) \;=\; \inf_{x\in X}\varphi_1(x) \;=\; f^*. \tag{3.12}$$

Relations (3.10)–(3.12) imply (3.8).

Let $\lambda^{(1)} \geq \lambda$ (i.e. $\lambda_i^{(1)} \geq \lambda_i \quad \forall i \in I$). Then

$$\Phi(x,\lambda^{(1)}) \;=\; \Phi(x,\lambda) \;+\; \sum_{i\in I}(\lambda_i^{(1)} - \lambda_i)h_i^+(x) \;\geq\; \Phi(x,\lambda). \tag{3.13}$$

Relation (3.13) yields

$$\psi_1(\lambda^{(1)}) = \inf_{x\in X}\Phi(x,\lambda^{(1)}) \geq \psi_1(\lambda) \quad \forall\lambda^{(1)} \geq \lambda. \tag{3.14}$$

THEOREM 3.1 *Let all the hypotheses of Lemma 3.1 hold. If*

$$\limsup_{\varepsilon\downarrow 0}\frac{f^* - f_\varepsilon}{\varepsilon} \leq C < \infty \tag{3.15}$$

where

$$f_\varepsilon = \inf_{x\in\Omega_\varepsilon} f(x), \quad \Omega_\varepsilon = \{x \in X \mid h_i(x) \leq \varepsilon \quad \forall i \in I\}, \quad \varepsilon \geq 0,$$

then there exist $\lambda^* \in \Lambda$ and $x^* \in \Omega$ such that

$$\psi_1(A) \;=\; \inf_{x\in X}\Phi(x,\lambda) \;=\; \Phi(x^*,\lambda) \;=\; f(x^*) \;=\; f^* \quad \forall\lambda \geq \lambda^*. \tag{3.16}$$

Proof. It has already been observed that if condition (3.15) holds then the calmness condition (2.20) is also satisfied. By Theorem 2.1 there exist $A^* < \infty$ and $x^* \in \Omega$ such that (2.21) takes place, i.e.

$$f^* \;=\; F(x^*,A) \;=\; \inf_{x\in X}F(x,A) \;=\; f(x^*) \quad \forall A \geq A^*.$$

Thus, for $\lambda_A = (A, ..., A)$ we get

$$f^* \;= \Phi(x^*,\lambda_A) \;=\; F(x^*,A) \;=\; \inf_{x\in X}F(x,A) \;=\; \inf_{x\in X}\Phi(x,\lambda_A). \tag{3.17}$$

By (3.14) and (3.17)

$$\psi_1(\lambda) = \inf_{x\in X}\Phi(x,\lambda) \;\geq\; \inf_{x\in X}\Phi(x,\lambda_A) \;=\; f^* \quad \forall\lambda \geq \lambda_A. \tag{3.18}$$

Relations (3.8) and (3.18) mean that

$$\psi_1(\lambda) \;=\; f^* \quad \forall\lambda \geq \lambda_A$$

which is equivalent to (3.17) with $\lambda^* = \lambda_A$.

Remark 3.2 Theorem 3.1 states that if the calmness condition (3.15) holds then the problem of minimizing a function f on a set Ω described by (3.1) is equivalent to that of minimizing the function $\Phi(x, \lambda)$ (see (3.7)) on the set X if the vector λ is sufficiently large (i.e., all the components of $\lambda = (\lambda_1, ..., \lambda_N)$ are large enough). In [4, 5] the same result is established under different conditions.

Note that the function Φ is essentially nonsmooth even if all the functions f and h_i's are smooth. Let us consider this case in detail.

4. Lagrange multipliers

Now let us consider problem (P) where

$$\Omega = \{x \in X \mid h_i(x) \le 0 \quad \forall i \in I\}$$

and functions f and h_i's are smooth on $\mathbb{R}^n$. Assume that $\Omega \ne \emptyset$ and the calmness condition (3.15) and the conditions of Lemma 3.1 hold. Then by Theorem 3.1 for λ sufficiently large we have

$$\min_{x \in X} \Phi(x, \lambda) \;=\; \Phi(x^*, \lambda) \;=\; f^* \;=\; \min_{x \in \mathbb{R}^n} f(x) \;=\; f(x^*), \tag{4.1}$$

i.e. the constrained minimization problem (P) is reduced to the problem of minimizing the function $\Phi(x, \lambda)$ on $\mathbb{R}^n$. The function

$$\Phi(x, \lambda) \;=\; f(x) + \sum_{i \in I} \lambda_i h_i^+(x) \;=\; f(x) + \sum_{i \in I} \lambda_i(x) \max\{0, h_i(x)\}$$

is nonsmooth in x. This function is subdifferentiable, i.e. it is directionally differentiable and its directional derivative $\Phi'(x, \lambda; g)$ is of the form

$$\Phi'(x, \lambda; g) \;=\; \lim_{\alpha \downarrow 0} \frac{1}{\alpha} \Big[\Phi(x + \alpha g, \lambda) - \Phi(x, \lambda) \Big] \;=\; \max_{v \in \partial\Phi(x, \lambda)} (v, g)$$

where $\partial\Phi(x, \lambda)$ is the subdifferential of Φ at x (see, e.g., [6, 7])

$$\partial\Phi(x, \lambda) \;=\; \{f'(x)\} + \sum_{i \in I} \lambda_i \partial h_i^+(x), \tag{4.2}$$

$$\partial h_i^+(x) \;=\; \begin{cases} \{0_n\}, & h_i(x) < 0; \\ h_i'(x), & h_i(x) > 0; \\ \mathrm{co}\,\{0_n, h_i'(x)\}, & h_i(x) = 0. \end{cases}$$

Here $h_i'(x)$ is the gradient of h_i at x.

Let $\lambda^* = (\lambda_1^*, ..., \lambda_N^*)$ be such that (4.1) holds for $\lambda = \lambda^*$. By the necessary condition for a minimum (see [6, 7])

$$0_n \in \partial\Phi(x^*, \lambda^*). \tag{4.3}$$

Condition (4.3) implies that there exist coefficients $\alpha_i \in [0,1]$ $(i \in I)$ such that

$$f'(x^*) + \sum_{i \in I} \lambda_i^* \alpha_i h_i'(x^*) = 0_n. \tag{4.4}$$

Put $\overline{\lambda}_i = \alpha_i \lambda_i^* \quad \forall i \in I$. Condition (4.4) means that

$$L_x'(x^*, \overline{\lambda}) = 0_n \tag{4.5}$$

where

$$L(x, \lambda) = f(x) + \sum_{i \in I} \lambda_i h_i(x), \qquad \overline{\lambda} = (\overline{\lambda}_1, ..., \overline{\lambda}_N). \tag{4.6}$$

The function $L(x, \lambda)$ defined by (4.6) is called the *Lagrange function* related to problem (P) (or just the *Lagrangian*). Condition (4.5) implies that there exist coefficients (called *Lagrange multipliers*) $\overline{\lambda}_1, ..., \overline{\lambda}_N$ such that the minimizer x^* of f on Ω is a stationary point of $L(x, \overline{\lambda})$ (to be more precise, x^* is an inf-stationary point of $L(x, \overline{\lambda})$ at x^*, i.e. at x^* the necessary condition for a minimum of $L(x, \overline{\lambda})$ holds, but since L is smooth, the necessary conditions for a minimum and a maximum coincide). The following example shows that x^* may be not a minimizer of L but its maximizer.

EXAMPLE 4.1 Let

$$x \in \mathbb{R}, \ f(x) = \sqrt{x^2 + a}, \ a > 0, \ \Omega = \{x \in \mathbb{R} \mid h(x) = 1 - x^2 \leq 0\}.$$

Clearly,

$$\min_{x \in \Omega} f(x) = f(x^*) = \sqrt{1 + a}$$

where $x^* = +1$ or $x^* = -1$.

Let

$$\Omega_\varepsilon = \{x \in \mathbb{R} \mid h(x) \leq \varepsilon\}.$$

It is easy to see that

$$f_\varepsilon = \min_{x \in \Omega_\varepsilon} f(x) = \sqrt{1 + a - \varepsilon}.$$

Since

$$\frac{f^* - f_\varepsilon}{\varepsilon} = \frac{\sqrt{1+a} - \sqrt{1+a-\varepsilon}}{\varepsilon} \underset{\varepsilon \to 0}{\longrightarrow} \frac{1}{2(1+a)} < \infty$$

then the calmness condition (3.15) is satisfied and Theorem 3.1 holds. Therefore for any sufficiently large A

$$\min_{x\in\mathbb{R}} \Phi(x, A) \;=\; \min_{x\in\mathbb{R}} [f(x) + A\max\{0, h(x)\}] \;=\; \Phi(x^*, A).$$

Take the minimizer $x^* = +1$. We have

$$\partial\Phi(x^*, A) \;=\; \{f'(x^*)\} + A\, co\{0, h'(x^*)\} \;=\; \{\frac{1}{\sqrt{1+a}} + A\, co\{0, -2\}\}.$$

It follows from (4.3) that

$$0 \;=\; f'(x^*) + \frac{1}{2\sqrt{1+a}} h'(x^*),$$

i.e. $\lambda = \frac{1}{2\sqrt{1+a}}$. Thus, the Lagrangian is

$$L(x, \lambda) \;=\; f(x) + \frac{1}{2\sqrt{1+a}} h(x) = \sqrt{x^2+a} + \frac{1}{2\sqrt{1+a}}(1 - x^2).$$

The minimizer $x^* = -1$ provides the same Lagrangian. (In general it is not true: different minimizers produce different Lagrangians.)

For stationary points of $L(x, \lambda)$:

$$L'_x(x, \lambda) \;=\; \frac{x}{\sqrt{x^2+a}} - \frac{x}{\sqrt{1+a}} = x[\sqrt{1+a} - \sqrt{x^2+a}].$$

The equation $L'_x(x, \lambda) \;=\; 0$ has the following solutions:

$$x_1 = 0, \quad x_2 = +1, \quad x_3 = -1.$$

The point $x_1 = 0$ doesn't belong to Ω, the points $x_2 = +1$ and $x_3 = -1$ are maximizers of $L(x, \lambda)$, not minimizers.

Remark 4.1 Thus, if the functions f and h_i's are smooth on $\mathbb{R}^n$ and the abstract constraint is absent (i.e. $X = \mathbb{R}^n$) then there exist coefficients $\lambda_1, ..., \lambda_N$ such that $\lambda_i \geq 0\ \forall i \in I$ and the minimizer x^* of the problem (P) (to be more precise, the minimizer x^* of the function f on the set Ω) is a stationary point of the Lagrangian $L(x, \lambda) = f(x) + \sum \lambda_i h_i$ (and the function $L(x, \lambda)$ is smooth in x).

Let us consider problem (P) where Ω is given by (3.1) and the functions f and h_i's are again smooth on X. Assume also that $\Omega \neq \emptyset$ and the calmness condition (3.15) holds. By Theorem 3.1 if coefficients $\lambda_1, ..., \lambda_N$ are large enough then the constrained minimization problem (P) is reduced to the problem of minimizing the function $\Phi(x, \lambda) =$

$f(x)+\sum \lambda_i h_i^+(x)$ on the set X. The function $\Phi(x,\lambda)$ is subdifferentiable with the subdifferential $\partial\Phi(x,\lambda)$ (see (4.2)). Let $x^* \in \Omega$ be a minimizer of f on Ω and λ^* be "large enough". Then, by Theorem 3.1, x^* is a minimizer of $\Phi(x,\lambda^*)$ on X. By the necessary condition for a minimum

$$\Phi'(x,\lambda^*;g) \geq 0 \quad \forall g \in \Gamma(x^*) \tag{4.7}$$

where $\Gamma(x^*)$ is the Bouligand cone to X at x^*:

$$\Gamma(x^*) = \{v = \mu g \mid \exists \{x_i\} : x_i \in X, x_i \neq x_i^*,\ g_i = \frac{x_i - x^*}{||x_i - x^*||} \to g,\ \mu \geq 0\}.$$

A point x^* satisfying (4.7) is called an *inf-stationary point* of Φ on X. The cone $\Gamma(x^*)$ can always be represented in the form

$$\Gamma(x^*) = \bigcup_{A \in \mathcal{A}(x^*)} A \tag{4.8}$$

where

$$0_n \in A,\ A \subset \Gamma(x^*) \quad \forall A \in \mathcal{A}(x^*), \tag{4.9}$$

$\mathcal{A}(x^*)$ is a family of convex closed sets. Such a family always exists. For example, one can take

$$\mathcal{A}(x^*) = \{l \in \mathbb{R}^n \mid l = \{g = \mu g_0 | \mu \geq 0\},\ g_0 \in \Gamma(x^*),\ ||g_0|| = 1\}.$$

The condition (4.7) is equivalent to (see [6], Lemma V.2.2)

$$\partial\Phi(x^*,\lambda^*) \bigcap A^+ \neq \emptyset \quad \forall A \in \mathcal{A}(x^*) \tag{4.10}$$

where A^+ is the cone conjugate to A:

$$A^+ = \{v \in \mathbb{R}^n \mid (v,g) \geq 0 \quad \forall g \in A\}.$$

It follows from (4.2) and (4.10) that for every $A \in \mathcal{A}(x^*)$ there exists a set of coefficients $\bar{\lambda}_1(A), ..., \bar{\lambda}_N(A)$ (depending on A) such that

$$f'(x) + \sum_{i \in I} \bar{\lambda}_i(A) h_i'(x^*) \in A^+. \tag{4.11}$$

It means that the point x^* is an inf-stationary point of the function $L(x,\bar{\lambda}(A)) = f(x) + \sum_{i\in I} \bar{\lambda}_i(A) h_i(x)$ on the set $x^* + A$. If X is a convex set then $\Gamma(x^*)$ is convex and (4.7) is equivalent to the condition

$$\partial\Phi(x^*,\lambda^*) \bigcap \Gamma^+(x^*) \neq \emptyset.$$

In this case x^* is an inf-stationary point of $L(x, \bar{\lambda}(A))$ on $x^* + \Gamma(x^*)$ or, what is equivalent, on X.

Thus, if X is convex, there exists a Lagrangian $L(x, \bar{\lambda})$ such that the minimizer x^* of f on Ω is an inf-stationary point of the function $L(x, \bar{\lambda})$ on X. If X is not convex, there exist a family of convex closed cones $\mathcal{A}(x^*)$ and a bundle of Lagrangians $L(x, \bar{\lambda}(A))$ such that for every $A \in \mathcal{A}(x^*)$ the point x^* is an inf-stationary point of the function $L(x, \bar{\lambda}(A))$ on the set $x^* + A$.

If $M(A)$ is the set of inf-stationary points of $L(x, \bar{\lambda}(A))$ on $x^* + A$ then

$$x^* \in \bigcap_{A \in \mathcal{A}(x^*)} M(A). \tag{4.12}$$

It will be shown in Example 4.2 that, in general, there exists no Lagrangian $L(x, \lambda)$ such that x^* is an inf-stationary point of the function $L(x, \lambda)$ on X.

The representation (4.8) is not unique. From the practical standpoint we are interested to have in $\mathcal{A}(x^*)$ (see (4.8)) as few elements as possible.

EXAMPLE 4.2 Let $x = (x_1, x_2) \in \mathbb{R}^2$,

$$X = \{x \in \mathbb{R}^2 \mid x \notin X_1\}, \ X_1 = \{x \in \mathbb{R}^2 \mid x_1 > 0, \ x_2 > 0\},$$

$$\Omega \ = \ \{x \in X \mid x_1 - x_2 = 0\}, \quad f(x) = -x_1 - x_2.$$

It is clear that

$$\Omega \ = \ \{x \in X \mid h_1(x) = x_1 - x_2 \leq 0, \ h_2(x) = x_2 - x_1 \leq 0\}.$$

It is easy to see that

$$\min_{x \in \Omega} f(x) = f(x^*) = f^* = 0$$

where $x^* = (0, 0)$. Let us check condition (3.15). We have

$$\Omega_\varepsilon \ = \ \{x \in X \mid h_1(x) \leq \varepsilon, \ h_2(x) \leq \varepsilon\},$$

$$f_\varepsilon = \min_{x \in \Omega_\varepsilon} f(x) = f(x_\varepsilon) = -\varepsilon, \quad x_\varepsilon = (0, \varepsilon), \quad \frac{f^* - f_\varepsilon}{\varepsilon} = 1.$$

Thus, (3.15) holds, therefore Theorem 3.1 is valid. In our case $\Gamma(x^*) = X$. Clearly, $\Gamma(x^*) = A_1 \cup A_2$ where

$$A_1 = \{x = (x_1, x_2) \mid x_1 \leq 0\}, \quad A_2 = \{x = (x_1, x_2) \mid x_2 \leq 0\}.$$

The cones A_1 and A_2 are convex and closed and

$$A_1^+ = \{v = (-\lambda, 0) \mid \lambda \geq 0\}, \quad A_2^+ = \{v = (0, -\lambda) \mid \lambda \geq 0\}.$$

Take A_1 and find coefficients $\bar{\lambda}_1$ and $\bar{\lambda}_2$ satisfying (4.11). Since

$$f'(x^*) = (-1, -1),\ h_1'(x^*) = (1, -1),\ h_2'(x^*) = (-1, 1)$$

then (4.11) is equivalent to

$$f'(x^*) + \lambda_1 h_1'(x^*) + \lambda_2 h_2'(x^*) \in A_1^+ \ = \ \{v = (-\lambda, 0) \mid \lambda > 0\}. \quad (4.13)$$

Find λ, λ_1, and λ_2 satisfying (4.13):

$$\begin{cases} -1 + \lambda_1 - \lambda_2 = -\lambda, \\ -1 - \lambda_1 + \lambda_2 = 0. \end{cases}$$

This system (w.r. to λ, λ_1, and λ_2) has the following set of solutions:

$$\{(\lambda, \lambda_1, \lambda_2) \mid \lambda = 2, \lambda_2 = \lambda_1 + 1\}. \quad (4.14)$$

The corresponding Lagrangian is (since $h_2(x) = -h_1(x)$)

$$\begin{aligned} L_1(x) &= f(x) + \lambda_1 h_1(x) + \lambda_2 h_2(x) = f(x) - h_1(x) = \\ &= -x_1 - x_2 - (x_1 - x_2) = -2x_1 . \end{aligned}$$

Since $L_1(x)$ is convex, the set of inf-stationary points of the function $L_1(x)$ on the set $x^* + A_1 = A_1$ coincides with the set M_1 of minimizers of $L_1(x)$ on A_1:

$$M_1 \ = \ \{(0, x_2) \mid x_2 \in \mathbb{R}\}.$$

Analogously, for A_2 we find that (4.11) is equivalent to

$$f'(x^*) + \lambda_1 h_1'(x^*) + \lambda_2 h_2'(x^*) \in A_2^+ \ = \ \{v = (0, -\lambda) \mid \lambda > 0\}. \quad (4.15)$$

Find λ, λ_1, and λ_2 satisfying (4.13):

$$\begin{cases} -1 + \lambda_1 - \lambda_2 = 0, \\ -1 - \lambda_1 + \lambda_2 = -\lambda. \end{cases}$$

The set of solutions of this system is:

$$\{(\lambda, \lambda_1, \lambda_2) \mid \lambda = 2, \lambda_2 = \lambda_1 - 1\}. \quad (4.16)$$

The corresponding Lagrangian is

$$L_2(x) \ = \ f(x) + \lambda_1 h_1(x) + \lambda_2 h_2(x) \ = \ f(x) + h_1(x) = -2x_2.$$

Since $L_2(x)$ is convex, the set of inf-stationary points of the function $L_2(x)$ on the set $x^* + A_2 = A_2$ coincides with the set M_2 of minimizers of $L_2(x)$ on A_2:

$$M_2 = \{(x_1, 0) \mid x_1 \in \mathbb{R}\}.$$

Applying (4.12) we conclude that

$$x^* \in M_1 \cap M_2 = \{(0,0)\}.$$

Now let us try to find a Lagrangian

$$L(x, \lambda) = f(x) + \lambda_1 h_1(x) + \lambda_2 h_2(x)$$

such that x^* is an inf-stationary point of $L(x, \lambda)$ on X. By (4.11) it should be

$$L'(x^*, \lambda) \in A_1^+, \tag{4:17}$$

$$L'(x^*, \lambda) \in A_2^+. \tag{4.18}$$

Inclusions (4.17) and (4.18) imply (see (4.14) and (4.16)) that λ_1 and λ_2 must satisfy the relations

$$\begin{cases} \lambda_2 = \lambda_1 + 1, \\ \lambda_2 = \lambda_1 - 1 \end{cases}$$

which are inconsistent.

Thus, there exists no Lagrangian $L(x, \lambda)$ such that x^* is an inf-stationary point of $L(x, \lambda)$ on X.

Remark 4.2 Above we considered the function $L(x, \lambda)$ which is a linear function of f and h_i's (see (4.6)). A question may be raised whether a smooth function

$$H(f, h_1, ..., h_N) = f + H_1(h_1, ..., h_N) \tag{4.19}$$

exists such that the minimizer x^* of f on the set Ω (given by (3.1)) is an inf-stationary point of the function $\mathcal{H}(x) = H(f(x), h_1(x), ..., h_N(x))$ on the set X. Let us try to find such a function for the problem discussed in Example 4.2. We have $\mathcal{H}(x) = f(x) + H_1(f(x), h_1(x), ..., h_N(x))$ where $f(x) = -x_1 - x_2$, $h_1(x) = x_1 - x_2$, $h_2(x) = x_2 - x_1$. Arguing as above (see (4.13) and (4.15)) one concludes that

$$\mathcal{H}'(x^*) \in A_1^+, \tag{4.20}$$

$$\mathcal{H}'(x^*) \in A_2^+. \tag{4.21}$$

Here $\mathcal{H}'(x^*) = f'(x^*) + a_1 h_1'(x^*) + a_2 h_2'(x^*)$ where

$$f'(x^*) = (-1, -1), \ h_1'(x^*) = (1, -1), \ h_2'(x^*) = (-1, 1), \tag{4.22}$$

$$a_1 = \frac{\partial H_1(h_1(x^*), h_2(x^*))}{\partial h_1}, \quad a_2 = \frac{\partial H_1(h_1(x^*), h_2(x^*))}{\partial h_2}.$$

Inclusion (4.20) implies that there exists $\lambda_1 \geq 0$ such that

$$\mathcal{H}'(x^*) = f'(x^*) + a_1 h_1'(x^*) + a_2 h_2'(x^*) = (-\lambda_1, 0).$$

Taking into account (4.22) we get the system

$$\begin{cases} -1 + a_1 - a_2 = -\lambda_1, \\ -1 - a_1 + a_2 = 0 \end{cases}$$

which yields

$$\lambda_1 = 2, \ a_1 - a_2 = -1. \tag{4.23}$$

Analogously, inclusion (4.21) implies that there exists $\lambda_2 \geq 0$ such that $\mathcal{H}'(x^*) = (0, -\lambda_2)$ and we get the system

$$\begin{cases} -1 + a_1 - a_2 = 0, \\ -1 - a_1 + a_2 = -\lambda_2 \end{cases}$$

which yields

$$\lambda_2 = 2, \ a_1 - a_2 = 1. \tag{4.24}$$

The relations (4.23) and (4.24) are incompatible.

This example demonstrates that, in general, there exists no smooth function $H(f, h_1, ..., h_N)$ of the form (4.19) such that the minimizer of f on Ω is an inf-stationary point of the function

$$\mathcal{H}(x) = H(f(x), h_1(x), \ldots, h_N(x))$$

on the set X.

Remark 4.3 A detailed discussion of smooth Mathematical Programming Problems with smooth "functional" constraint and convex "abstract" constraints is presented in [1] where an extended Bibliography is also attached. Nonlinear Lagrangians in Nonsmooth Mathematical Programming are studied in [2] where different duality results are stated as well.

5. Acknowledgment

The author is grateful to B. D. Craven and A. M. Rubinov for helpful discussions in St. Petersburg (Russia) and Ballarat (Australia).

References

1. Rockafellar R. T. (1993), Lagrange multipliers and Optimality. *SIAM Review*, vol. 35, No. 2, pp. 183-238.
2. Rubinov A. M., Glover B. M., Yang X. Q. (1997), Extended Lagrange and Penalty Functions in Continuous Optimization. *Research Report. University of Ballarat, School of Information Technology and Math Sciences. Australia* No. 19/97, 20 pp.
3. Demyanov V. F. (1994), Exact penalty functions in problems of nonsmooth optimization. *Vestnik of St. Petersburg University*, Ser.1, No 4(22), pp. 21-27.
4. Demyanov V. F., Di Pillo G. and Facchinei F.(1998), Exact penalization via Dini and Hadamard conditional derivatives. *Optimization Methods and Software*, Vol. 9, pp. 19-36.
5. Di Pillo G., Facchinei F., Exact penalty functions for nondifferentiable programming problems. *Eds. F.H. Clarke, V.F. Demyanov and F. Giannessi: Nonsmooth Optimization and Related Topics*. N. Y.: Plenum, 1989. pp. 89-107.
6. Demyanov V. F. and Rubinov A. M. (1995), *Constructive Nonsmooth Analysis*. Frankfurt a/M, Verlag Peter Lang.
7. Rockafellar R. T. (1970), *Convex Analysis*. Princeton, N. J., Princeton University Press.

A Nontopological Two-function Minimax Theorem with Monotone Transformations of the Functional Values

Ferenc Forgó
Budapest University of Economic Sciences
Department of Operations Research
H-1828 Budapest, Hungary
Email: forgo@nero.bke.hu

Abstract. A two-function minimax inequality with the following characteristics is proved: (i) the function on the less-side of the inequality is generalized concave in its first variable while the function on the greater-side is generalized convex in its second variable, (ii) generalized convexity/concavity is defined in terms of averages of monotone transformations of values of both functions, (iii) the proof is elementary, neither separation nor fixed-point theorems are used.

1. Introduction

Let X and Y be nonempty sets, $f, g : X \times Y \longrightarrow \mathbb{R}$ functions with $f \le g$, f bounded. A nontopological two-function minimax theorem asserts that the inequality

$$\inf_Y \max_F f(x,y) \le \sup_X \inf_Y g(x,y) \tag{1}$$

holds for any finite set $F \subset X$.

Inequality (1) has been proved under various conditions each of which is some generalization of convexity and/or (nontopological) connectedness. Good markers in the jungle of different results are Simons' papers [5], [6], [7]. In [6], Simons suggests that the character of theorems and their proofs is inherently different depending on the geometry of function f and g. If f is (generalized) concave in its first variable and g is (generalized) convex in its second variable, then fixed-point type methods seem to be needed for the proof while the opposite case when f is convex in its second variable and g is concave in its first variable calls for Hahn-Banach type separation. Lin and Quan [4] were the first to prove (1) in the first case without recourse to any kind of fixed-point related ideas. The price to pay for being able to do so is to "mix up" the functional values of f and g in the (generalized) convexity/concavity conditions designed to make (1) hold. Lin and Quan's result [4] was generalized by Forgó and Joó [1], allowing Irle's [3] averaging functions to replace the special average Lin and Quan [4] used in their conditions.

F. Giannessi et al. (eds.), Optimization Theory, 99–110.

In this paper, the main result of Forgó and Joó [1] will be further generalized in a new direction. Our minimax theorem remains to be given in terms of functions f and g but the (generalized) convexity/concavity conditions will also involve monotone transforms of f and g. Mixing values of f and g in the conditions is still required but the freedom to choose from a large family of monotone transformations greatly extends the power of the theorem.

The proof of the theorem goes along the lines of the proof in [1]. Since great care must be taken where and when formulas and/or argumentation have to be changed or adjusted to the new conditions, we present the proof in its entirety. Apology is due for the unavoidable duplications of certain parts of the proof in [1].

Two-function minimax inequalities can be used for proving the existence of solutions to variational inequalities, the existence of Nash equilibria in normal-form noncooperative games and fixed-point theorems. Simons' paper [6] is an excellent reference in this regard. Some new ideas on the classification and applicability of two-function minimax theorems can also be found in Forgó [2].

2. The main result

Averaging functions introduced by Irle [3] will play an important role in the two-function minimax theorem we are going to prove. A continuous function $\rho : \mathbb{R}^2 \longrightarrow \mathbb{R}$ is said to be an averaging function if it is increasing in each variable and

$$\rho(\lambda,\lambda)=\lambda, \quad \lambda\neq\mu \Longrightarrow \min\{\lambda,\mu\}<\rho(\lambda,\mu)<\max\{\lambda,\mu\}. \tag{2}$$

Just as in [1], the proof of (1) will be based on a lemma.

LEMMA 1. *Let ρ_1, ρ_2 be two averaging functions, X, Y nonempty sets, $f, g : X \times Y \longrightarrow \mathbb{R}$ such that*

$$f \le g \quad \text{on} \quad X\times Y, \quad f \text{ is bounded}. \tag{3}$$

Let furthermore

$$\beta := \sup_X \inf_Y g(x,y),$$

$\varphi, \psi : \mathbb{R} \longrightarrow \mathbb{R}$ monotone increasing continuous functions which satisfy

$$\varphi(c) \ge c, \quad \text{for all} \quad c \ge \beta \tag{4}$$

$$\psi(d) \le d, \quad \text{for all} \quad d \ge \beta. \tag{5}$$

If for any $x_1, x_2 \in X$*, there exists* $x_3 \in X$ *to satisfy*

$$y \in Y \Longrightarrow \psi(f(x_3, y)) \geq \rho_1(\psi(f(x_1, y)), g(x_2, y)) \tag{6}$$

and for any $y_1, y_2 \in Y$*, there exists* $y_3 \in Y$ *to satisfy*

$$x \in X \Longrightarrow \varphi(g(x, y_3)) \leq \rho_2(f(x, y_1), \varphi(g(x, y_2)), \tag{7}$$

then for any $x, w \in X$*, we have*

$$\inf_Y \max \{f(x, y), g(w, y)\} \leq \sup_X \inf_Y g(x, y). \tag{8}$$

Proof. By way of contradiction, assume that for some $x_1, w_1 \in X$

$$\alpha := \inf_Y \max \{f(x_1, y), g(w_1, y)\} > \sup_X \inf_Y g(x, y). \tag{9}$$

Starting with (x_1, w_1), define the sequence $\{(x_n, w_n)\}$ inductively as follows. Assume that x_n and w_n have already been constructed and

$$\inf_Y \max \{f(x_n, y), g(w_n, y)\} \geq \alpha. \tag{10}$$

Define x^* by

$$\psi(f(x^*, y)) \geq \rho_1(\psi(f(x_n, y)), g(w_n, y)). \tag{11}$$

By assumption (6) such an x^* exists. Let γ be any number to satisfy

$$\alpha < \gamma < \beta. \tag{12}$$

Introduce the notation

$$N_\gamma f(x) := \{y \in Y : f(x, y) \leq \gamma\}. \tag{13}$$

First we show that

$$\text{either} \quad N_\gamma g(x^*) \subset N_\gamma f(x_n) \quad \text{or} \quad N_\gamma g(x^*) \subset N_\gamma g(w_n). \tag{14}$$

Since $\gamma > \beta$, we have $N_\gamma g(x) \neq \emptyset$ for any $x \in X$ which implies by $g \geq f$ that $N_\gamma f(x) \neq \emptyset$ for all $x \in X$. From $\alpha > \gamma$ and (10), it follows that

$$N_\gamma f(x_n) \cap N_\gamma g(w_n) = \emptyset. \tag{15}$$

Now we prove that

$$N_\gamma g(x^*) \subset N_\gamma f(x^*) \subset N_\gamma f(x_n) \cup N_\gamma g(w_n). \tag{16}$$

The first inclusion follows from $g \geq f$. Suppose that the second inclusion does not hold, that is, there is a $y \in Y$ for which

$$f(x^*, y) \leq \gamma\,, \quad f(x_n, y) > \gamma > \beta\,, \quad g(w_n, y) > \gamma > \beta\,. \tag{17}$$

By (5) and (11), and since ρ_1 is an averaging function, we get

$$\psi(f(x^*, y)) \geq \rho_1(\psi(f(x_n, y)), g(w_n, y)) \geq \rho_1(\psi(f(x_n, y)), \psi(g(w_n, y)))\,.$$

This means that $\psi(f(x^*, y)) < \psi(f(x_n, y))$ and $\psi(f(x^*, y)) < \psi(g(w_n, y))$ cannot both hold, or equivalently $f(x^*, y) < f(x_n, y)$ and $f(x^*, y) < g(w_n, y)$ cannot both hold which contradicts (17).

Suppose now that (14) does not hold. Then, by (16), there exist z_1 and z_2 such that

$$z_1 \in N_\gamma g(x^*) \cap N_\gamma f(x_n)\,, \quad z_2 \in N_\gamma g(x^*) \cap N_\gamma g(w_n)\,. \tag{18}$$

Beginning with $m = 3$, we define z_m for all $m \geq 3$ by induction, as follows.

Denote i_m (respectively j_m) the largest index less than m for which $z_{i_m} \in N_\gamma f(x_n)$, (respectively $z_{j_m} \in N_\gamma g(w_n)$). Define z_m by (7), that is, for all $x \in X$ we have

$$\varphi(g(x, z_m)) \leq \rho_2\left(f(x, z_{i_m}), \varphi(g(x, z_{j_m}))\right)\,. \tag{19}$$

Because $\gamma > \beta$, by condition (4) and the monotonicity of φ and ρ_2, we have for all $x \in X$

$$\varphi(g(x, z_m)) \leq \rho_2\left(f\left(x, z_{i_m}\right), \varphi\left(g(x, z_{j_m})\right)\right) \leq \rho_2(\gamma, \varphi(\gamma)) \leq \varphi(\gamma)$$

which implies $g(x, z_m) \leq \gamma$. Thus $g(x^*, z_m) \leq \gamma$, that is, $z_m \in N_\gamma g(x^*)$ for all m. Then, by (14) and (15)

$$\text{either} \quad z_m \in N_\gamma f(x_n) \quad \text{or} \quad z_m \in N_\gamma g(w_n)\,. \tag{20}$$

Suppose first that there are infinitely many m with $z_m \in N_\gamma f(x_n)$. Denote $\{z_{m_k}\}$ the subsequence of these. From $f\left(x_n, z_{m_{k+1}}\right) \leq \gamma < \alpha$, taking into account inequalities (10) and (19), we get by the monotonicity of φ and ρ_2 that

$$\alpha \leq g\left(w_n, z_{m_{k+1}}\right) \leq \varphi^{-1}\left(\rho_2\left(f\left(w_n, z_{m_k}\right), \varphi\left(g\left(w_n, z_{j_{m_{k+1}}}\right)\right)\right)\right)$$

and consequently

$$\varphi(\alpha) \leq \varphi\left(g\left(w_n, z_{m_{k+1}}\right)\right) \leq \rho_2\left(f\left(w_n, z_{m_k}\right), \varphi\left(g\left(w_n, z_{j_{m_{k+1}}}\right)\right)\right)\,.$$

Since $\varphi\left(g\left(w_n, z_{j_{m_{k+1}}}\right)\right) \leq \varphi(\gamma)$, therefore

$$f(w_n, z_{m_k}) \geq \varphi\left(g\left(w_n, z_{m_{k+1}}\right)\right)$$

which implies by assumption (4) that

$$f(w_n, z_{m_k}) \geq g\left(w_n, z_{m_{k+1}}\right) \geq f\left(w_n, z_{m_{k+1}}\right),$$

that is, the sequence $\{f(w_n, z_{m_k})\}$ is monotone decreasing as $k \to \infty$ and is bounded from below by $\varphi(\alpha)$. Denote

$$A := \lim_{k\to\infty} f(w_n, z_{m_k}) \geq \varphi(\alpha) \geq \alpha\,.$$

Then, we have

$$\begin{aligned}
&f\left(w_n, z_{m_{k+1}}\right) \leq g\left(w_n, z_{m_{k+1}}\right) \leq \\
&\varphi^{-1}\left(\rho_2\left(f(w_n, z_{m_k}), \varphi\left(g\left(w_n, z_{j_{m_{k+1}}}\right)\right)\right)\right) \leq \\
&\varphi^{-1}\left(\rho_2\left(f(w_n, z_{m_k}), \varphi(\gamma)\right)\right).
\end{aligned}$$

Hence

$$\varphi\left(f\left(w_n, z_{m_{k+1}}\right)\right) \leq \rho_2\left(f(w_n, z_{m_k}), \varphi(\gamma)\right) \leq \rho_2\left(\varphi\left(f(w_n, z_{m_k})\right), \varphi(\gamma)\right).$$

Taking the limit as $k \to \infty$, we get

$$\varphi(A) \leq \rho_2(\varphi(A), \varphi(\gamma))$$

which is a contradiction since $\varphi(A) \geq \varphi(\alpha) > \varphi(\gamma)$ and ρ_2 is an averaging function.

This shows that there are no infinitely many $z_m \in N_\gamma f(x_n)$. Hence, for some M, for all $m \geq M$, we have $z_m \in N_\gamma g(w_n)$ and for some $m_0 < M$ (m_0 is the largest index for which $f(x_n, z_{m_0}) \leq \gamma$)

$$\varphi(g(x, z_{m+1})) \leq \rho_2\left(f(x, z_{m_0}), \varphi(g(x, z_m))\right)$$

holds for all $x \in X$.

Now $g(w_n, z_m) \leq \gamma$ implies $f(x_n, z_m) \geq \alpha$ by (10) which, in turn, results in

$$\varphi(\alpha) \leq \varphi(g(x_n, z_{m+1})) \leq \rho_2(\gamma, \varphi(g(x_n, z_m)))$$

for all $m \geq M$. Hence $g(x_n, z_{m+1}) \leq g(x_n, z_m)$, that is, the sequence $\{g(x_n, z_m)\}$ is decreasing as $m \to \infty$ and is bounded from below by $\varphi(\alpha)$. This results in a contradiction in the same way as it did in the case of the sequence $\{f(w_n, z_{m_k})\}$.

Under the hypothesis that (14) is not true, the generation of the (not necessarily distinct) points z_m would go indefinitely. But we have just shown that this is impossible, therefore (14) must hold true.

Secondly, based on (14), we will show that

$$\text{either} \quad N_\gamma f(x^*) \subset N_\gamma f(x_n) \quad \text{or} \quad N_\gamma g(x^*) \subset N_\gamma g(w_n) . \tag{21}$$

For this, it is enough to prove $N_\gamma g(x^*) \subset N_\gamma f(x_n) \Longrightarrow N_\gamma f(x^*) \subset N_\gamma f(x_n)$ since the reverse implication is always true by $g \geq f$. Since, by (16), $N_\gamma f(x^*) \subset N_\gamma f(x_n) \cup N_\gamma g(w_n)$, the indirect assumption $N_\gamma f(x^*) \not\subset N_\gamma f(x_n)$ implies the existence of a $z_2 \in N_\gamma f(x^*) \cap N_\gamma g(w_n)$.

Take any element $z_1 \in N_\gamma g(x^*) \subset N_\gamma f(x_n)$ and define the sequence $\{z_m\}$ by induction, using (7), as follows:

$$\varphi(g(x, z_3)) \leq \rho_2(f(x, z_2), \varphi(g(x, z_1))) ,$$
$$\varphi(g(x, z_{m+1})) \leq \rho_2(f(x, z_2) , \varphi(g(x, z_m))) , \quad (m \geq 3)$$

for all $x \in X$.

Since $f(x^*, z_2) \leq \gamma$ and $g(x^*, z_1) \leq \gamma$, we get

$$\varphi(g(x^*, z_3)) \leq \rho_2(\gamma, \varphi(\gamma)) \leq \varphi(\gamma)$$

implying $g(x^*, z_3) \leq \gamma$, and by induction on m, $g(x^*, z_{m+1}) \leq \gamma$. This means that $z_{m+1} \in N_\gamma g(x^*) \subset N_\gamma f(x_n)$ and then, by (10), we have $g(w_n, z_{m+1}) \geq \alpha$.

Hence

$$\varphi(\alpha) \leq \varphi(g(w_n, z_{m+1})) \leq \rho_2(f(w_n, z_2), \varphi(g(w_n, z_m))) , \quad (m \geq 3) . \tag{22}$$

Since $z_2 \in N_\gamma g(w_n)$, we have $f(w_n, z_2) \leq g(w_n, z_2) \leq \gamma$ and then from (22) we deduce that $\{g(w_n, z_2)\}$ is a decreasing sequence for $m \geq 4$. Let $\alpha \leq B := \lim_{m \to \infty} g(w_n, z_m)$. Then, from (22) we get

$$\varphi(B) \leq \rho_2(\gamma, \varphi(B)) \leq \rho_2(\varphi(\gamma), \varphi(B)) ,$$

a contradiction. Hence (21) is true.

Now we turn to the third and final part of the proof. If

$$N_\gamma f(x^*) \subset N_\gamma f(x_n) ,$$

then let $(x_{n+1}, w_{n+1}) := (x^*, w_n)$, and if

$$N_\gamma g(x^*) \subset N_\gamma g(w_n) ,$$

then let $(x_{n+1}, w_{n+1}) := (x_n, x^*)$.

In the first case, $y \in N_\gamma f(x^*)$ implies $y \in N_\gamma f(x_n)$, thus $g(w_n, y) \geq \alpha$ by (10). Hence

$$\inf_Y \max\{f(x^*, y), g(w_n, y)\} = \inf_Y \max\{f(x_{n+1}, y), g(w_{n+1}, y)\} \geq \gamma. \tag{23}$$

Since the choice of x^* does not depend on γ and (23) holds for any $\gamma < \alpha$, we have

$$\inf_Y \max\{f(x_{n+1}, y), g(w_{n+1}, y)\} \geq \alpha.$$

Similar reasoning leads us to the above inequality in the second case, as well.

By the construction of the sequence $\{x_n, w_n\}$, we obtain from (21)

$$N_\gamma f(x_{n+1}) \subset N_\gamma f(x_n), \quad N_\gamma g(w_{n+1}) \subset N_\gamma g(w_n), \quad (n \geq 1).$$

If $N_\gamma f(x^*) \subset N_\gamma f(x_n)$, then $\psi(f(x_{n+1}, y)) \geq \rho_1(\psi(f(x_n, y)), g(w_n, y))$ for all $y \in Y$.

Since $y \in N_\gamma f(x_{n+1}) \Longrightarrow y \in N_\gamma f(x_n) \Longrightarrow g(w_n, y) \geq \alpha$, therefore

$$\psi(f(x_{n+1}, y)) \geq \rho_1(\psi(f(x_n, y)), \alpha) \geq \rho_1(\psi(f(x_n, y), \psi(\alpha))$$

for all $y \in N_\gamma f(x_{n+1})$. Hence

$$\inf_Y f(x_{n+1}, y) = \inf_{y \in N_\gamma f(x_{n+1})} \psi^{-1}(\psi(f(x_{n+1}, y))) \geq \tag{24}$$

$$\inf_Y \psi^{-1}\left(\rho_1\left(\psi\left(f(x_n, y)\right), \psi(\alpha)\right)\right) = \psi^{-1}\left(\rho_1\left(\psi\left(\inf_Y f(x_n, y)\right), \psi(\alpha)\right)\right)$$

by the continuity and monotonicity of ψ and ρ_1.

If $N_\gamma g(x^*) \subset N_\gamma g(w_n)$, then we can similarly obtain

$$\inf_Y g(w_{n+1}, y) \geq \psi^{-1}\left(\rho_1\left(\psi(\alpha), \psi\left(\inf_Y g(w_n, y)\right)\right)\right). \tag{25}$$

Define the (infinite) sequences $\{c_n\}$ and $\{d_n\}$ by

$$c_n := \inf_Y f(x_n, y), \quad d_n := \inf_Y g(w_n, y), \quad (n \geq 1).$$

Let $\{c_{n_k}\}$ be the subsequence of $\{c_n\}$ where $N_\gamma f(x_{n_k+1}) \subset N_\gamma f(x_{n_k})$ for all $k \geq 1$ and $\{d_{n_\ell}\}$ the subsequence of $\{d_n\}$ where $N_\gamma g(w_{n_\ell+1}) \subset N_\gamma g(w_{n_\ell})$ for all $\ell \geq 1$. Clearly, at least one of $\{c_{n_k}\}$ and $\{d_{n_\ell}\}$ must be infinite.

When $\{c_{n_k}\}$ is infinite, then

$$c_{n_k} \leq \beta < \alpha, \quad c_{n_k+1} \geq \psi^{-1}\left(\rho_1\left(\psi\left(c_{n_k}\right), \psi(\alpha)\right)\right)$$

or equivalently

$$\psi(c_{n_k}) < \psi(\alpha),$$
$$\psi(c_{n_k+1}) \geq \rho_1(\psi(c_{n_k}), \psi(\alpha)) > \psi(c_{n_k}), \quad (k \geq 1). \qquad (26)$$

Taking the limit and denoting $c^* := \lim_{k\to\infty} c_{n_k} = \lim_{k\to\infty} c_{n_k+1}$, we obtain from (26)

$$\psi(c^*) \geq \rho_1(\psi(c^*), \psi(\alpha)) > \psi(c^*)$$

which is a contradiction. If $\{d_{n_\ell}\}$ is infinite, then we arrive at a contradiction in the same way. Hence, (9) is false and the lemma is proved. □

Now we are ready to state and prove the main result of the paper.

THEOREM 2. *Under the conditions of Lemma 1, for any finite set* $F \subset X$

$$\inf_Y \max_F f(x,y) \leq \sup_X \inf_Y g(x,y). \qquad (27)$$

Proof. (27) trivially holds for $|F| = 1$. The proof goes by induction on $|F|$. In particular, we will prove that for any finite set $F \subset X$ and any set $U \subset Y$ which fulfills assumptions (6) and (7) with U in place of Y, we have

$$\inf_U \max_F f(x,y) \leq \sup_X \inf_U g(x,y). \qquad (28)$$

Assume that (28) holds for any F with $|F| \leq n$. Take

$$G := \{x_1, \ldots, x_n, x_{n+1}\}$$

and $U \subset Y$ for which (6) and (7) hold. Since f is bounded,

$$\inf_U \max_G f(x,y) > -\infty.$$

Suppose indirectly that

$$\inf_U \max_G f(x,y) > \sup_X \inf_U g(x,y) \geq \sup_X \inf_Y g(x,y) = \beta. \qquad (29)$$

Then, there are real numbers γ, λ, α to satisfy

$$\inf_U \max_G f(x,y) > \gamma > \lambda > \alpha > \sup_X \inf_U g(x,y) \geq \beta. \qquad (30)$$

Define

$$V := \{y \in U : g(x_{n+1}, y) < \gamma\},$$
$$\overline{V} := U \setminus V = \{y \in U : g(x_{n+1}, y) \geq \gamma\}.$$

If $V = \emptyset$, then

$$\gamma \leq \inf_U g(x_{n+1}, y) \leq \sup_X \inf_U g(x, y)$$

which contradicts (30). Thus we may suppose that $V \neq \emptyset$. Let $y_1, y_2 \in V$ be arbitrary. Then

$$\max\{g(x_{n+1}, y_1), g(x_{n+1}, y_2)\} < \gamma .$$

By (4) and (7), there is a $y_3 \in V$, such that

$$\begin{aligned}&\varphi(g(x_{n+1}, y_3)) \leq \rho_2(f(x_{n+1}, y_1), \varphi(g(x_{n+1}, y_2))) \leq \\ &\max\{f(x_{n+1}, y_1), \varphi(g(x_{n+1}, y_2))\} \leq \\ &\max\{g(x_{n+1}, y_1), \varphi(g(x_{n+1}, y_2))\} < \max\{\gamma, \varphi(\gamma)\} = \varphi(\gamma)\end{aligned}$$

implying

$$g(x_{n+1}, y_3) < \gamma \tag{31}$$

which means that $y_3 \in V$ and V (just as U) meets condition (7) and obviously (6), too.

Let $F := G \setminus \{x_{n+1}\}$. Thus, by the definition of V and since

$$f(x_{n+1}, y) \leq g(x_{n+1}, y) < \gamma \qquad \text{and} \qquad \max_G f(x, y) > \gamma$$

for all $y \in Y$, we have

$$\max_G f(x, y) = \max_F f(x, y) .$$

Therefore

$$\inf_V \max_G f(x, y) = \inf_V \max_F f(x, y) .$$

By induction,

$$\inf_V \max_F f(x, y) \leq \sup_X \inf_V g(x, y) .$$

If $\overline{V} = \emptyset$, then $V = U$ and

$$\inf_U \max_G f(x, y) = \inf_V \max_G f(x, y) = \inf_V \max_F f(x, y) \leq \sup_X \inf_U g(x, y) \tag{32}$$

contradicting (29) .

Assume now that $\overline{V} \neq \emptyset$. We will prove that in this case there is an $x' \in X$ for which

$$f(x', y) > \alpha \tag{33}$$

for all $y \in \overline{V}$.

Let $z_0 := x_{n+1}$. If for some $k = 0, 1, \ldots f(z_k, y) \geq \lambda$ for all $y \in \overline{V}$, then set $x' := z_k$. Otherwise, define $z_k \in X$, using (6), by

$$\psi(f(z_k, y)) \geq \rho_1(\psi(f(z_{k-1}, y)), g(z_0, y)) \quad \text{for all} \quad y \in \overline{V}.$$

Define $\overline{V}' := \{y \in \overline{V} : f(z_{k-1}, y) < \gamma\}$. Clearly, $\overline{V}' \neq \emptyset$. By (5) and the monotonicity of ψ, we have

$$\psi(f(z_k, y)) \geq \rho_1(\psi(f(z_{k-1}, y)), g(z_0, y)) > \psi(f(z_{k-1}), y))$$

for all $y \in \overline{V}'$, which implies

$$f(z_k, y) > f(z_{k-1}, y) \tag{34}$$

for all $y \in \overline{V}'$. Similarly, for any $y \in \overline{V} \setminus \overline{V}'$ we have

$$f(z_k, y) \geq \gamma. \tag{35}$$

If the sequence $\{z_k\}$ is finite, then (33) immediately follows. So we assume that $\{z_k\}$ is infinite. Define for any $k = 0, 1, \ldots$.

$$r_k := \inf_{y \in \overline{V}} f(z_k, y).$$

Then $r_k < \lambda$ for any $k = 0, 1, \ldots$ and the sequence $\{r_k\}$ is increasing by (34) and (35). Thus $\{r_k\}$ converges and $\delta := \lim_{k \to \infty} r_k \leq \lambda$. If $\delta < \lambda$, then by the continuity of ρ_1 and ψ and by the definition of z_{k+1}

$$\psi(\delta) = \psi\left(\lim_{k\to\infty} r_k\right) \geq \rho_1\left(\psi\left(\lim_{k\to\infty} r_k\right), \lambda\right) \geq \rho_1(\psi(\delta), \psi(\lambda)) > \psi(\delta)$$

which is impossible. Therefore $\delta = \lambda$ and thus for a large enough k_0,

$$\inf_{y \in \overline{V}} f(z_{k_0}, y) > \alpha$$

and we set $x' := z_{k_0}$.

If $\overline{V} = \emptyset$, then (32) implies the existence of $x'' \in X$ such that

$$g(x'', y) > \alpha$$

holds for all $y \in V$. Hence

$$\alpha < \inf_U \max\{f(x', y), g(x'', y)\}.$$

Then, by Lemma 1

$$\alpha < \inf_U \max\{f(x', y), g(x'', y)\} \leq \sup_X \inf_U g(x, y)$$

which contradicts (30). □

3. Remarks

(i) If φ and ψ are the identity functions, then Theorem 1 reduces to the two-function minimax inequality of [1].

(ii) It is not hard to find nontrivial examples of functions φ and ψ satisfying conditions (3) and (4). For example, if $f \geq 0$, $g \geq 0$, $\beta = 1$, then $\varphi, \psi : \mathbb{R}^+ \longrightarrow \mathbb{R}$, $\varphi(x) := x^a$, $0 < a \leq 1$, $\psi(x) := x^b$, $(b \geq 1)$ meet conditions (3) and (4). Note that in this case, the (generalized) concavity condition involves f^b, the (generalized) convexity condition involves g^a, yet the assertion of the theorem is about f and g.

(iii) Generally, for a given averaging function ρ_1, the condition

$$\psi(f(x_3, y)) \geq \rho_1(\psi(f(x_1, y)), g(x_2, y)) \tag{36}$$

neither implies nor is implied by

$$f(x_3, y) \geq \rho_1(f(x_1, y), g(x_2, y)),$$

the condition used in [1]. In general, it is also impossible to find an averaging function σ_1, for which (36) is equivalent to

$$f(x_3, y) \geq \sigma_1(f(x_1, y), g(x_2, y)).$$

So, the obvious shortcuts to reduce Theorem 1 to the main result of [1] do not work. It seems that one has to go the hard way: redo the proof in [1] (just as we have done it in this paper) carefully making changes where conditions (6) and (7) are involved.

4. Acknowledgements

Research was supported by grants OTKA T023881 and FKFP 59/1997.

References

1. Forgó, F. and Joó, I., *A general nontopological two-function minimax theorem*, Arch. Math., **71** (1998), 376–383.
2. Forgó, F., *On two-function minimax inequalities: a guided tour*, to appear in SZIGMA.
3. Irle, I., *A general minimax theorem*, Z. Oper. Res., **29** (1985), 229–247.
4. Lin, B.-L. and Quan, X.-C., *A two functions symmetric nonlinear minimax theorem*, Arch. Math., **57** (1991), 75–79.

5. Simons, S., *Minimax and variational inequalities, are they of fixed point or Hahn-Banach type?* In: Game theory and mathematical economics, O. Moeschlin and D. Pallaschke, eds., 379–388. North Holland, Amsterdam 1981.
6. Simons, S., *Two-function minimax theorems and variational inequalities for functions on compact and noncompact sets, with some comments on fixed-point theorems*, Proceedings of Symposia in Pure Mathematics, **45** (1986), Part 2, 377–392.
7. Simons, S., *Minimax theorems and their proofs*, in: Minimax and Applications ed. Ding-Zhu Du and Panos M. Pardalos, Kluwer Academic Publishers, Boston, 1995, 1–23.

Non-smooth Optimization with Randomization

László Gerencsér
Laboratory of Applied Mathematics,
Computer and Automation Institute, Hungarian Academy of Sciences,
H-1518 Budapest, P.O.Box 63 Hungary
e-mail: gerencser@sztaki.hu

Zsuzsanna Vágó
Laboratory of Operation Research and Decision Systems,
Computer and Automation Institute, Hungarian Academy of Sciences,
H-1518 Budapest, P.O.Box 63 Hungary
e-mail: vago@oplab.sztaki.hu

Abstract. The purpose of the paper is to describe a modification of Spall's simultaneous perturbation stochastic approximation (SPSA) algorithm for non-smooth functions. The method is particularly useful when it is hard or expensive to calculate the value of the function.

Key words: Stochastic approximation, gradient method, non-smooth optimization

Introduction

The simultaneous perturbation stochastic approximation (SPSA) method developed by J. C. Spall in [4] is considered an effective tool for solving difficult optimization problems. SPSA is essentially a randomized Kiefer-Wolfowitz method in which the gradient estimation is based on only two random measurements per iteration. The method is particularly useful in high dimension and in problems, where the cost function can be computed only by expensive simulations. In its original form, SPSA is applicable to functions which are three times continuously differentiable. The purpose of the paper is to extend SPSA to non-smooth optimization.

1. Non-smooth optimization

Let $L(\theta), \theta \in \mathbb{R}^p$ be a cost-function, which is assumed to be convex and Lipschitz-continuous, but non-differentiable. On the contrary, we assume that $L(\theta)$ has a *strictly non-smooth minimum* at θ^* in the following sense: for some $c > 0$

$$L(\theta) - L(\theta^*) \geq c|\theta - \theta^*|.$$

F. Giannessi et al. (eds.), Optimization Theory, 111–117.

A strictly non-smooth minimum may arise for problems defined by

$$L(\theta) = \max_{i=1,\dots,m} g_i(\theta),$$

where $g_i(\theta)$ themselves are Lipschitz-continuous, convex functions (cf. (2.2) of [3]).

For the minimization of $L(\cdot)$ a number of methods are available in the theory of non-smooth optimization (cf. [3]). In all the methods suggested there, it is assumed that the directional derivatives

$$L'(\theta, d) = \lim_{t \searrow 0} \frac{L(\theta + td) - L(\theta)}{t}$$

exist in every direction d. The proposed methods include the subgradient methods and the bundle methods. The efficiency of classical non-smooth optimization techniques depend considerably on the complexity of the function $L(\theta)$ itself. In this paper, we develop a new method for non-smooth optimization which seems to be applicable for problems of high complexity. It is based on a randomization technique by which we get a sufficiently smooth approximation of $L(\cdot)$. Thus, we get a suboptimal solution, but this is acceptable in many applications.

We proceed as follows: first we generate a sufficiently smooth approximation $\overline{L}(\theta)$ to $L(\theta)$ by locally *averaging* it which is implemented by a suitable randomization technique. Since the value or even an unbiased estimate for the gradient of the approximating function is not directly available, we use the simultaneous perturbation method (cf. [4]) to approximate the gradient, followed by a stochastic gradient method. In short, the minimization of $\overline{L}(\theta)$ will be performed by the SPSA method (cf. [4]).

2. Smooth approximation

Let $h(\theta)$ be a sufficiently smooth density function with bounded support, say $h(\theta) = 0$ for $|\theta_i| \geq \delta_m$ for all $i = 1, ..., p$, with some $\delta_m > 0$. The approximation of $L(\theta)$ will be defined as

$$\overline{L}(\theta) = \int_{\mathbf{R}} L(\theta + \delta) h(\delta)\, d\delta. \tag{2.1}$$

Obviously, $\overline{L}(\cdot)$ is convex. The value of the approximating function can be estimated for each θ as follows. Let $\delta\theta$ be a random vector having density function $h(\theta)$. Then, we can write

$$\overline{L}(\theta) = \mathrm{E}L(\theta + \delta\theta).$$

If δ_m is small, then the mass distribution defined by $h(\theta)$ is concentrated around zero, and hence, $\overline{L}(\theta)$ will be a good approximation of $L(\theta)$. Indeed, we have

$$L(\theta) - \overline{L}(\theta) = \int_{\mathbf{R}} (L(\theta) - L(\theta+\delta))h(\delta)\, d\delta,$$

and from here,

$$|L(\theta) - \overline{L}(\theta)| \leq K\delta_m,$$

where K is the Lipschitz constant of $L(\cdot)$. On the other hand, since $L(\theta)$ has a strictly non-smooth minimum at θ^*, it follows that for some $c > 0$, we have

$$\overline{L}(\theta) - \overline{L}(\theta^*) \geq L(\theta) - L(\theta^*) - 2K\delta_m \geq c|\theta - \theta^*| - 2K\delta_m.$$

Now, fix any small $\delta_0 > 0$. Then, for $|\theta - \theta^*| > \delta_0$, we get

$$\overline{L}(\theta) - \overline{L}(\theta^*) \geq c\delta_0 - 2K\delta_m > 0$$

for sufficiently small δ_m. Thus, the approximating function $\overline{L}(\theta)$ has its minimum inside the sphere of radius δ_0 around θ^*, if δ_m is sufficiently small.

The function $\overline{L}(\theta)$ is not computable explicitly, however, for the purpose of iterative minimization we can easily generate a Monte-Carlo estimate $\hat{L}_n(\theta)$ of $\overline{L}(\theta)$ for any fixed θ at any time n by generating a random perturbation $\delta\theta_n$ with density function $h(\theta)$ and setting

$$\hat{L}_n(\theta) = L(\theta + \delta\theta_n). \tag{2.2}$$

Thus, $\hat{L}_n(\theta)$ is an unbiased estimation of $\overline{L}(\theta)$. Technically speaking, the measurement error is

$$\varepsilon_n = \hat{L}_n(\theta) - \overline{L}(\theta) = L(\theta + \delta\theta_n) - \overline{L}(\theta) \tag{2.3}$$

and this depends on θ, i.e., we have a *state-dependent noise sequence.*

The difficulty in minimizing $\overline{L}(\theta)$ is that there is no direct way to estimate the gradient of $\overline{L}(\theta)$. Thus, we have to resort to numerical differentiation using the noise-corrupted function values $\hat{L}_n(\theta)$. This can be done in a very effective manner, using simultaneous random perturbation of the parameters (cf. [4]), and will be described in the next section.

3. The SPSA method

In this section, we summarize the basic ingredients of the SPSA method that was developed in [4]. The modification of the SPSA algorithm that is presented here was developed in [1, 2] where the assumed boundedness condition of [4] is replaced by a resetting mechanism, enforcing the estimator sequence to stay in a bounded domain. In the cited papers, a rate of convergence result is derived for the moments of the estimation error.

Let k denote the iteration time. At time k, we take a random perturbation vector

$$\Delta_k = (\Delta_{k1}, ..., \Delta_{kp})^T,$$

where Δ_{ki} is a double sequence of independent identically and symmetrically distributed, bounded random variables satisfying certain technical conditions. (cf. Section III of [4]). A standard perturbation that will be used is the double sequence Δ_{ki}

$$P'(\Delta_{ki} = +1) = 1/2 \qquad P'(\Delta_{ki} = -1) = 1/2,$$

where P' denotes the probability-measure associated with the random variable Δ_{ki}. Define the random vector

$$\Delta_k^{-1} = (\Delta_{k1}^{-1}, ..., \Delta_{kp}^{-1})^T.$$

Now, let $0 < c_k \leq 1$ be a fixed sequence of positive numbers defined as $c_k = c/k^\gamma$ with some $0 < \gamma < 1/2$. Let $D_0 \subset \text{int } D$ be a compact, convex truncation domain, the technical conditions of which are specified in Condition 2.3 of [2]. Among other things we require that $\theta^* \in \text{int } D_0$.

For each $\theta \in D_0$, we take two measurements

$$\hat{L}_k^+(\theta) = \hat{L}_{2k-1}(\theta + c_k\Delta_k) = L(\theta + \delta_{2k-1} + c_k\Delta_k)$$

and

$$\hat{L}_k^-(\theta) = \hat{L}_{2k}(\theta - c_k\Delta_k) = L(\theta + \delta_{2k} - c_k\Delta_k).$$

Then, the i-th component of the gradient estimator at time k and $\theta \in D_0$ is

$$H_i(k, \theta) = \frac{\hat{L}_k^+(\theta) - \hat{L}_k^-(\theta)}{2c_k\Delta_{ki}}. \tag{3.1}$$

The SPSA method is essentially a stochastic gradient methods using the gradient estimate given above. Let a_k be a fixed sequence of positive numbers with a_k denoting the stepsize at time k. A standard choice for it is $a_k = a/k^\delta$ with $0 < \delta \leq 1$, and $a > 0$. We start with an initial estimate $\widehat{\theta}_0$, and then, a sequence of estimated parameters, denoted by $\widehat{\theta}_{k+1} =$, $k = 0, 1, \ldots$ is generated recursively:

$$\widehat{\theta}_{k+1} = \widehat{\theta}_k - a_{k+1} H(k+1, \widehat{\theta}_k). \tag{3.2}$$

To ensure convergence we add a resetting mechanism so that the sequence of the estimators stay within the region D_0. For this purpose, we redefine $\widehat{\theta}_{k+1}$ as follows:

$$\widehat{\theta}_{k+1} := \widehat{\theta}_0 \qquad \text{when} \quad \widehat{\theta}_{k+1} \notin D_0.$$

The convergence properties of the above algorithm were proved in [2] for three times continuously differentiable functions.

4. Computational experiments

We present simulation results for two examples. These examples were taken from [3] and are considered typical hard non-smooth problems.

Example 1. The cost function is defined as

$$L(\theta) = \max_{k=1,\ldots d} |\lambda_k^T \theta|, \quad \theta \in \mathbb{R}^d,$$

where the components of the vectors λ_k are chosen randomly to be $+1$ or -1 with equal probability, for $k = 1, \ldots, d$. The minimum of the above function is trivially 0 with a minimizing point at $\theta^* = 0$. In the simulation, we had $d = 20$, and we made 5 000 iterations. The initial point was $\widehat{\theta}_0 = 0.5 \cdot (1, \ldots, 1)$.

To see the effect of the smoothing parameter we compared the values of the cost function along the iteration by using three different δ_m-s. In *Figure 1*, we plot the values of the cost function $L(\widehat{\theta}_k)$ vs. the iteration time for $\delta_m = 0.001$ (solid line), $\delta_m = 0.1$ (dotted line) and $\delta_m = 1$ (dash-dotted line). Clearly, the iteration converges in all three cases, but the rate of convergence is much better for small smoothing parameters. We increased the dimension up to $d = 100$.

Example 2. We considered

$$L(\theta) = \sum_{i=1}^{n} \lambda_i |\theta_i - \theta_i^*|,$$

with $\theta \in \mathbb{R}^n$, where $\theta^* = (\theta_1^*, \ldots, \theta_n^*)$ was chosen randomly, uniformly in the n-dimensional unit cube, and the weight coefficients λ_i, $i = 1, \ldots, n$ were chosen randomly, uniformly in $[1, 2]$. The minimum of

$L(\cdot)$ is 0 and it is achieved at $\theta = \theta^*$. In our example, the dimension was 20, and we made 5 000 iterations. We plotted the value of $L(\widehat{\theta}_k)$ against the iteration time in *Figure 2*. In this figure, we plotted the result with different smoothing parameters, $\delta_m = 0.001$ (solid line), $\delta_m = 0.1$ (dotted line) and $\delta_m = 1$ (dash-dotted line).

Acknowledgements

This research was supported by Hungarian Research Grant OTKA T 032932 and Bolyai János Research Fellowship.

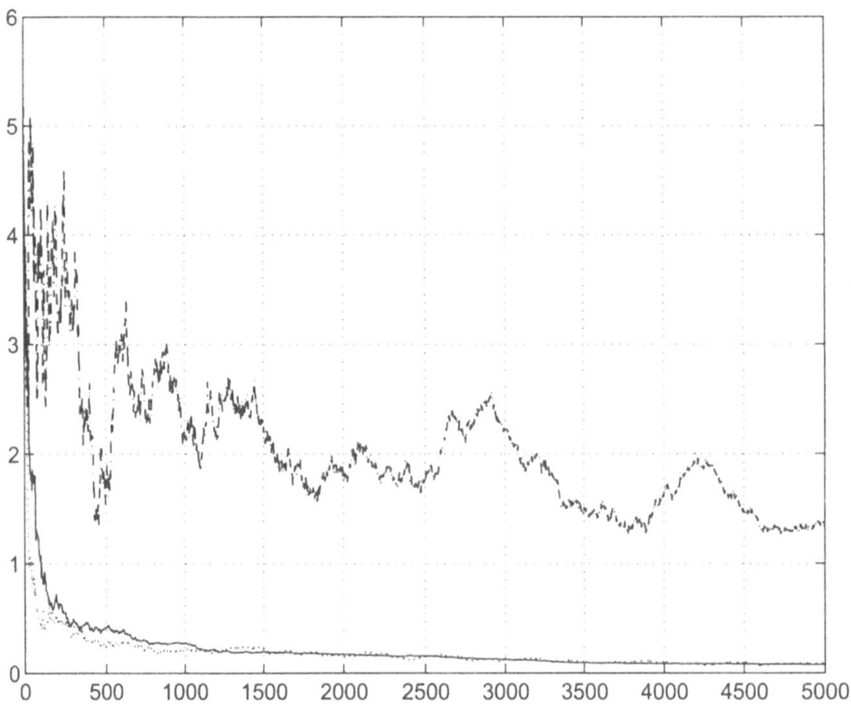

Figure 1. The value of $L(\widehat{\theta}_k)$ with different smoothing parameters

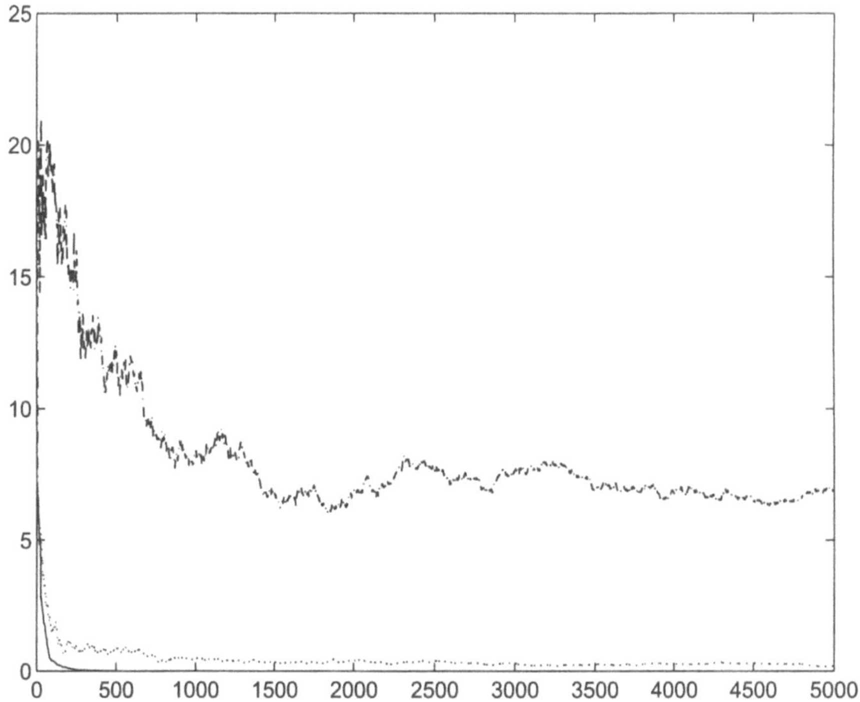

Figure 2. The value of $L(\widehat{\theta}_k)$ with different smoothing parameters

References

1. L. Gerencsér. Rate of convergence of a simultaneuous perturbation stochastic approximation method for function minimization. In I. Csiszár and Gy. Michaletzky, editors, *Proc. of the Bernoulli World Congress' Satellite Conference on Stochastic Differential and Difference Equations*, pages 67–75. Győr, August 21-24, 1996, Birkhauser, 1997.
2. L. Gerencsér. Rate of convergence of moments for a simultaneuous perturbation stochastic approximation method for function minimization. *IEEE Trans. Automat. Contr.*, 44:894–906, 1999.
3. C. Lemarechal. Nondifferentiable optimization. In G.L. Nemhauser et al., editor, *Handbooks in Operations Research and Management Science*, pages 529–572. Elsevier Science Publishers B.V. North-Holland, 1989.
4. J.C. Spall. Multivariate stochastic approximation using a simultaneous perturbation gradient approximation. *IEEE Trans. Automat. Contr.*, 37:332–341, 1992.

The Sherman-Morrison Formula for the Determinant and its Application for Optimizing Quadratic Functions on Condition Sets Given by Extreme Generators

Gerzson Kéri
Computer and Automation Research Institute
Hungarian Academy of Sciences
e-mail: keri@oplab.sztaki.hu

Abstract. First a short survey is made of formulas, which deal with either the inverse, or the determinant of perturbed matrices, when a given matrix is modified with a scalar multiple of a dyad or a finite sum of dyads. By applying these formulas, an algorithmic solution will be developed for optimizing general (i. e. nonconcave, nonconvex) quadratic functions on condition sets given by extreme generators. (In other words: the condition set is given by its internal representation.) The main idea of our algorithm is testing copositivity of parametral matrices.

Key words: copositive matrix, determinant calculus, extreme generator, matrix inversion, matrix perturbation, optimization, parametral matrix, quadratic programming.

1. Introduction

If $A(h)$ is a parametral real symmetric matrix, we may be interested to seek the smallest value of h — provided that there is one — for which $v^T A(h)v \geq 0$ holds for all $v \geq 0$. Optimization of quadratic functions may lead to this type of problem, if the condition set is given by its internal representation.

Two special cases of parametral matrices are of special interest to us. The first is

$$A(h) = A + huu^T, \tag{1.1}$$

where A is a constant real symmetric matrix and u is a given vector. In this case $A(h)$ is obtained by adding a dyad to A.

The second case is the generalization of the first one, when the modification of a given matrix with a finite sum of dyads is under consideration. Then we have

$$A(h) = A + h \cdot \sum_{i=1}^{k} u_i u_i^T = A + hUU^T. \tag{1.2}$$

In the present paper some methods will be worked out for the above mentioned problem, where — as it will be seen — the determinant, as

F. Giannessi et al. (eds.), Optimization Theory, 119–138.

well as the adjoint of the parametral matrix $A(h)$ will play an important role. This is why our problem has a close connection to methods — known from the literature as the Sherman-Morrison formula or the Sherman-Morrison-Woodbury formula — which deal with the inversion of perturbed matrices of type $A(h) = A + uv^T$ or $A(h) = A + UV^T$. The probably first appearance of an identity of this type was found by Duncan [8] in 1944. Sherman and Morrison published their work [14] in 1949, Woodbury [17] in 1950.

Different formulations of this matrix identity have been cited e. g. by Householder [10] in 1964, by Ortega and Rheinboldt [13] in 1970. A special case of the formula is contained in Bodewig's book [1], 1956. This list of occurrences of the formula is far not complete. A simpler identity of this type is often mentioned in connection with basis changes in linear programming.

We shall deduce formulas not only for the inverse, but also for the determinant of parametral matrices given in the form (1.1) or (1.2). These formulas will be applied; however, for quadratics and not for LP.

Methods developed for testing copositivity of constant matrices will be applied for the determination of the critical value of h. The probably best applicable results towards this direction were achieved by Cottle, Habetler, and Lemke [4] in 1970. For a copositivity test it is enough to deal with determinants and adjoints of different principal submatrices, see Theorems 3.1 and 4.2 (Keller's theorem) in [4]. Later Hadeler [9] and Väliaho [15]-[16] gave some other criteria for the copositivity of quadratic matrices. For our purpose – i.e. for copositivity test of parametral matrices – the test involved in Keller's theorem seems to be the most applicable. As checking copositivity is an NP-hard problem (See e. g. Murty and Kabadi [12]), so it would be a vain hope to reduce the number of steps to polynomial in our algorithms; some kinds of reductions are possible, however.

Standardization of QP and using copositivity for checking global optimality in general QP problems occur e. g. in the papers of Bomze [2], Danninger [6], Bomze and Danninger [3], [7]. The same kind of standardization (without using this word) is also applied in a previous publication of the author of the present paper [11].

2. Definitions, notations, and abbreviations

Let R^r denote the Euclidean r-space, R^r_+ its nonnegative orthant, and C^r the Euclidean complex r-space.

DEFINITION 2.1 *A real symmetric $r \cdot r$ matrix A is copositive (abbreviated CP), if $x^T A x \geq 0$ holds for every $x \in R^r_+$.*

The abbreviation *NCP* will be used for "noncopositive".

DEFINITION 2.2 *A real symmetric matrix A is copositive of order k (abbreviated kCP), if every $k \cdot k$ principal submatrix of A is CP.*

DEFINITION 2.3 *A real symmetric $r \cdot r$ matrix A is copositive of exact order k, if it is kCP, but not $(k+1)CP$.*

An important special case for *"copositive of exact order k"* comes forward in the following definition.

DEFINITION 2.4 *A real symmetric $r \cdot r$ matrix A is almost copositive if it is $(r-1)CP$, but not CP — provided that $r > 1$. A real symmetric matrix $A = [a]$ of order 1 is almost copositive if $a < 0$.*

The determinant of a quadratic matrix A will be denoted by $|A|$, its adjoint by $\operatorname{adj} A$. The element of the transposed matrix of $\operatorname{adj} A$ with coordinates (i, j) will be denoted by A_{ij}; this is the algebraic complement (otherwise co-factor) belonging to the entry a_{ij} of A. The adjoint of $A = [a]$ is defined the matrix $[1]$, regardless of the value of a.

When 0 or 1 will be used as vectors, that means they contain all zeros, or all ones, respectively, as their components.

We note that the concept of *"copositive matrix"* was first introduced by Motzkin in 1952, while *"almost copositive matrix"* by Väliaho in 1989.

3. The Sherman-Morrison formula and related topics

Throughout this whole section, let A, B denote real or complex $r \cdot r$ matrices, u, v denote r-vectors, U, V denote $r \cdot k$ matrices, $I^{(r)}$ denote the unit matrix of order r, and h denote an arbitrary real or complex number.

The Sherman-Morrison formula applies to inverting the modification of a given matrix with a dyad. If not a single dyad, but a finite sum of dyads (i. e. a matrix-product UV^T) is added to A, then the Sherman-Morrison-Woodbury formula is applicable as a more general matrix equality. Further, it is not very difficult to give similar formulas for determinants rather than inverses, which we shall call "the Sherman-Morrison formula for the determinant", and the "Sherman-Morrison-Woodbury formula for the determinant".

PROPOSITION 3.1 (Sherman-Morrison formula). *If A is invertible and $1 + hv^TA^{-1}u \neq 0$, then*

$$(A + huv^T)^{-1} = A^{-1} - h \cdot \frac{A^{-1}uv^TA^{-1}}{1 + hv^TA^{-1}u}. \tag{3.1}$$

We shall prove a more general formula that will occur in the next statement.

PROPOSITION 3.2 (Sherman-Morrison-Woodbury formula). *If* A *and* $D^{(k)}(h) = I^k + hV^TA^{-1}U$ *are invertible, then*

$$(A + hUV^T)^{-1} = A^{-1} - hA^{-1}U[D^{(k)}(h)]^{-1}V^TA^{-1}. \tag{3.2}$$

Proof.

$$(A+hUV^T)\{A^{-1} - hA^{-1}U[D^{(k)}(h)]^{-1}V^TA^{-1}\} = AA^{-1} + hUV^TA^{-1} -$$
$$-hAA^{-1}U[D^{(k)}(h)]^{-1}V^TA^{-1} - h^2 \cdot UV^TA^{-1}U[D^{(k)}(h)]^{-1}V^TA^{-1} =$$
$$= I^{(r)} + hUV^TA^{-1} - hU(I^{(k)} + hV^TA^{-1}U)[D^{(k)}(h)]^{-1}V^TA^{-1} =$$
$$= I^{(r)} + hUV^TA^{-1} - hU[D^{(k)}(h)][D^{(k)}(h)]^{-1}V^TA^{-1} = I^{(r)}.$$

PROPOSITION 3.3 *For arbitrary square matrices* A *and* B,

$$|A+B| = \sum_{i_1=0}^{1}\sum_{i_2=0}^{1}\cdots\sum_{i_r=0}^{1}|C^{(i_1i_2\ldots i_r)}|, \tag{3.3}$$

where the j*-th column of* $C^{(i_1i_2\ldots i_r)}$ *is defined by*

$$c_j^{(i_1i_2\ldots i_r)} = \begin{cases} a_j\,, & \text{if } i_j = 0 \\ b_j\,, & \text{if } i_j = 1 \end{cases} \quad (j = 1, 2, \ldots, r).$$

We omit the proof, which is elementary and is known from the determinant calculus.

PROPOSITION 3.4 *If* $\operatorname{rank} B \le 1$, *then*

$$|A+B| = |A| + \sum_{i=1}^{r}\sum_{j=1}^{r} b_{ij}A_{ij}. \tag{3.4}$$

Proof. Let us consider the matrices $C^{(i_1i_2\ldots i_r)}$ introduced in the previous proposition. If $i_1 + i_2 + \cdots + i_r \ge 2$, then $C^{(i_1i_2\ldots i_r)}$ contains at least two columns selected from B, and therefore the assumption about the rank of B implies that

$$|C^{(i_1i_2\ldots i_r)}| = 0.$$

If $i_j = 1$ and $i_t = 0$ for $t \in \{1, 2, \ldots, r\} \setminus \{j\}$, then

$$|C^{(i_1i_2\ldots i_r)}| = |a_1\; a_2\; \ldots\; a_{j-1}\; b_j\; a_{j+1}\; \ldots\; a_r| = \sum_{i=1}^{r} b_{ij}A_{ij}$$

according to the determinant expansion theorem. Now we can write

$$|A+B| = |A| + \sum_{i_1+i_2+\cdots+i_r=1} |C^{(i_1 i_2 \ldots i_r)}| = |A| + \sum_{j=1}^{r}\left(\sum_{i=1}^{r} b_{ij}A_{ij}\right).$$

COROLLARY 3.5 (Sherman-Morrison formula for the determinant).

$$|A + huv^T| = |A| + hv^T(\operatorname{adj} A)u. \tag{3.5}$$

Proof. The dyad huv^T is an $r \cdot r$ matrix with rank equal to 1. According to Proposition 3.4, we have

$$|A + huv^T| = |A| + \sum_{i=1}^{r}\sum_{j=1}^{r} hu_i v_j A_{ij} = |A| + hv^T(\operatorname{adj} A)u.$$

COROLLARY 3.6 *If A is invertible, then*

$$|A + huv^T| = |A| \cdot (1 + hv^T A^{-1}u). \tag{3.6}$$

The proof is obvious, if we take into account that $\operatorname{adj} A = |A| \cdot A^{-1}$.

COROLLARY 3.7 *If A and $A + huv^T$ are invertible, then*

$$1 + hv^T A^{-1} u \neq 0.$$

The proof is evident from (3.6).

For the more general case, where the vectors u and v are replaced by the matrices U and V, statements, parallel to Corollaries 3.6–3.7 will also be formulated. Before doing this, we state the following lemma, which will help us to prove the generalization of Corollary 3.6.

LEMMA 3.8 *Let $D = D^{(k)} = [d_{ij}]_{i,j=1,2,\ldots,k}$ denote a real or complex $k \cdot k$ matrix, and denote*

$$D^{(t)} = [d_{ij}]_{i,j=1,2,\ldots,t}$$

the leading principal submatrices of D. Then we state the following:

$$|D^{(k)}| = d_{kk} \cdot |D^{(k-1)}| - \sum_{i=1}^{k-1}\sum_{j=1}^{k-1} (\operatorname{adj} D^{(k-1)})_{ij} d_{ki} d_{jk}. \tag{3.7}$$

Proof. Expand the determinant $|D^{(k)}|$ according to its last row, and then expand the given smaller determinants — with the exception of $|D^{(k-1)}|$ — according to their last columns. Then we obtain the sum standing on the right hand side of (3.7).

THEOREM 3.9 (Sherman-Morrison-Woodbury formula for the determinant). *If A is invertible, then*

$$|A + hUV^T| = |A| \cdot |D^{(k)}(h)|, \tag{3.8}$$

where

$$D^{(k)}(h) = I^{(k)} + hV^T A^{-1} U. \tag{3.9}$$

Proof. We prove the theorem by induction on k. If $k = 1$, then the assertion reduces to the statement of Corollary 3.6.

Consider the modified matrix $A + hUV^T$ in the form

$$A + h \cdot \sum_{i=1}^{k} u_i v_i^T$$

where u_i and v_i are r-vectors, and assume that the assertion of the theorem is true for $k = 1$, i. e.

$$|A + h \cdot \sum_{i=1}^{k-1} u_i v_i^T| = |A| \cdot |D^{(k-1)}(h)|.$$

We suppose temporarily, that $D^{(k-1)}(h)$ is invertible. Then, according to Corollary 3.6,

$$|A + hUV^T| = |(A + h \cdot \sum_{i=1}^{k-1} u_i v_i^T) + hu_k v_k^T| =$$

$$= |A + h \cdot \sum_{i=1}^{k-1} u_i v_i^T| \cdot \{1 + hv_k^T (A + h \cdot \sum_{i=1}^{k-1} u_i v_i^T)^{-1} u_k\} =$$

$$= |A| \cdot |D^{(k-1)}(h)| \cdot \{1 + hv_k^T (A + h \cdot \sum_{i=1}^{k-1} u_i v_i^T)^{-1} u_k\}.$$

According to Proposition 3.2 we can obtain, that

$$(A + h \cdot \sum_{i=1}^{k-1} u_i v_i^T)^{-1} = A^{-1} - hA^{-1}[U^{(k-1)}][D^{(k-1)}(h)]^{-1}[V^{(k-1)}]^T A^{-1}$$

where

$$U^{(k-1)} = [u_1\ u_2\ \dots\ u_{k-1}]$$

and similarly

$$V^{(k-1)} = [v_1\ v_2\ \dots\ v_{k-1}].$$

Thus

$$hv_k^T(A + h \cdot \sum_{i=1}^{k-1} u_i v_i^T)^{-1} u_k =$$

$$= hv_k^T A^{-1} u_k - h^2 \cdot v_k^T A^{-1} [U^{(k-1)}][D^{(k-1)}(h)]^{-1} [V^{(k-1)}]^T A^{-1} u_k =$$

$$= hv_k^T A^{-1} u_k - \sum_{i=1}^{k-1} \sum_{j=1}^{k-1} (hv_k^T A^{-1} u_i)[D^{(k-1)}(h)]_{ij}^{-1} (hv_j^T A^{-1} u_k) =$$

$$= hv_k^T A^{-1} u_k - \sum_{i=1}^{k-1} \sum_{j=1}^{k-1} d_{ki}(h) \cdot \frac{(\text{adj}\, D^{(k-1)}(h))_{ij}}{|D^{(k-1)}(h)|} \cdot d_{jk}(h).$$

Now we can write

$$|A + hUV^T| = |A| \cdot |D^{(k-1)}(h)| \cdot \{1 + hv_k^T A^{-1} u_k -$$

$$- \sum_{i=1}^{k-1} \sum_{j=1}^{k-1} \frac{(\text{adj}\, D^{(k-1)}(h))_{ij}}{|D^{(k-1)}(h)|} \cdot d_{ki}(h) d_{jk}(h)\} =$$

$$= |A| \cdot \{d_{kk}(h) \cdot |D^{(k-1)}(h)| - \sum_{i=1}^{k-1} \sum_{j=1}^{k-1} (\text{adj}\, D^{(k-1)}(h))_{ij} d_{ki}(h) d_{jk}(h)\}.$$

Finally, by applying Lemma 3.8 for $D^{(k)}(h)$, we obtain that (3.8) is valid, provided that $D^{(k-1)}(h)$ is invertible. As clearly $D^{(k-1)}(0) = I^{(k-1)}$, we can conclude that (3.8) is valid, if h is in a neighbourhood of zero. Since the expression

$$|A + hUV^T| - |A| \cdot |D^{(k)}(h)|$$

is a polynomial, i. e. analytical function of h, thus if it vanishes in a neighbourhood of zero, then it does so for any complex h. By this the proof is complete.

COROLLARY 3.10 *If both A and $A + hUV^T$ are invertible, then $I^{(k)} + hV^T A^{-1} U$ is invertible as well.*

Remark 3.11 The assertions of Proposition 3.2 and Corollary 3.10 can be united in the following way: Suppose that A is invertible. In this case $A + hUV^T$ is invertible if and only if $I^{(k)} + hV^T A^{-1} U$ is invertible, and then

$$(A + hUV^T)^{-1} = A^{-1} - hA^{-1} U (I^{(k)} + hV^T A^{-1} U)^{-1} V^T A^{-1}.$$

The Sherman-Morrison-Woodbury formula is cited in this form in [13], where $k \leq r$ is assumed. As we have seen, the restriction $k \leq r$ is not necessary to suppose.

4. Domain of copositivity of a parametral matrix

For the discussion of this section $A(h)$ will denote a real parametral $r \cdot r$ matrix, having the properties, we specify in the following definition.

DEFINITION 4.1 *A real parametral $r \cdot r$ matrix is proper parametral matrix, if it has the following features:*

(i) $A(h)$ *is symmetric for every real* h.

(ii) $A(h)$ *is a continuous function of* h.

(iii) For any principal submatrix $B(h)$ *of* $A(h)$ *it is true that if* $B(h_1)$ *is CP and* $h_2 > h_1$, *then* $B(h_2)$ *is also CP.*

(iv) There exists some $h \in R$ *such that* $A(h)$ *is NCP.*

The next statement contains examples for proper parametral matrices.

PROPOSITION 4.2 *If A is a real symmetric $r \cdot r$ matrix and U is a real nonzero $r \cdot k$ matrix, then $A(h) = A + hUU^T$ is a proper parametral matrix. As a special case, for a nonzero r-vector u, $A(h) = A + huu^T$ is a proper parametral matrix.*

Proof. The validity of (i) and (ii) is obvious. If $h_2 > h_1$, then $(h_2 - h_1)UU^T$ is positive semidefinite, therefore its principal submatrices are all *CP*. The sum of two *CP* matrices is also *CP*. This way, feature (iii) is established.

For the proof of feature (iv) it is enough to show that h may assume such values, for which $A(h)$ has a negative entry in its main diagonal. Indeed, let $\overline{u}_i$ be a nonzero row vector of U, then $a_{ii}(h) = a_{ii} + h\overline{u}_i\overline{u}_i^T < 0$ holds for any such h for which $h < -\frac{a_{ii}}{\overline{u}_i\overline{u}_i^T}$.

DEFINITION 4.3 *The domain of copositivity of a proper parametral matrix $A(h)$ is the set*

$$G = \{\, h : \ A(h) \text{ is } CP \,\}. \tag{4.1}$$

DEFINITION 4.4 *The threshold number of copositivity (briefly threshold) of a proper parametral matrix $A(h)$ is the number*

$$h^* = \inf G = \inf\{\, h : \ A(h) \text{ is } CP \,\}. \tag{4.2}$$

Remark 4.5 Because of the assumption involved in feature (iv) of Definition 4.1, only the following two possibilities may occur:

a) G is empty, and then $h^* = +\infty$;

b) G is a nonempty closed convex set, and then h^* is finite. Further, in this case $G = [h^*, +\infty)$.

5. Connection between quadratic optimization and copositivity of parametral matrices

Let us consider the following optimization problem:

$$\begin{array}{c} \text{minimize } f(x) = x^T Dx + 2g^T x \\ \text{subject to } x \in K = \{Qy + Sz : y \geq 0, z \geq 0, 1^T z = 1\}, \end{array} \tag{5.1}$$

where D is a given real symmetric $n \cdot n$ matrix, g is a given n-vector, Q is a given $n \cdot r_1$ matrix, S is a given $n \cdot r_2$ matrix ($r_1 \geq 0$ and $r_2 \geq 1$).

On the matrix D, which determines the type of the objective function, there are no further conditions. That means that the objective function $f(x)$ can be nonconvex and also nonconcave.

The matrices Q and S contain the column vectors q_i and s_i respectively.

Now let us apply the following notations:

$$A = \begin{bmatrix} Q^T & 0 \\ S^T & 1 \end{bmatrix} \begin{bmatrix} D & g \\ g^T & 0 \end{bmatrix} \begin{bmatrix} Q & S \\ 0^T & 1^T \end{bmatrix} =$$

$$= \begin{bmatrix} Q^T DQ & Q^T DS + Q^T g 1^T \\ S^T DQ + 1g^T Q & S^T DS + 1g^T S + S^T g 1^T \end{bmatrix} \tag{5.2}$$

and

$$A(h) = \begin{bmatrix} Q^T & 0 \\ S^T & 1 \end{bmatrix} \begin{bmatrix} D & g \\ g^T & h \end{bmatrix} \begin{bmatrix} Q & S \\ 0^T & 1^T \end{bmatrix}. \tag{5.3}$$

Here A is a real symmetric $r \cdot r$ matrix with $r = r_1 + r_2$. Denoting by I the set, whose elements are the last r_2 row indices of matrix A, i. e.

$$I = \{r_1 + 1, r_1 + 2, \ldots, r_1 + r_2\}, \tag{5.4}$$

then the entries of the parametral matrix $A(h)$ are the following:

$$a_{ij}(h) = \begin{cases} a_{ij} & \text{if } i \notin I \text{ or } j \notin I \\ a_{ij} + h & \text{if } i \in I \text{ and } j \in I. \end{cases} \tag{5.5}$$

According to Proposition 4.2, $A(h)$ is a proper parametral matrix. With the newly introduced notations the objective function increased by the additive constant h can be written as follows:

$$f(x)+h = \begin{bmatrix} x^T & 1 \end{bmatrix} \begin{bmatrix} D & g \\ g^T & h \end{bmatrix} \begin{bmatrix} x \\ 1 \end{bmatrix} =$$

$$= \begin{bmatrix} y^T Q^T + z^T S^T & 1 \end{bmatrix} \begin{bmatrix} D & g \\ g^T & h \end{bmatrix} \begin{bmatrix} Qy + Sz \\ 1 \end{bmatrix} =$$

$$= \begin{bmatrix} y^T & z^T \end{bmatrix} \begin{bmatrix} Q^T & 0 \\ S^T & 1 \end{bmatrix} \begin{bmatrix} D & g \\ g^T & h \end{bmatrix} \begin{bmatrix} Q & S \\ 0^T & 1^T \end{bmatrix} \begin{bmatrix} y \\ z \end{bmatrix} = \tag{5.6}$$

$$= \begin{bmatrix} y^T & z^T \end{bmatrix} A(h) \begin{bmatrix} y \\ z \end{bmatrix} = v^T A(h) v.$$

At the last phase of the transformation we applied the notation

$$v = \begin{bmatrix} y \\ z \end{bmatrix}. \tag{5.7}$$

On the basis of equality (5.6) an interesting relationship between the optimum value of problem (5.1) and the domain of copositivity of $A(h)$ can be recognized. Namely, if h^* is the smallest value, for which $f(x) + h^* \geq 0$ for every $x \in K$, i. e. the optimum value of (5.1) is $-h^*$, then according to (5.6), h^* is also the smallest value for which the matrix $A(h^*)$ is copositive, i. e. the threshold of $A(h)$ is h^*.

Our last observation will be precisely described in the following two statements, which include also the case of the unbounded objective function:

PROPOSITION 5.1 *For the parametral matrix $A(h)$ specified at (5.3), its threshold is finite if and only if the function $f(x)$ is bounded from below on K.*

PROPOSITION 5.2 *If for the parametral matrix $A(h)$ specified at (5.3), its threshold is finite and equals h^*, then the optimum value of problem (5.1) is $-h^*$.*

On the basis of the observations made so far we can go even further and derive a conclusion for the optimal solutions of (5.1). Namely, a certain vector $x^* \in K$ is an optimal solution to problem (5.1) if and only if $f(x^*) + h^* = 0$, i. e. for the corresponding v^* the equation $v^{*T} A(h^*) v^* = 0$ holds. So the following statement is also true.

PROPOSITION 5.3 *If for the parametral matrix $A(h)$ specified at (5.3), its threshold is finite and equals h^*, moreover for a certain vector* $v^* = \begin{bmatrix} y^* \\ z^* \end{bmatrix}$ *conditions*

$$\begin{aligned} &v^* \in R^r_+, \\ &\sum_{i \in I} v_i^* = 1, \\ &v^{*T} A(h^*) v^* = 0 \end{aligned} \tag{5.8}$$

hold, then the vector

$$x^* = Qy^* + Sz^* \tag{5.9}$$

is an optimal solution of problem (5.1). Conversely, if the vector produced in the form of (5.9) is an optimal solution of (5.1), then (5.8) holds for $v^* = \begin{bmatrix} y^* \\ z^* \end{bmatrix}$.

From the above statements it follows that the quadratic programming problem (5.1) is equivalent to solving the following *derived problem* — a) and b) — for a proper parametral matrix.

a) Determine the h^* threshold of $A(h)$. (More accurately: first decide whether it is finite or not. If it is finite, then compute its value.)

b) In the case when h^* is finite, find a vector v^* satisfying conditions (5.8).

Remark 5.4 It can never occur that $h^* = -\infty$, because $A(-a_{ii} - 1)$ is NCP for any $i \in I$ (cf. Proposition 6.6).

On the contrary, $h^* = +\infty$ can occur, and according to Proposition 5.1 it does occur in every case when $f(x)$ is unbounded from below on set K. Now let us consider the following example:

$$\min\{x_1^2 - 2x_2 : x_1 \geq 0, x_2 \geq 0\}.$$

The application of (5.3) for this special case gives the matrix

$$A(h) = \begin{bmatrix} 1 & 0 & 0 \\ 0 & 0 & -1 \\ 0 & -1 & h \end{bmatrix}.$$

This matrix cannot be *CP* for any real h, because in the case of non-positive h, taking the vector $v^T = \begin{bmatrix} 0 & 1 & 1 \end{bmatrix}$, and substituting it into the quadratic form with $A(h)$ results in

$$v^T A(h) v = -2 + h < 0,$$

while in the case of a positive h, taking $v^T = \begin{bmatrix} 0 & h & 1 \end{bmatrix}$, we obtain that

$$v^T A(h) v = -h < 0.$$

Remark 5.5 If h^* is finite, then $A(h^*)$ is CP (cf. Proposition 6.4).

6. Algorithms for proper parametral matrices

Our next purpose is to develop algorithmic procedures for finding the threshold number h^* of a proper parametral matrix. In view of the definition of almost copositive matrices, h^* is the smallest such value, for which none of the principal submatrices of $A(h)$ is almost copositive. This fact suggests the following scheme as an algorithm skeleton (Algorithm I.).

Denote by $B_1(h), B_2(h), \dots, B_N(h)$ the principal submatrices of $A(h)$ in a nondecreasing order of their size; i. e. the first r such submatrices $B_1(h), B_2(h), \dots, B_r(h)$ are of order 1, while the last such submatrix $B_N(h)$ is identical to $A(h)$. The sequential order of submatrices of same algebraic order is optional.

Algorithm I.

Initialization:

A1) Let $t = 1$.
A2) Let h_0 be any real value for which $A(h_0)$ is NCP.

Main cycle of the algorithm: While $t \leq N$, perform steps B1-B4.

B1) If $B_t(h_{t-1})$ is not almost copositive, then let $h_t = h_{t-1}$ and go to B4, else continue at B2.
B2) If the set $\{ h : B_t(h)$ is not almost copositive, $h > h_{t-1} \}$ is empty, then stop. (In this case h^* is unbounded.)
B3) Let $h_t = \inf \{ h : B_t(h)$ is not almost copositive, $h > h_{t-1} \}$.
B4) Let $t = t + 1$.

Normal termination

C1) Let $h^* = h_N$.
C2) Stop.

This formulation of the algorithm does not contain a copositivity test. It will be built into the algorithm after the proof of this simpler algorithm version, which can be done with the help of the following definition and lemma.

DEFINITION 6.1 *Let A denote a real symmetric $r \cdot r$ matrix, and $\beta = (B_1, B_2, \ldots, B_N)$ denote the ordered set of all principal submatrices of A in a nondecreasing order of their size. Then A is defined to be β-copositive of order t, if all of $B_1, B_2, \ldots, B_t$ are CP.*

Clearly, A is $1CP$ if and only if A is β-copositive of order r, A is $2CP$ if and only if A is β-copositive of order $r + \binom{r}{2}$, etc.

LEMMA 6.2 *If $A(h)$ is a proper parametral $r \cdot r$ matrix, h_1 is a real number such that $A(h_1)$ is β-copositive of order $t-1$, but not of order t, where $1 \leq t \leq N$, and*

$$h_2 = \inf\{h : B_t(h) \text{ is not almost copositive, } h > h_1\}$$

is finite, then $A(h_2)$ is β-copositive of order t.

Proof. The copositivity of $B_1(h_2), B_2(h_2), \ldots, B_{t-1}(h_2)$ follows from feature (iii) of proper parametral matrices (Definition 4.1), while the copositivity of $B_t(h_2)$ follows from the definition of h_2.

Proof of Algorithm I. If the set, appearing at Step B2 is empty, then clearly $A(h)$ is NCP for any real h. Now let us suppose, that the algorithm ends with a normal termination. Then $B_1(h_1)$ is CP according to the definition of h_1 given in step B3 of the algorithm. Applying the assertion of Lemma 6.2, it follows by induction, that $A(h_t)$ is β-copositive of order t if $t \in \{1, 2, \ldots, N\}$. For $t = N$ this means that $A(h_N)$ is CP. Because $A(h_0)$ is NCP, there should be a $t \in \{1, 2, \ldots, N\}$, for which

$$h_{t-1} < h_t = h_N.$$

Then $B_t(h)$ is NCP for any $h < h_N$, and consequently so is $A(h)$. We have seen; however, that $A(h_N)$ is CP, thus h_N is equal to the threshold number of copositivity h^* of $A(h)$.

It is not very nice in the description of Algorithm I, that it contains certain clumsy details such as

- to determine if a submatrix is almost copositive or not;
- to find the smallest value of the parameter h, for which a given submatrix is not almost copositive.

Now we are going to make these details clearer. For this purpose, the first — and probably the most important — tool is the following theorem.

THEOREM 6.3 (Cottle-Habetler-Lemke's theorem). *Suppose that a real symmetric $r \cdot r$ matrix A is $(r-1)CP$. In this case A is NCP if and only if the following two conditions hold:*
(i) $|A| < 0$;
(ii) $\operatorname{adj} A \geq 0$.

For the proof of this theorem see [4], Theorem 3.1.

Next some properties of *CP* and almost copositive matrices as well as statements on the adjoint of real quadratic matrices will be listed. All of these can be proved very simply, in an elementary way, so their proofs will be omitted.

PROPOSITION 6.4 *The set of all CP matrices of order r is a closed convex cone.*

PROPOSITION 6.5 *Adding a nonnegative matrix to a CP matrix results in another CP matrix.*

PROPOSITION 6.6 *Every principal submatrix of a CP matrix is also CP. (As a special case, the elements of its main diagonal are nonnegative.)*

PROPOSITION 6.7 *An almost copositive matrix has at least one negative entry in each of its rows.*

PROPOSITION 6.8 *If a symmetrical row-column permutation is performed for a CP (or almost copositive) matrix, the result is also a CP (or almost copositive) matrix.*

PROPOSITION 6.9 *The product of a singular matrix and its adjoint always yields the zero matrix.*

PROPOSITION 6.10 *The rank of the adjoint of a singular matrix cannot be greater than 1. In other words, if A is a singular matrix, then all columns of* $\operatorname{adj} A$ *are proportional to each other.*

Now we are going to reformulate our general algorithm and build a copositivity test into it. Before doing that, we state two more lemmas.

LEMMA 6.11 *If $A(h)$ is a proper parametral matrix, then*

$$H = \{h : A(h) \text{ is almost copositive}\} \tag{6.1}$$

is convex (i. e. it is either finite or infinite interval).

Proof. Let us consider the sets

$$H_1 = \{h : A(h) \text{ is } (r-1)\,CP\} \tag{6.2}$$

and

$$H_2 = \{h : A(h) \text{ is } CP\}. \tag{6.3}$$

By the definition of almost copositive matrices,

$$H = H_1 \setminus H_2. \tag{6.4}$$

Obviously H_1 and H_2 are intervals on the real axis, both of them being infinite from right. From this it follows that H is also an interval.

LEMMA 6.12 *Let us suppose that for a proper parametral matrix $A(h)$, the set specified in (6.1) as set H is not empty, and $h_0 = \sup H$ has a finite value. Then $A(h_0)$ has the following two features:*

$$|A(h_0)| = 0,$$

$$\operatorname{adj} A(h_0) \geq 0. \tag{6.5}$$

Proof. Denote by h_1 an arbitrary element of H. According to Lemma 6.11, $A(h)$ is almost copositive for $h \in [h_1, h_0)$; therefore, according to Theorem 6.3, $|A(h)| < 0$ and $\operatorname{adj} A(h) \geq 0$ are also satisfied if $h \in [h_1, h_0)$. From this it follows by continuity that the second line of (6.5) is true and

$$|A(h_0)| \leq 0. \tag{6.6}$$

If $h > h_0$, then $A(h)$ is obviously $(r-1)\,CP$, but not almost copositive, so it must be CP. Then $A(h_0)$ is also CP, and therefore $|A(h_0)| < 0$ is not possible, according to Theorem 6.3, i. e. the first line of (6.5) holds as well.

By the help of Theorem 6.3 and Lemma 6.12 we can adjust the general Algorithm I into the following form (where a copositivity test is already built into the algorithm):

Algorithm I'.

Initialization:

A1) Let $t = 1$.

A2) Let h_0 be any real value for which $A(h_0)$ is NCP.

Main cycle of the algorithm: While $t \leq N$, perform steps B1-B4.

B1) If $|B_t(h_{t-1})| \geq 0$ or $\operatorname{adj} B_t(h_{t-1})$ has a negative entry, then let $h_t = h_{t-1}$ and go to B4, else continue at B2.

B2) If the set $\{h : |B_t(h)| = 0, h > h_{t-1}\}$ is empty, then stop. (In this case h^* is unbounded.)

B3) Let $h_t = \min\{h : |B_t(h)| = 0, h > h_{t-1}\}$.

B4) Let $t = t + 1$.

Normal termination

C1) Let $h^* = h_N$.

C2) Stop.

In the next part of this section we turn our attention to two special cases, when $A(h) = A + hUU^T$ or $A(h) = A + huu^T$. In the first, more general case let V_t denote the submatrix of U consisting of all such row vectors of U, which correspond to a given submatrix B_t of A. With this notation

$$B_t(h) = B_t + hV_tV_t^T, \tag{6.7}$$

and Steps B2-B3 of Algorithm I' require the solution of the equation

$$|B_t + hV_tV_t^T| = 0. \tag{6.8}$$

This is an algebraic equation of order not greater than the rank of V_t. If B_t is invertible, then — according to Theorem 3.9 — (6.8) is equivalent to

$$|I + hV_t^TB_t^{-1}V_t| = 0, \tag{6.9}$$

i. e. (6.8) leads to seek the eigenvalues of $V_t^TB_t^{-1}V_t$. We shall prove the following assertion:

THEOREM 6.13 *Suppose, that A is a real symmetric $r \cdot r$ matrix, U is a real $r \cdot k$ matrix, and let $A(h) = A + hUU^T$. Further, suppose, that $A(\hat{h})$ is almost copositive for some $\hat{h} \in R$. Denote by H the set*

$$H = \{h : |A(h)| = 0, h > \hat{h}\}, \tag{6.10}$$

and by E the spectrum of $U^T[A(\hat{h})]^{-1}U$, i. e.

$$E = \{\lambda : |U^T[A(\hat{h})]^{-1}U - \lambda I^{(k)}| = 0\}. \tag{6.11}$$

Then we state the following:

$\inf H$ is finite if and only if there exists a $\lambda \in E$ such that $\lambda < 0$, and then

$$\inf H = \min\{\hat{h} - \frac{1}{\lambda} : \lambda \in E, \lambda < 0\}. \tag{6.12}$$

Proof will be given by referring to the following lemma.

LEMMA 6.14 *With the assumption of Theorem 6.13, we state that $|A(h)| = 0$ if and only if $h \neq \hat{h}$ and $\frac{1}{\hat{h}-h} \in E$.*

Proof. The almost copositive matrix $A(\hat{h})$ is invertible according to Theorem 6.3. Therefore Theorem 3.9 can be applied in the following way:

$$|A(h)| = |A + hUU^T| = |A(\hat{h}) + (h - \hat{h})UU^T| =$$

$$= |A(\hat{h})||I^{(k)} + (h - \hat{h})U^T[A(\hat{h})]^{-1}U|.$$

From this it follows that $|A(h)| = 0$ if and only if $h \neq \hat{h}$, and

$$\left|U^T[A(\hat{h})]^{-1}U - \frac{1}{\hat{h} - h}I^{(k)}\right| = 0.$$

Proof of Theorem 6.13

The value of $\inf H$ is finite if and only if H is nonempty. According to Lemma 6.14, H is nonempty if and only if $\frac{1}{\hat{h}-h} \in E$ for some $h > \hat{h}$, i. e. if E contains one or more negative elements. The roots of equation $|A(h)| = 0$ are the elements of the set

$$\left\{h : \frac{1}{\hat{h} - h} = \lambda, \lambda \in E\right\} = \left\{\hat{h} - \frac{1}{\lambda} : \lambda \in E\right\}$$

from which follows the validity of (6.12).

Now we are ready to concretize Algorithm I' for the special case where $A(h) = A + hUU^T$.

Algorithm II.

Replace steps B2)-B3) in Algorithm I' by the following:

B2) Let E_t denote the spectrum of $V_t^T[B_t(h_{t-1})]^{-1}V_t$. If E_t contains only nonnegative elements, then stop. (In this case h^* is unbounded.)

B3) Let $h_t = h_{t-1} - \frac{1}{\min\{\lambda:\lambda\in E_t,\lambda<0\}}$.

(All other steps are the same as in Algorithm I'.)

Now let us consider the simpler special case where $A(h) = A + huu^T$. In this case $|A(h)| = 0$ is a linear equation, the solution of which can be given explicitly by the help of Corollary 3.5. Moreover, in this case the set E introduced in (6.11) has a single element, namely

$$E = \{u^T[A(\hat{h})]^{-1}u\} = \left\{\frac{u^T \operatorname{adj} A(\hat{h})u}{|A(\hat{h})|}\right\}. \tag{6.13}$$

These observations make it possible, that Steps B2 and B3 of the algorithm become much simpler. We shall choose, however, a different approach, and apply the following assertion.

LEMMA 6.15 *Consider the proper parametral matrix* $A(h) = A + huu^T$, *and suppose that* $A(\hat{h})$ *is almost copositive for some* $\hat{h} \in R$. *Then we state the following:*

a) $|A(h)| = 0$ *if and only if*

$$u^T \operatorname{adj} A(\hat{h})u \neq 0 \tag{6.14}$$

and

$$h = \hat{h} - \frac{|A(\hat{h})|}{u^T \operatorname{adj} A(\hat{h})u}. \tag{6.15}$$

b) Let $\overline{h}$ *be an arbitrary real number. Then* $|A(h)| = 0$ *if and only if*

$$|A(\overline{h}+1)| - |A(\overline{h})| \neq 0 \tag{6.16}$$

and

$$h = \hat{h} - \frac{|A(\hat{h})|}{|A(\overline{h}+1)| - |A(\overline{h})|}. \tag{6.17}$$

c) For the solution of $|A(h)| = 0$, $h > \hat{h}$ *if and only if*

$$|A(\overline{h}+1)| - |A(\overline{h})| > 0. \tag{6.18}$$

d) If (6.18) holds, then denoting by h^* *the value obtained according to the right hand side expression of (6.15), and by* v^* *an arbitrary nonzero column vector of* adj $A(h^*)$, *the following are true:*

$$A(h^*)v^* = 0, v^* \geq 0, \text{ and } u^T v^* \neq 0.$$

Proof. By applying Corollary 3.5 we get

$$|A(h)| = |A(\hat{h})| + (h - \hat{h})u^T \operatorname{adj} A(\hat{h})u \tag{6.19}$$

where

$$|A(\hat{h})| < 0 \tag{6.20}$$

according to Theorem 6.3. This implies assertion a).

From (6.19) it follows that

$$|A(\overline{h}+1)| - |A(\overline{h})| = u^T \operatorname{adj} A(\hat{h}) u \tag{6.21}$$

which validates assertion b). Assertion c) follows from (6.17) and (6.20). To prove assertion d), using the matrix identity

$$A \operatorname{adj} A = |A| \cdot I,$$

and applying Lemma 6.12, one can take the conclusion that $A(h^*)v^* = 0$ and $v^* \geq 0$. Clearly, (6.21) remains true when we substitute $\hat{h}$ by an arbitrary real number. Thus

$$u^T \operatorname{adj} A(h^*)\, u = |A(\overline{h}+1)| - |A(\overline{h})| > 0.$$

From this it follows that $u^T \operatorname{adj} A(h^*) \neq 0$. Because the rank of the adjoint of a singular matrix cannot be greater than 1, $u^T v^* = 0$ would imply that $u^T \operatorname{adj} A(h^*) = 0$. As this is not the case, consequently $u^T v^* \neq 0$ is true.

Finally we remark that the difference of two determinants standing on the left of (6.21) is generally easier to handle than a quadratic form like the expression of the right hand side of (6.21).

The assertions stated in Lemma 6.15 can be used for the adaptation of Algorithm I' to the case $A(h) = A + huu^T$:

Algorithm III.

Replace steps B2)-B3) in Algorithm I' by the following:

B2) If $|B_t(h_{t-1}+1)| - |B_t(h_{t-1})| \leq 0$, then stop. (In this case h^* is unbounded.)

B3) Let

$$h_t = h_{t-1} - \frac{|B_t(h_{t-1})|}{|B_t(h_{t-1}+1)| - |B_t(h_{t-1})|}.$$

(All other steps are the same as in Algorithm I'.)

References

1. E. Bodewig, *Matrix calculus*, North-Holland Publishing Co., Amsterdam, 1956.
2. I. M. Bomze, On standard quadratic optimization problems, *J. Global Optimiz.*, 13:369–387, 1998.

3. I. M. Bomze, and G. Danninger, A finite algorithm for solving general quadratic problems, *J. Global Optimiz.*, 4:1–16, 1994.
4. R. W. Cottle, G. J. Habetler, and C. E. Lemke, On classes of copositive matrices, *Linear Algebra and its Applications*, 3:295–310, 1970.
5. R. W. Cottle, J. S. Pang, and R. E. Stone, *The linear complementarity problem*, Academic Press, 1992.
6. G. Danninger, Role of copositivity in optimality criteria for nonconvex optimization problems, *J. Optim. Theo. Appl.*, 75:535–558, 1992.
7. G. Danninger, and I. M. Bomze, Using copositivity for global optimality criteria in concave quadratic programming problems, *Math. Programming*, 62:575–580, 1993.
8. W. J. Duncan, Some devices for the solution of large sets of simultaneous linear equations, *Phil. Mag.*, 35:660–670, 1944.
9. K. P. Hadeler, On Copositive Matrices, *Linear Algebra and its Applications*, 49:79–89, 1983.
10. A. Householder, *The theory of matrices in numerical analysis*, Blaisdell Publishing Co., Boston, Massachusets, 1964.
11. G. Kéri, On a class of quadratic forms, in: A. Prékopa, ed., Survey of mathematical programming, 231-247, Akadémiai Kiadó, Budapest, 1979.
12. K. G. Murty, and S. N. Kabadi, Some NP-complete problems in quadratic and linear programming, *Math. Programming*, 39:117–129, 1987.
13. J. M. Ortega and W. C. Rheinboldt, *Iterative solution of nonlinear equations in several variables*, McGraw-Hill, New York, 1970.
14. J. Sherman and W. J. Morrison, Adjustment of an inverse matrix corresponding to changes in the elements of a given column or a given row of the original matrix, *Ann. Math. Statist.*, 20:621, 1949.
15. H. Väliaho, Criteria for Copositive Matrices, *Linear Algebra and its Applications*, 81:19–34, 1986.
16. H. Väliaho, Almost Copositive Matrices, *Linear Algebra and its Applications*, 116:121–134, 1989.
17. M. A. Woodbury, *Inverting modified matrices*, Statist. Res. Group, Mem. Rep., No. 42., Princeton Univ., Princeton, N. J., 1950.

Duality For D.C. Optimization Over Compact Sets

Juan-Enrique Martinez-Legaz
CODE and Departament d'Economia i d'Història Econòmica
Universitat Autònoma de Barcelona
08193 Bellaterra, Spain

Michel Volle
Département de Mathématiques
Université d'Avignon
33, rue Louis Pasteur, 84000 Avignon, France

1. Introduction

Convex duality theory has been successfully used for different optimization problems dealing with differences of convex functions (see for instance [10, 8, 3, 12, 11, 4] and references therein). This paper contains an alternative approach to the duality theory for d.c. programming problems developed in [6]. That theory associates to the general problem of minimizing a d.c. function under several d.c. constraints on an arbitrary locally convex space a dual problem defined in terms of conjugate functions, in such a way that, under suitable constraint qualifications, a strong duality theorem holds. These constraint qualifications are satisfied in specific situations considered in [6]. However they are in general difficult to verify.

In this work we follow a quite different strategy: we only consider problems over compact convex sets, on which the objective and constraint functions take finite values. Thanks to these simplifying assumptions we get a strong duality theorem without any constraint qualification. Another advantage of this new approach is that, unlike in [6], here we obtain an (essentially) unconstrained dual problem. However, contrary to [6], no multipliers appear in the formulas given in the present paper.

We consider two applications: the minimization of a d.c. function on a (not necessary convex) compact set, and the general linear programming problem with 0,1 variables. None of these problems can be analyzed in the framework of [6], since, in both cases, the needed constraint qualifications do not hold.

[1] This work was partially supported by DGICYT (Spain), Project PB98-0867, and by the Comissionat per Universitats i Recerca de la Generalitat de Catalunya. Grant 1998SGR-00062.

F. Giannessi et al. (eds.), Optimization Theory, 139–146.

2. The problem

Given a real topological vector space X with topological dual X^*, we are concerned with the problem

$$(\mathcal{P}): \quad \begin{array}{ll} \text{minimize} & g(x) - h(x) \\ \text{subject to} & g_i(x) - h_i(x) \leq 0 \qquad \forall i \in I_m := \{1, \cdots, m\} \\ & x \in K\,, \end{array}$$

where K is a (nonempty) compact convex subset of X and $g, h, g_i, h_i : X \to \mathbb{R} \cup \{+\infty\}$ are extended real-valued functions taking finite values on K. We assume that the functions g, g_i ($i \in I_m$) are lower semicontinuous and convex on K. The assumptions concerning the functions h, h_i will be made precise later on.

We denote by $F(\mathcal{P})$ (resp. $v(\mathcal{P})$) the feasible set (resp. the optimal value) of problem $(\mathcal{P})$. Observe that $F(\mathcal{P})$ may be empty: we set, as usual, $\inf \emptyset = +\infty$.

To each function $f : X \to \mathbb{R} \cup \{+\infty\}$ corresponds its Legendre-Fenchel conjugate f^*, which is defined on X^* by

$$f^*(x^*) := \sup_{x \in \operatorname{dom} f} \{\langle x, x^* \rangle - f(x)\}\,.$$

Here $\operatorname{dom} f := \{x \in X : f(x) < +\infty\}$ denotes the effective domain of f and, as usual, $\langle\ ,\ \rangle$ is the bilinear pairing between X and X^*. The biconjugate f^{**} of f is defined on X by

$$f^{**}(x) := \sup_{x^* \in \operatorname{dom} f^*} \{\langle x, x^* \rangle - f^*(x^*)\}\,.$$

One always has the Fenchel inequality

$$\langle x, x^* \rangle - f(x) \leq f^*(x^*)\,;$$

equality holds at $x \in \operatorname{dom} f$ iff x^* belongs to the subdifferential of f at x, which is defined by

$$\partial f(x) := \{x^* \in X^* : f(u) - f(x) \geq \langle u - x, x^* \rangle \quad \forall u \in X\}\,.$$

The function f is said to be subdifferentiable at x if $\partial f(x) \neq \emptyset$. We denote by δ_K the indicator function of K: $\delta_K(x) = 0$ if $x \in K$, $\delta_K(x) = +\infty$ if $x \in X \backslash K$. Let us begin by rewriting the feasible set of $\mathcal{P}$ (compare with [6] Lemma 2.1).

LEMMA 1. *Assume that the h_i's are subdifferentiable on $F(\mathcal{P})$, then*

$$F(\mathcal{P}) = \bigcup_{(x_1^*, \cdots, x_m^*) \in \prod_{i=1}^m \operatorname{dom} h_i^*} \{x \in K : g_i(x) - \langle x, x_i^* \rangle + h_i^*(x_i^*) \leq 0 \ \forall i \in I_m\}\,.$$

Proof. Let $(x_1^*, \cdots, x_m^*) \in \prod_{i=1}^{m} \text{dom } h_i^*$ and $x \in K$ be such that, for each $i \in I_m$, $g_i(x) - \langle x, x_i^* \rangle + h_i^*(x_i^*) \leq 0$; from the Fenchel's inequality it follows that $g_i(x) - h_i(x) \leq 0$, so that $x \in F(\mathcal{P})$: inclusion $\supset$ holds in Lemma 1. Conversely, let $x \in F(\mathcal{P})$; for each $i \in I_m$ choose $x_i^* \in \partial h_i(x)$; by the Fenchel's equality we then have $x_i^* \in \text{dom } h_i^*$ and $0 \geq g_i(x) - h_i(x) = g_i(x) - \langle x, x_i^* \rangle + h_i(x_i^*)$ for any $i \in I_m$ and the proof is complete. □

We also need another lemma from convex analysis.

LEMMA 2. *Let $f, f_1, \cdots, f_m$ be real-valued lower semicontinuous (l.s.c.) convex functions on the compact convex set K; then*

$$\inf \{f(x) : x \in K, f_i(x) \leq 0 \;\; \forall i \in I_m\} = \sup_{\lambda \in \mathbb{R}_+^m} \inf_{x \in K} \{f(x) + \sum_{i=1}^{m} \lambda_i f_i(x)\}.$$

Proof. It is well known that the left-hand member above coincides with

$$\inf_{x \in K} \sup_{\lambda \in \mathbb{R}_+^m} \{f(x) + \sum_{i=1}^{m} \lambda_i f_i(x)\}.$$

Now equality directly comes from Sion's minimax theorem ([9] Theorem 4.2). □

We are now in a position to state the main result of this paper, in which we denote by Δ the convex set

$$\Delta := \text{dom } h^* \times \text{dom } h_1^* \times \cdots \times \text{dom } h_m^*,$$

and adopt the convention $0 \cdot (+\infty) = +\infty$.

THEOREM 1. *Let g, g_i, h, h_i be as in the beginning of the section; assume moreover that the h_i's are subdifferentiable on $F(\mathcal{P})$ and that $h(x) = h^{**}(x)$ for any $x \in F(\mathcal{P})$; then*

$$v(\mathcal{P}) = \inf_{(x^*, x_1^*, \cdots, x_m^*) \in \Delta} \; \sup_{\lambda = (\lambda_1, \cdots, \lambda_m) \in \mathbb{R}_+^m} \{h^*(x^*) + \sum_{i=1}^{m} \lambda_i h_i^*(x_i^*) - $$
$$- (g + \delta_K + \sum_{i=1}^{m} \lambda_i g_i)^*(x^* + \sum_{i=1}^{m} \lambda_i x_i^*)\}.$$

Proof. As $h(x) = h^{**}(x)$ for any $x \in F(\mathcal{P})$, Lemma 1 entails easily

$$v(\mathcal{P}) = \inf_{(x^*, x_1^*, \cdots, x_m^*) \in \Delta} \inf_{x \in K} \{g(x) - \langle x, x^* \rangle + h^*(x^*) :$$
$$g_i(x) - \langle x, x_i^* \rangle + h_i^*(x_i^*) \leq 0 \quad \forall i \in I_m\}.$$

Now observe that for any $(x^*, x_1^*, \cdots, x_m^*) \in \Delta$, $h^*(x^*)$, $h_1^*(x_1^*)$, $\cdots$, $h_m^*(x_m^*)$ are real numbers (otherwise the functions h^*, h_1^*, $\cdots$, h_m^* would be identically equal to $+\infty$). Applying Lemma 2 with

$$f(x) = g(x) - \langle x, x^* \rangle + h^*(x^*), \qquad f_i(x) = g_i(x) - \langle x, x_i^* \rangle + h_i^*(x_i^*),$$

we obtain

$$v(\mathcal{P}) = \inf_{(x^*, x_1^*, \cdots, x_m^*) \in \Delta} \sup_{\lambda \in \mathbb{R}_+^m} \inf_{x \in K} \{ g(x) - \langle x, x^* \rangle + h^*(x^*) + \\ + \sum_{i=1}^m \lambda_i (g_i(x) - \langle x, x_i^* \rangle + h_i^*(x_i^*)) \}.$$

We conclude the proof by using the very definition of the Legendre-Fenchel conjugate of the function

$$g + \delta_K + \sum_{i=1}^m \lambda_i \, g_i .$$

□

Let us pay a special attention to the case when $h = 0$.

COROLLARY 1. *Let g, g_i, h_i be as in Theorem 1 and $h = 0$; then*

$$v(\mathcal{P}) = \inf \{ g(x) : x \in K, \; g_i(x) - h_i(x) \leq 0 \quad \forall i \in I_m \} =$$

$$\inf_{(x_1^*, \cdots, x_m^*) \in \prod_{i=1}^m \operatorname{dom} h_i^*} \sup_{\lambda \in \mathbb{R}_+^m} \{ \sum_{i=1}^m \lambda_i h_i^*(x_i^*) - (g + \delta_K + \sum_{i=1}^m \lambda_i g_i)^* (\sum_{i=1}^m \lambda_i x_i^*) \}.$$

Moreover, if $\bar{x}$ is an optimal solution of $(\mathcal{P})$ then the infimum on the right-hand side is attained at any

$$(x_1^*, \cdots, x_m^*) \in \prod_{i=1}^m \partial h_i(\bar{x}).$$

Proof. Since $h = 0$ one has $h^* = \delta_{\{0\}}$, so that dom $h^* = \{0\}$ and the formula comes directly from Theorem 1. Assume now that $x_i^* \in \partial h_i(\bar{x})$ for any $i \in I_m$; then, using Fenchel equality and Fenchel inequality,

$$\sum_{i=1}^m \lambda_i h_i^*(x_i^*) - (g + \delta_K + \sum_{i=1}^m \lambda_i g_i)^* (\sum_{i=1}^m \lambda_i x_i^*) \leq$$

$$\sum_{i=1}^m \lambda_i (\langle \bar{x}, x_i^* \rangle - h_i(\bar{x})) - \langle \bar{x} \, , \sum_{i=1}^m \lambda_i x_i^* \rangle + g(\bar{x}) + \sum_{i=1}^m \lambda_i g_i(\bar{x}) =$$

$$g(\bar{x}) + \sum_{i=1}^m \lambda_i (g_i(\bar{x}) - h_i(\bar{x})) \leq g(\bar{x}) = v(\mathcal{P}),$$

which completes the proof. □

Remark 1. Observe that problem $(\mathcal{P})$ does admit an optimal solution if $F(\mathcal{P}) \neq \emptyset$, and h, h_i are upper semicontinuous on K.

In the case where X is a Hausdorff locally convex topological vector space one can say more. First of all it suffices to assume that the convex set K is compact with respect to the weak topology $\sigma(X, X^*)$. This is due to the fact that g and the $g_i's$ being l.s.c. and convex, these functions are automatically l.s.c. with respect to $\sigma(X, X^*)$. Moreover, K being $\sigma(X, X^*)$-compact, the support function δ_K^* is finite-valued and continuous on X^* with respect to the Mackey topology $\tau(X^*, X)$. It then follows from [7] Proposition 9b that

$$(g + \sum_{i=1}^{m} \lambda_i g_i + \delta_K)^* = (g + \sum_{i=1}^{m} \lambda_i g_i)^* \square \delta_K^*,$$

where the symbol $\square$ denotes the infimal convolution operator: the infimal convolution of two functions $\varphi, \psi : X^* \to \mathbb{R} \cup \{+\infty\}$ is given by

$$(\varphi \square \psi)(x^*) = \inf_{u^* \in X^*} \{\varphi(u^*) + \psi(x^* - u^*)\} .$$

Consequently we can state

THEOREM 2. *Assume that K is a $\sigma(X, X^*)$-compact convex subset of the Hausdorff locally convex space X and let g, g_i, h, h_i be as in Theorem 1; then*

$$v(\mathcal{P}) = \inf_{(x^*, x_1^*, \cdots, x_m^*) \in \Delta} \sup_{u^* \in X^*} \sup_{\lambda \in \mathbb{R}_+^m} \{h^*(x^*) + \sum_{i=1}^{m} \lambda_i h_i^*(x_i^*) -$$
$$((g + \sum_{i=1}^{m} \lambda_i g_i)^*(u^*) + \delta_K^*(x^* - u^* + \sum_{i=1}^{m} \lambda_i x_i^*))\} .$$

Remark 2. In the case where the convex functions $g_1, \cdots, g_m$ are continuous at a common point of K (where g is finite) one has moreover ([7])

$$(g + \sum_{i=1}^{m} \lambda_i g_i)^*(u^*) = \min \{g^*(u_0^*) + \sum_{i=1}^{m} (\lambda_i g_i)^*(u_i^*) : \sum_{j=0}^{m} u_j^* = u^*\} .$$

Observe that $\lambda_i g_i = \delta_{\text{dom } g_i}$ when $\lambda_i = 0$. According to Theorem 2, the value of problem $(\mathcal{P})$ is then entirely expressed in terms of the Legendre-Fenchel conjugate of the data: $g^*, g_i^*, h^*, h_i^*, \delta_K^*, \delta_{\text{dom } g_i}^*$.

3. Applications

3.1. Minimizing a d.c. function over a compact set

Here X is a Hilbert space and C is a compact subset of X. We consider the problem

$$(\mathcal{P}_1): \quad \text{minimize } g(x) - h(x) \qquad \text{subject to } x \in C,$$

where $g, h : X \to \mathbb{R} \cup \{+\infty\}$ are finite on the (compact) closed convex hull $K = \overline{\text{co}}\, C$ of C. We assume that g is l.s.c. and convex on K and that $h(x) = h^{**}(x)$ for any $x \in C$.

Problem $(\mathcal{P}_1)$ can be clearly reformulated as

$$\text{minimize } g(x) - h(x) \qquad \text{subject to } \frac{1}{2}\, d^2(x, C) \leq 0,$$

where

$$d(x, C) = \inf_{u \in C} \|x - u\|$$

denotes the distance between $x \in X$ and C. By using Asplund formula ([1])

$$\frac{1}{2}\, d^2(x, C) = \frac{1}{2}\, \|x\|^2 - (\frac{1}{2}\, \|\ \|^2 + \delta_C)^*(x)\,,$$

problem $(\mathcal{P}_1)$ can be rewritten as

$$\begin{aligned} &\text{minimize} && g(x) - h(x) \\ &\text{subject to} && \tfrac{1}{2}\|x\|^2 - (\tfrac{1}{2}\, \|\ \|^2 + \delta_C)^*(x) \leq 0 \\ &&& x \in \overline{\text{co}}\, C\,. \end{aligned}$$

In order to apply Theorem 1 to problem $(\mathcal{P}_1)$ let us observe that $h_1 := (\frac{1}{2}\, \|\ \|^2 + \delta_C)^*$ is a l.s.c. convex function which, C being compact, is finite everywhere. Therefore h_1 is continuous and subdifferentiable on the whole Hilbert space X ([2]). Moreover, the l.s.c. convex function $\frac{1}{2}\, \|\ \|^2 + \delta_{\overline{\text{co}}\, C}$ is a minorant of $\frac{1}{2}\, \|\ \|^2 + \delta_C$, so that

$$\frac{1}{2}\, \|\ \|^2 + \delta_{\overline{\text{co}}\, C} \;\leq\; (\frac{1}{2}\, \|\ \|^2 + \delta_C)^{**} = h_1^* \;\leq\; \frac{1}{2}\, \|\ \|^2 + \delta_C\,.$$

It follows that

$$C \subset \text{dom}\ h_1^* \subset \overline{\text{co}}\, C$$

and, as dom h_1^* is convex, $\overline{\text{dom}\ h_1^*} = \overline{\text{co}}\, C$. From Theorem 1 we then have

COROLLARY 2.

$$v(\mathcal{P}_1) = \inf_{(x^*,x_1^*)\in \text{dom } h^* \times \overline{\text{co}}\, C} \sup_{\lambda \geq 0} \{h^*(x^*) + \lambda(\frac{1}{2} \|\ \|^2 + \delta_C)^{**}(x_1^*) - (g + \delta_{\overline{\text{co}}\, C} + \frac{\lambda}{2} \|\ \|^2)^*(x^* + \lambda x_1^*)\} .$$

Of course Corollary 2 specializes to the minimization of a convex (resp. concave) function over a compact set. We omit the corresponding formulas. Observe also that Corollary 1 can be applied to problem $(\mathcal{P}_1)$.

3.2. 0–1 LINEAR PROGRAMMING

Given $c \in \mathbb{R}^n$, $a_j \in \mathbb{R}^n$, $b_j \in \mathbb{R}$ $(j \in I_p)$, let us consider the general linear programming problem with 0,1 variables:

$$\begin{array}{lll} (\mathcal{P}_2): & \text{minimize} & \langle c, x\rangle \\ & \text{subject to} & b_j - \langle a_j, x\rangle \leq 0\,,\ j \in I_p \\ & & x \in \{0,1\}^n . \end{array}$$

This problem can be easily reformulated (with $e := (1, \cdots, 1)$) as

$$\begin{array}{ll} \text{minimize} & \langle c, x\rangle \\ \text{subject to} & b_j - \langle a_j, x\rangle \leq 0\,,\ j \in I_p \\ & \frac{1}{2}\langle e, x\rangle - \frac{1}{2}\|x\|^2 \leq 0 \\ & x \in \{0,1\}^n . \end{array}$$

At this stage one can apply Corollary 1 by taking $m = p+1$, $g = \langle c, \cdot\rangle$, $g_j = b_j - \langle a_j, \cdot\rangle$ and $h_j = 0$ for $j \in I_p$, $g_m = \frac{1}{2}\ \langle e, \cdot\rangle$, $h_m = \frac{1}{2}\|\ \|^2$ and $K = [0,1]^n$. In this way we get the following formula, where $r^+ = \max\{r, 0\}$ denotes the positive part of a real number r and x_k^*, c_k, a_{jk} $(k \in I_n,\ j \in I_p)$ are the components of x^*, c, a_j, respectively.

COROLLARY 3.

$$v(\mathcal{P}_2) = \inf_{x^* \in \mathbb{R}^n} \sup_{\lambda \in \mathbb{R}_+^{p+1}} \{\frac{\lambda_{p+1}}{2} \|x^*\|^2 - \\ - \sum_{k=1}^{n} (\lambda_{p+1}(x_k^* - \frac{1}{2}) - c_k + \sum_{j=1}^{p} \lambda_j a_{jk})^+ + \sum_{j=1}^{p} \lambda_j b_j\} .$$

Remark 3. According to Corollary 1 and denoting by $\bar{x}$ an optimal solution of problem $(\mathcal{P}_2)$ (which exists whenever $(\mathcal{P}_2)$ is feasible), the infimum on $x^* \in \mathbb{R}^n$ in Corollary 3 is attained at $x^* \in \partial h_m(\bar{x}) = \{\bar{x}\}$.

4. Acknowledgement

The contribution of the first author was completed while he was visiting the Laboratoire d'Analyse Appliquée et Optimisation of the Université de Bourgogne. He is grateful to Abderrahim Jourani for his invitation and to him and the other members of the Laboratoire by their kind hospitality during the visit.

References

1. E. Asplund, "Differentiability of the metric projection in finite-dimensional Euclidean space", *Proc. Amer. Math. Soc.*, Vol. 38 (1973), 218-219.
2. I. Ekeland, R. Temam, *"Analyse convexe et problèmes variationnels"*, Dunod, Gauthier-Villars, 1974.
3. J.-B. Hiriart-Urruty, "Generalized differentiability, duality and optimization for problems dealing with differences of convex functions". *Convexity and duality in optimization* (Groningen, 1984), 37-70, Lecture Notes in Econom. and Math. Systems, 256, Springer, Berlin-New York, 1985.
4. R. Horst, N. V. Thoai, "DC programming: overview", *J. Optim. Theory Appl.*, Vol. 103 (1999), no. 1, 1-43.
5. B. Lemaire, "Duality in reverse convex optimization", *SIAM J. Optim.*, Vol. 8 (1998), no. 4, 1029-1037.
6. J.-E. Martinez-Legaz, M. Volle, "Duality in d.c. programming : the case of several d.c. constraints", *J. Math. Anal. Appl.*, Vol. 237 (1999), 657-671.
7. J.-J. Moreau, *"Fonctionnelles convexes"*, Séminaire sur les équations aux dérivées partielles, Collège de France, 1966.
8. I. Singer, "A Fenchel-Rockafellar type duality theorem for maximization", *Bull. Austral. Math. Soc.*, Vol. 20 (1979), no. 2, 193–198.
9. M. Sion, "On general minimax theorems", *Pacific J. Math.*, Vol. 8 (1958), 171-176.
10. J. F. Toland, "Duality in nonconvex optimization", *J. Math. Anal. Appl.*, Vol. 66 (1978), no. 2, 399-415.
11. H. Tuy, "D.C. optimization: theory, methods and algorithms", *Handbook of global optimization,* 149-216, Nonconvex Optim. Appl., 2, Kluwer Acad. Publ., Dordrecht, 1995.
12. M. Volle, "Concave duality: application to problems dealing with difference of functions", *Math. Programming*, Vol. 41 (1988), no. 2 (Ser. A), 261-278.

Characterization of Monotone Operators by Using a Special Class of Preinvex Functions

S. Z. Németh
Laboratory of Operations Research and Decision Systems,
Computer and Automation Institute, Hungarian Academy of Sciences,
H-1518 Budapest, P.O.Box 63 Hungary
e-mail: snemeth@oplab.sztaki.hu

Abstract. For every operator $F : \mathbf{R}^n \to \mathbf{R}^n$ the scalar increment $\Diamond F(x) : \mathbf{R}^n \to \mathbf{R}$ will be introduced at every point $x \in \mathbf{R}^n$, by the relation

$$\Diamond F(x)(u) = \langle F(x+u) - F(x), u \rangle,$$

where $\langle . , . \rangle$ denotes the canonical scalar product of $\mathbf{R}^n$. If F is hemicontinuous, it will be proved that F is monotone (with respect to duality) if and only if its scalar increment is preinvex in every point of $\mathbf{R}^n$. A special class of preinvex functions, called starvex functions, will be introduced. In the differentiable case this class of preinvex functions coincide with the corresponding class of invex functions. In the noncontinuous case it will be shown that the monotonicity of F coincides with the starvexity of its scalar increment at every point of $\mathbf{R}^n$. If F is differentiable the monotonicity of F will be proved to be equivalent with the invexities of all its scalar increments. Finally, a characterization for strictly monotone operators will be given.

Key words: monotone operator, invex set, preinvex function, invex function, local minimum, global minimum, stationary point

1. Introduction

Since M.A. Hanson wrote his famous article [3], more than 100 articles have been published in the subject of invexity. Everything started from the simple but very useful observation that for a differentiable function $f : \mathbf{R}^n \to \mathbf{R}$, the existence of a vector function η defined on $\mathbf{R}^n \times \mathbf{R}^n$ such that the inequality

$$f(v) - f(u) \geq \langle \nabla f(u), \eta(v,u) \rangle$$

holds for all $u, v \in \mathbf{R}^n$, implies that the stationary points of f are global minima. (Here $\langle . , . \rangle$ denotes the canonical scalar product of $\mathbf{R}^n$.) B. D. Craven proposed the term of *invex* for such functions [1] and proved that the class of invex functions is precisely the class of functions whose stationary points are global minima [2]. If f is invex with respect to a given η, then f is called η-*invex*. In the nondifferentiable case, we cannot speak about the gradient of a function. In this case, V. Jeyakumar proposed another generalization for convex functions, the preinvex functions [4]. First of all, a set $A \subset \mathbf{R}^n$ is called

F. Giannessi et al. (eds.), Optimization Theory, 147–157.

invex with respect to a vector function $\eta : \mathbf{R}^n \times \mathbf{R}^n \to \mathbf{R}^n$ (or in short η-*invex*) if for all u, $v \in \mathbf{R}^n$ and all $t \in [0,1]$

$$u + t\eta(v,u) \in A.$$

If $A \subset \mathbf{R}^n$ is η-invex and $f : A \to \mathbf{R}$, then f is called η-*preinvex* if the inequality

$$f(u + t\eta(v,u)) \leq (1-t)f(u) + tf(v)$$

holds for all u, $v \in A$ and all $t \in [0,1]$. If there is an η so that f is η-preinvex, then f is called *preinvex*. It can be easily proved that the local minima of such a function are global minima (see e.g. R. Pini [9]). R. Pini shows that in the differentiable case η-preinvexity implies η-invexity [9]. By a counterexample she proves that the converse is not true [9]. However, imposing restrictions on η, η-invexity implies η-preinvexity (see Condition C. [5]). We also remark R. Pini's article [8] where she defines the notion of invex functions on manifolds and relates it to geodesic convexity [10].

We study a class of invex functions, namely, $-pr_2$-invex functions where $pr_2(v,u) = u$. We call the $-pr_2$-preinvex functions *starvex-*functions. This term came from the very simple observation that the $-pr_2$-invex sets are exactly the starlike sets with respect to the origin of $\mathbf{R}^n$. In spite of the failure of Condition C [5] we prove that $-pr_2$-invexity implies $-pr_2$-preinvexity (if the domain of definition is starlike with respect to the origin of $\mathbf{R}^n$). Hence the $-pr_2$-invex functions which are defined on starlike domains with respect to the origin of $\mathbf{R}^n$ are exactly the differentiable starvex functions. The starvex functions are strictly related to monotone operators (i.e., the operators $F : \mathbf{R}^n \supset G \to \mathbf{R}^n$ which satisfy the inequality $\langle F(y) - F(x), y - x \rangle \geq 0$, for all x, $y \in G$). We call the *scalar increment* at $x \in \mathbf{R}^n$ of an operator $F : \mathbf{R}^n \to \mathbf{R}^n$ the function $\Diamond F(x) : \mathbf{R}^n \to \mathbf{R}$ defined by

$$\Diamond F(x)(u) = \langle F(x+u) - F(x), u \rangle.$$

We prove that F is monotone exactly when all its scalar increments are starvex. More generally if F is hemicontinuous, we prove that the monotonicity of F is equivalent with the preinvexity of $\Diamond F(x)$, for all $x \in \mathbf{R}^n$. In the differentiable case, we prove that the above result holds if preinvexity is replaced by invexity. If F is arbitrary, we prove that F is strictly monotone (i.e., F is monotone and $\Diamond F(x)(h) = 0$ implies $h = 0$) if and only if 0 is the only local minimum of $\Diamond F(x)$ for all $x \in \mathbf{R}^n$. We remark that for a fixed η, finding an invex function is a difficult problem. Moreover, for many η there do not exist nontrivial preinvex functions.

If, e.g., $\eta(v,u) = Av + Bu$, where A and B are linear operators and -1 is not an eigenvalue of B, it can be relatively easily proved that the η-preinvex functions $f : \mathbf{R}^n \to \mathbf{R}$ are the constant functions. To show this it is enough to take $t = 1$ in the defining inequality of the preinvexity and use the surjectivity of $B + I$, where I is the identical operator. Our paper outlines a nontrivial class of preinvex functions. Using monotonicity, many preinvex (and invex) functions can and will be generated.

Having fixed the notions of monotone and strictly monotone operator, invex set, invex and preinvex functions, in the next sections, we give the new definitions and results only.

2. Preliminary results

Let $pr_2 : \mathbf{R}^n \times \mathbf{R}^n \to \mathbf{R}^n$ be the projection function on the second component defined by

$$pr_2(v,u) = u,$$

for all u, $v \in \mathbf{R}^n$.

Recall that a set $A \subset \mathbf{R}^n$ is called starlike with respect to $x_0 \in \mathbf{R}^n$ if $(1-t)x_0 + tx \in \mathbf{R}^n$ for all $t \in [0,1]$ and $x \in A$.

We trivially have as follows:

LEMMA 2.1 *The set $A \subset \mathbf{R}^n$ is $-pr_2$-invex if and only if it is starlike with respect to the origin 0 of $\mathbf{R}^n$.*

This lemma motivates the following definition:

DEFINITION 2.1 *The $-pr_2$-preinvex functions are called* starvex.

First of all, let us see which functions defined on $\mathbf{R}$ are starvex.

THEOREM 2.1 *Let $f : \mathbf{R} \to \mathbf{R}$; $f(0) = 0$. Then, f is starvex if and only if $f(u) \geq 0$, for all $u \in \mathbf{R}$ and $g : \mathbf{R} \setminus \{0\} \to \mathbf{R}$ defined by*

$$g(u) = \frac{f(u)}{u}$$

is monotone increasing.

Proof. Suppose that $f : \mathbf{R} \to \mathbf{R}$; $f(0) = 0$, is a starvex function. We have

$$f(tu) \leq tf(u) + (1-t)f(v), \tag{2.1}$$

$\forall u,\, v \in \mathbf{R}$ and $\forall t \in [0,1]$. Inserting $t = 0$ into (2.1), we have $f(v) \geq 0$, $\forall v \in \mathbf{R}^n$. Replacing v with 0 in (2.1), we obtain

$$f(tu) \leq tf(u), \tag{2.2}$$

$\forall u \in \mathbf{R}$ and $\forall t \in [0,1]$.

Let $p,\, q \in \mathbf{R}$; $p \geq q$. We have to prove that

$$\frac{f(p)}{p} \geq \frac{f(q)}{q}. \tag{2.3}$$

If $p > 0 > q$, the inequality (2.3) is trivial (since f is nonnegative). If $p \geq q > 0$ $(0 > p \geq q)$, then $0 \leq q/p \leq 1$ $(0 \leq p/q \leq 1)$. Using inequality (2.2) with $t = q/p$ and $u = p$ $(t = p/q$ and $u = q)$, we obtain (2.3).

Conversely, suppose that f takes nonnegative values only and g is monotone increasing. We have to prove that

$$f(tu) \leq tf(u) + (1-t)f(v),$$

$\forall u,\, v \in \mathbf{R}$ and $\forall t \in [0,1]$. Since f takes nonnegative values only, it is sufficient to prove that

$$f(tu) \leq tf(u), \tag{2.4}$$

$\forall u \in \mathbf{R}$ and $\forall t \in [0,1]$. For $t = 0$, inequality (2.4) is trivial. Hence, we can suppose that $t \in (0,1]$. Then, $0 < tu \leq u$ $(0 > tu \geq u)$ if $u > 0$ $(u < 0)$. In both of the cases, the monotonicity of g implies (2.4). □

Our next aim is to see which functions defined on $\mathbf{R}^n$ are starvex.

THEOREM 2.2 *Let $f : \mathbf{R}^n \to \mathbf{R}$; $f(0) = 0$. Then, f is starvex if and only if $f(u) \geq 0$, for all $u \in \mathbf{R}^n$ and $g_u : (0,1] \to \mathbf{R}$ defined by*

$$g_u(t) = \frac{f(tu)}{t}$$

is monotone increasing for all $u \in \mathbf{R}^n$.

Proof. Suppose that f is starvex. Similarly to the proof of Theorem 2.1, it can be shown that $f(v) \geq 0$ and

$$f(tv) \leq tf(v), \tag{2.5}$$

$\forall v \in \mathbf{R}^n$ and $\forall t \in (0,1]$. Let $0 < p \leq q \leq 1$. We have to prove that

$$\frac{f(pu)}{p} \leq \frac{f(qu)}{q}, \tag{2.6}$$

$\forall u \in \mathbf{R}^n$. Since $p/q \in (0,1]$, we can apply inequality (2.5) with $t = p/q$ and $v = qu$, thus we obtain inequality (2.6).

Conversely, suppose that f is nonnegative and g_u is monotone for all $u \in \mathbf{R}^n$. Similarly to the proof of Theorem 2.1, it is sufficient to prove that

$$f(tu) \leq tf(u), \tag{2.7}$$

$\forall u \in \mathbf{R}^n$ and $\forall t \in [0,1]$. For $t = 0$, inequality (2.7) is trivial. If $0 < t \leq 1$, then (2.7) follows from the monotonicity of g_u. □

In the proof of Theorem 3.1, we shall use as follows:

LEMMA 2.2 *Let* $f : \mathbf{R}^n \to \mathbf{R}$; $f(0) = 0$, *be a preinvex function so that*

$$\limsup_{t \downarrow 0} \frac{f(tu)}{t} \geq 0, \tag{2.8}$$

for all $u \in \mathbf{R}^n$. *Then,* $f(u) \geq 0$ *for all* $u \in \mathbf{R}^n$.

Proof. Since f is preinvex, there is an $\eta : \mathbf{R}^n \times \mathbf{R}^n \to \mathbf{R}^n$ so that

$$f(u + t\eta(v,u)) \leq (1-t)f(u) + tf(v), \tag{2.9}$$

$\forall u$, $v \in \mathbf{R}$ and $\forall t \in (0,1]$. Inserting $u = 0$ into (2.9), dividing by t and after that tending with t to 0, (2.8) implies that f is nonnegative. □

Consider the differentiable case. We want to show that $-pr_2$-invexity is equivalent to $-pr_2$-preinvexity.

THEOREM 2.3 *Let* $f : \mathbf{R}^n \to \mathbf{R}$ *be a differentiable function with* $f(0) = 0$. *Then, the following assertions are equivalent:*

(i) $\langle \nabla f(u), u \rangle \geq f(u) \geq 0, \quad \forall u \in \mathbf{R}^n$;
(ii) f is starvex (i.e., $-pr_2$-preinvex);
(iii) f is $-pr_2$-invex.

Proof. (ii)⇒(iii) it is a well-known implication and can be found e.g., in [9].

(iii)⇒(i) Since f is pr_2-invex, we have

$$f(v) - f(u) \geq \langle \nabla f(u), -u \rangle, \tag{2.10}$$

$\forall u$, $v \in \mathbf{R}^n$. Inserting $u = 0$ into (2.10), we obtain $f(v) \geq 0$, $\forall v \in \mathbf{R}^n$. Replacing v with 0 in (2.10), we get $\langle \nabla f(u), u \rangle \geq f(u)$, $\forall u \in \mathbf{R}^n$.

(i)⇒(ii) Let $g_u : (0,1] \to \mathbf{R}$; $g_u(t) = f(tu)/t$. Since $\langle \nabla f(tu), tu \rangle \geq f(tu)$ for all $u \in \mathbf{R}^n$ and all $t \in (0,1)$, we obtain

$$g_u'(t) = \frac{\langle \nabla f(tu), u \rangle t - f(tu)}{t^2} \geq 0.$$

Hence, g_u is monotone increasing on (0,1). Since g_u is continuous, g_u is monotone increasing on (0,1]. Applying Theorem 2.2, it yields that f is starvex. □

DEFINITION 2.2 *A function* $f : \mathbf{R}^n \to \mathbf{R}$ *is called* starlike *with respect to* x_0 *if for all* $x \in \mathbf{R}^n$ *and all* $t \in [0,1]$ *we have*

$$f((1-t)x_0 + tx_0) \leq (1-t)x_0 + tx.$$

Using Theorem 2.2, it is easy to prove as follows:

THEOREM 2.4 *A function* $f : \mathbf{R}^n \to \mathbf{R}$ *is starvex if and only if it is starlike with respect to the origin of* $\mathbf{R}^n$ *and bounded below by f(0). Particularly, if f is convex and bounded below by f(0), then f is starvex.*

It is easy to construct convex functions which are not bounded below (for example, $f : \mathbf{R} \to \mathbf{R}$ given by $f(x) = x + e^x$). By the above theorem, these functions are not starvex. In the last section, we shall give examples of starvex functions which are not convex.

We remark that the restriction $f(0) = 0$ imposed several times throughout this section does not restrict the generality, since the starvexity of f is equivalent to the starvexity of $f - f(0)$. The only reason for imposing this condition is that the functions considered in the next section are of this type.

3. Monotone operators and starvex functions

Let $F : \mathbf{R}^n \to \mathbf{R}^n$, $x \in \mathbf{R}^n$, and $\Diamond F(x) : \mathbf{R}^n \to \mathbf{R}$ be the function defined by

$$\Diamond F(x)(h) = \langle F(x+h) - F(x), h \rangle,$$

where $\langle .\,, . \rangle$ is the canonical scalar product of $\mathbf{R}^n$.

DEFINITION 3.1 $\Diamond F(x)$ *is called the* scalar increment *of* F *at* x.

Then, we have as follows:

THEOREM 3.1 *Suppose that* F *is hemicontinuous. Then,* F *is monotone if and only if its scalar increment* $\Diamond F(x)$ *is preinvex, for all* $x \in \mathbf{R}^n$.

Proof. Suppose that F is monotone. Then,

$$\diamondsuit F(x)(0) = 0 \tag{3.1}$$

and

$$\diamondsuit F(x)(u) \geq 0, \tag{3.2}$$

$\forall u,\ x \in \mathbf{R}^n$. Let $g_{xu} : (0, 1] \to \mathbf{R}$ be the function defined by

$$g_{xu}(t) = \frac{\diamondsuit F(x)(tu)}{t} = \langle F(x+tu) - F(x),\ u\rangle.$$

Let $t,\ s \in (0, 1]$; $t > s$. Then, we have

$$g_{xu}(t) - g_{xu}(s) = \frac{1}{t-s}\langle F(x+tu) - F(x+su),\ x+tu-(x+su)\rangle \geq 0,$$

since F is monotone. Hence,

$$g_{xu} \text{ is monotone.} \tag{3.3}$$

Bearing (3.1), (3.2) and (3.3) in mind, Theorem 2.2 implies that $\diamondsuit F(x)$ is starvex (hence preinvex) for all $x \in \mathbf{R}^n$.

Conversely, suppose that $\diamondsuit F(x)$ is preinvex for all $x \in \mathbf{R}^n$. Since F is hemicontinuous, we get

$$\limsup_{t \downarrow 0} \frac{\diamondsuit F(x)(tu)}{t} = \limsup_{t \to 0} \langle F(x+tu) - F(x),\ u\rangle = 0, \tag{3.4}$$

$\forall u,\ x \in \mathbf{R}^n$. Bearing (3.4) and $\diamondsuit F(x)(0) = 0$ in mind, $\forall x \in \mathbf{R}^n$, Lemma 2.2 implies

$$\diamondsuit F(x)(u) = \langle F(x+u) - F(x),\ u\rangle \geq 0,$$

$\forall u,\ x \in \mathbf{R}^n$. Hence, F is monotone. □

Without imposing restrictions on F, we have as follows:

THEOREM 3.2 *F is monotone if and only if its scalar increment $\diamondsuit F(x)$ is starvex for all $x \in \mathbf{R}^n$.*

Proof. The necessity can be proved similarly to Theorem 3.1.

Sufficiency. Since $\diamondsuit F(x)$ is starvex and $\diamondsuit F(x)(0) = 0$, $\forall x \in \mathbf{R}^n$, Theorem 2.2 implies

$$\diamondsuit F(x)(u) = \langle F(x+u) - F(x),\ u\rangle \geq 0,$$

$\forall u,\ x \in \mathbf{R}^n$. Hence, F is monotone. □

Using arguments similar to the above ones, the following can be proved:

THEOREM 3.3 *Let $A \subset \mathbf{R}^n$ be a starlike domain, with respect to the origin of $\mathbf{R}^n$. If $F : A \to \mathbf{R}^n$ is monotone, then $\diamondsuit F(0)$ is starvex.*

We remark that using Theorem 3.3 and the examples for monotone operators given in [6] and [7], it is possible to construct several examples of starvex functions. In the next section, we shall give some examples of starvex functions by using Theorems 2.2, 3.2 and 3.3.

In the differentiable case, we have as follows:

THEOREM 3.4 *Suppose that F is differentiable. Then, F is monotone if and only if its scalar increment $\diamondsuit F(x)$ is invex for all $x \in \mathbf{R}^n$.*

Proof. The necessity follows from Theorem 2.3 and Theorem 3.1.

Sufficiency. A simple computation yields the following formula:

$$d(\diamondsuit F(x))(u)(v) = \langle dF(x+u)(v),\, u\rangle + \langle F(x+u) - F(x),\, v\rangle, \quad (3.5)$$

$\forall x,\, u,\, v \in \mathbf{R}^n$. Inserting $u = 0$ into (3.5), we obtain

$$d(\diamondsuit F(x))(0)(v) = 0,$$

$\forall v,\, x \in \mathbf{R}^n$. Hence, $d(\diamondsuit F(x))(0) = 0$, $\forall x \in \mathbf{R}^n$. This means that 0 is a stationary point of $\diamondsuit F(x)$. Since $\diamondsuit F(x)$ is invex, 0 is a global minimum [2]. Hence, $\diamondsuit F(x)(u) \geq \diamondsuit F(x)(0) = 0$, for all $u,\, x \in \mathbf{R}^n$. Hence, F is monotone. □

If F is monotone, Theorem 3.1 implies that $\diamondsuit F(x)$ is starvex, hence, preinvex. Thus the local minima of $\diamondsuit F(x)$ are global minima. Suppose that u_0 is a global minimum of $\diamondsuit F(x)$. Hence,

$$\diamondsuit F(x)(u_0) \leq \diamondsuit F(x)(0) = 0\,.$$

On the other hand, the monotonicity of F implies that $\diamondsuit F(x)(u_0) \geq 0$. Hence, $\diamondsuit F(x)(u_0) = 0$. Conversely, if $u_0 \in \mathbf{R}^n$ so that $\diamondsuit F(x)(u_0) = 0$, the monotonicity of F implies that u_0 is a global minimum of $\diamondsuit F(x)$. Hence, if F is monotone, the set of global minima of $\diamondsuit F(x)$ is

$$M = \{u \in \mathbf{R}^n \mid \langle F(x+u) - F(x),\, u\rangle = 0\}. \quad (3.6)$$

This set is starlike with respect to the origin of $\mathbf{R}^n$, since it is a level set of the $-pr_2$-preinvex function $\diamondsuit F(x)$ and the $-pr_2$-invex sets are the starlike sets with respect to the origin of $\mathbf{R}^n$ (see Lemma 2.1 and Proposition 4.5 [9]). This assertion can be also verified directly without using Proposition 4.5 [9].

Similarly to the proof of Theorem 5.1 [6], it can be shown that if 0 is a local minimum of $\diamondsuit F(x)$ for all $x \in \mathbf{R}^n$, then F is monotone.

We have as follows:

THEOREM 3.5 *F is strictly monotone if and only if 0 is the only local minimum of $\diamondsuit F(x)$ for all $x \in \mathbf{R}^n$.*

Proof. *Necessity.* F is strictly monotone, hence, (3.6) implies that 0 is the only local minimum of $\diamondsuit F(x)$ for all $x \in \mathbf{R}^n$.

Sufficiency. Since 0 is a local minimum of $\diamondsuit F(x)$ for all $x \in \mathbf{R}^n$, F is monotone (as we remarked above). Suppose that F is not strictly monotone. Hence, there is an $x_0 \in \mathbf{R}^n$ and an $u_0 \in \mathbf{R}^n \setminus \{0\}$ so that

$$\diamondsuit F(x_0)(u_0) = 0. \tag{3.7}$$

By (3.6) and (3.7), $u_0 \neq 0$ is a local minimum of $\diamondsuit F(x_0)$. Hence, we obtained a contradiction. Thus, F is strictly monotone.

4. Examples

4.1. Example of a nonconvex starvex function

Consider the function $f : \mathbf{R} \to \mathbf{R}$ defined by

$$f(x) = ax^4 + bx^3 + cx^2.$$

If $a > 0$ and $(24/9)ac \leq b^2 \leq 3ac$, it is easy to verify, using Theorem 2.1, that f is starvex but not convex. Moreover, the set of starvex but not convex functions of the form $ax^4 + bx^3 + cx^2$ is given by the above inequalities. □

4.2. Example of a nonconvex starvex function

Let $F : \mathbf{R}^2 \to \mathbf{R}^2$ be defined by

$$F(x, y) = (x^2 - y^2, 2xy).$$

By using the computational methods of [7], it can be easily proved that F is monotone on

$$A = \{(x, y) \in \mathbf{R}^2 \mid x \geq 0\}.$$

On the other hand, $\diamondsuit F(0)(x, y) = x(x^2 + y^2)$. Hence, by Theorem 3.3, the function $f : A \to \mathbf{R}$ defined by

$$f(x, y) = x(x^2 + y^2)$$

is starvex. But F is not convex, since $\det Hf(1, 2) = -4 < 0$, where Hf is the Hessian of f. □

4.3. Example of a nonconvex starvex function

Similarly to the previous example, it can be proved that $F : \mathbf{R}^2 \to \mathbf{R}^2$ defined by

$$F(x, y) = (e^x \cos y - 1, e^x \sin y)$$

is monotone on

$$A = \{(x, y) \in \mathbf{R}^2 \mid y \in [-\pi/2, \pi/2]\}.$$

On the other hand, $\Diamond F(0)(x, y) = xe^x \cos y + ye^x \sin y - x$. Hence, by Theorem 3.3, the function $f : A \to \mathbf{R}^2$ defined by

$$f(x, y) = xe^x \cos y + ye^x \sin y - x$$

is starvex. But F is not convex, since

$$\frac{\partial^2 f}{\partial x^2}(-3 - \frac{\pi}{4}, \frac{\pi}{4}) = \frac{-\sqrt{2}}{2} e^{-3-\frac{\pi}{4}} < 0.$$

□

4.4. Example of a nonconvex starvex function of n variables

Consider the function $f : \mathbf{R}^n \to \mathbf{R}$ defined by

$$f(x) = \sum_{i=1}^{n} (a_i x_i^4 + b_i x_i^3 + c_i x_i^2),$$

where $x = (x_1, x_2, \ldots, x_n)$. Suppose that $a_i > 0$ and $b_i^2 \leq 3a_i c_i$ for all $i = 1, 2, \ldots, n$ and there is an $i_0 \in \{1, 2, \ldots, n\}$ such that $b_{i_0}^2 \geq (24/9) a_{i_0} c_{i_0}$. It is easy to verify, by using Theorem 3.2 that f is starvex but not convex. □

4.5. Example of a nonsmooth starvex function of n variables

Let F be a projection map onto a closed convex set C of $\mathbf{R}^n$ with nonempty and nonsmooth boundary. By [11], F is monotone. Hence, using Theorem 3.2, the function $f_x : \mathbf{R}^n \to \mathbf{R}$ given by

$$f_x(u) = \langle F(x + u) - F(x), u \rangle$$

is starvex for all $x \in \mathbf{R}^n$. Denote the boundary of C by ∂C. Then, f_x is not smooth on $\partial C - x$, since F is not smooth on ∂C. □

4.6. EXAMPLE OF A NONCONTINUOUS STARVEX FUNCTION

Let $g : \mathbf{R}^n \to \mathbf{R}_+$ be given by

$$g(u) = \|u\|f(\|u\|),$$

where $f : \mathbf{R}_+ \to \mathbf{R}_+$ is noncontinuous. Then, by using Theorem 2.2, it can be easily proved that g is starvex if and only if f is monotone. In this way, every noncontinuous monotone function generates a noncontinuous starvex function. □

Finally, we remark that a quadratic functional $f : \mathbf{R}^n \to \mathbf{R}$ given by $f(x) = \langle Ax, x\rangle$ (where $A : \mathbf{R}^n \to \mathbf{R}^n$ is a linear operator) is starvex if and only if it is convex. This can be easily verified by using Theorem 2.2.

Acknowledgements

This note was supported by Bolyai János Research Fellowship and Hungarian Research Grant OTKA T029572.

References

1. B. D. Craven, Invex functions and constrained local minima, *Bull. Austral. Math. Soc.* **24** (1981), 357-366.
2. B. D. Craven and B. M. Glover, Invex functions and duality, *J. Austral. Math. Soc.* (Series A) **39** (1985), 1-20.
3. M. A. Hanson, On Sufficiency of the Kuhn-Tucker Conditions, *Journal of Mathematical Analysis and Applications* **80** (1981), 545-550 .
4. V. Jeyakumar, Strong and weak invexity in mathematical programming, *Methods Oper. Res.* **55** (1985), 109-125.
5. S. R. Mohan and S. K. Neogy, On invex sets and preinvex functions, *Journal of Mathematical Analysis and Applications* **189** (1995), 901-908 .
6. S. Z. Németh, A scalar derivative for vector functions, *Rivista di Matematica Pura ed Applicata* N.10 (1992), 7-24.
7. S. Z. Németh, Scalar derivatives and spectral theory, *Mathematica*, Thome 35 (88) N.1, (1993), 49-58.
8. R. Pini, Convexity along curves and invexity, *Optimization* **29** (1994), 301-309.
9. R. Pini, Invexity and generalized convexity, *Optimization* **22** (1991)4, 513-525.
10. T. Rapcsák, *Smooth Nonlinear Optimization in* $\mathbf{R}^n$, Kluwer Academic Publishers (1997).
11. E. H. Zarantonello, Projections on convex sets in Hilbert space and spectral theory, I: Projections on convex sets, II: Spectral theory, *Contrib. Nonlin. Functional Analysis,* Proc. Sympos. Univ. Wisconsin, Madison (1971) 237-424

Two Approaches for Parallelizing the UEGO Algorithm

P. M. Ortigosa
Department of Computer Architecture and Electronics,
University of Almería, Spain
e-mail: pilar@ace.ual.es

I. García
Department of Computer Architecture and Electronics,
University of Almería, Spain
e-mail: inma@ace.ual.es

M. Jelásity
Research Group on Artificial Intelligence,
MTA-JATE, Szeged, Hungary
e-mail: jelasity@inf.u-szeged.hu

Abstract. In this work UEGO, a new stochastic optimization technique for accelerating and/or parallelizing existing search methods is described and parallelized. The skeleton of the algorithm is a parallel hillclimber. The separate hillclimbers work in restricted search regions (or clusters) of the search space. The volume of the clusters decreases as the search proceeds which results in a cooling effect similar to simulated annealing. Besides this, UEGO can be effectively parallelized because the amount of information exchanged among clusters (communication) is minimal. The purpose of this communication is to ensure that a hill is explored only by one hillclimber. UEGO makes periodic attempts to find new hills to climb. Two parallel algorithms of UEGO have been implemented and empirical results are also presented including an analysis of speedup.

Key words: Global Optimization, Stochastic Methods, Parallelism, Evolutionary Algorithms

1. Introduction

UEGO stands for *Universal Evolutionary Global Optimizer.* Though this method is not 'evolutionary' in the usual sense, we have kept the name for historical reasons. The predecessor of UEGO was GAS, a steady-state genetic algorithm with subpopulation support. GAS offers a solution to the so-called niche radius problem which is a common problem for many simple niching techniques such as *fitness sharing* [2, 3], *simple iteration* or the *sequential niching* [1]. This problem is related to functions that have multiple local optima and whose optima are unevenly spread throughout the search space. The solution of GAS involves a

[0] This work was supported by the Ministry of Education of Spain (CICYT TIC99-0361) and by the Tempus Structural Joint European Project.

F. Giannessi et al. (eds.), Optimization Theory, 159–177.

'cooling' technique which enables the search to focus on the promising regions of the space, starting off with a relatively large radius that decreases as the search proceeds. For more details on GAS the reader should consult [7].

In [6] an introduction to the history, motivation behind developing UEGO and its evaluation for binary problem is given.

The common part of UEGO with GAS is the species creation mechanism and the 'cooling' method. However, the species creation and 'cooling' mechanisms have been logically separated from the present optimization algorithm, so it is possible to implement any kind of optimizers that work 'inside a species'. This allows the adaptation of the method to a large number of possible search domains using existing domain specific optimizers while enjoying the advantages of the old GAS-style subpopulation approach. In this paper, an algorithm called SASS, proposed by Solis and Wets [10], was applied as the optimizer algorithm.

In the following sections the term *species* will be used instead of e.g. *cluster, zone, region, sub-domain,* etc. This may be strange, since in evolutionary computation this term normally refers to a population of similar individuals while here it denotes a subset of the search domain. This is not a major problem however; a species, in our sense, is nothing else but the set of all possible members that are similar according to some similarity measure which is application dependent. We think that the behavior of a species to be defined later has strong biological analogies.

1.1. Outline of the Paper

Section 2 describes UEGO; the basic concepts, the general algorithm and the theoretical tools used to set the parameters of the system based on a few user-given parameters. Sections 3 and 4 describe two parallel algorithms and show experimental results. Section 5 then provides a short summary.

2. Description of UEGO

In this section the basic concepts, the algorithm, and the setting of the parameters are outlined. In UEGO, a domain specific optimizer has to be implemented. Wherever we refer to 'the optimizer' in the paper, we mean this optimizer.

2.1. Basic Concepts

In the following part it will be assumed that the parameters of the function take values from the unit interval. This is easy to achieve for any function via normalization. With this assumption, some concepts (i.e. radius, attraction region) are easier to be understood.

A key notion in UEGO is that of a *species*. A species can be thought of as a window on the whole search space. This window is defined by its *centre* and a *radius*. The centre is a solution, and the radius is a positive number. Naturally, this definition assumes a *distance* defined over the search space. The role of this window is to 'localize' the optimizer, which is always called by a species and can 'see' only its window; so every new sample is taken from there. This means that the largest step made by the optimizer in a given species is no larger than the radius of the given species. If the value of a new solution is better than that of the old centre, the new solution becomes the centre and the window is moved.

The radius of a species is not arbitrary; it is taken from a list of decreasing radii, the *radius list*. The radii decrease in a regular fashion in geometrical progression. The first element of this list is always the diameter of the search space. This allows the largest species to always contain the whole space independently of its centre. The diameter is given by the largest distance between any of the two possible solutions according to the distance mentioned above. If the radius of a species is the *i-th* element of the list, then we say that the *level* of the species is i.

During the optimization process, a list of species is kept by UEGO. The algorithm is, in fact, a method for managing this *species-list* (i.e. creating, deleting and optimizing species); it will be described in Section 2.2.

2.2. The Algorithm

Firstly, some parameters of UEGO will be dealt with; more details can be found in Section 2.3.

As we mentioned earlier, every species has a fixed level during its lifetime. However, species-level operators are allowed to change this level as will be described. The maximal value for the level is given by a parameter called `levels`. Every valid level i (i.e. for levels from [1,`levels`]) has a corresponding radius value (r_i) and two integers (new_i, n_i) which denote the number of function evaluations. new_i is used when new species are created at a given level while n_i is applied when optimizing individual species. To define the algorithm fully, one more parameter is needed: the maximal length of the above species list (`max_spec_num`). The basic algorithm is shown in Algorithm 1. Now, the set of procedures called by UEGO will be described.

ALGORITHM 1. *Description of* UEGO *strategy*

Begin UEGO
  init_species_list
  optimize_species(n[1])
  for *i = 2 to levels*
    create_species(new[i]/length(species_list))
    fuse_species(r[i])
    shorten_species_list(max_spec_num)
    optimize_species(n[i]/max_spec_num)
    fuse_species(r[i])
  end *for*
End UEGO

`Init_species_list`. Create a species list consisting of a single species with a random centre at level 1.

`Create_species(evals)`. For every species on the list, create random pairs of solutions in the 'window' of the species, and for every such pair take the middle of the *section* connecting the pair. If the objective function value at the middle is worse than the values of the pair, then the members of the pair are inserted in the species list. Every new inserted species is assigned the actual level value (`i` in Algorithm 1).

The motivation behind this method is simple: to create species that are on different 'hills' so ensuring that there is a valley between the new species (see *Figure 1*).

Naturally this is a heuristic only. In higher dimensions it is possible (in fact typical) that many species are created even if the function is unimodal. This is an unlucky effect which is handled by the cooling process to ensure that at the beginning the algorithm does not create too many species capturing only the rough structure of the landscape. The parameter of this procedure is an upper bound for the number of function evaluations. Note that this algorithm needs a definition of *section* in the search space.

`Fuse_species(radius)`. If the centres of any pair of species from the species list are closer to each other than the given radius, the two species are fused. The centre of the new species will be the one with the best function value, while the level will be the minimum of the levels of the original species to be fused (see *Figure 2*). Naturally this method does not ensure that no species will overlap after fusion though the amount of the overlapping regions is typically highly decreased.

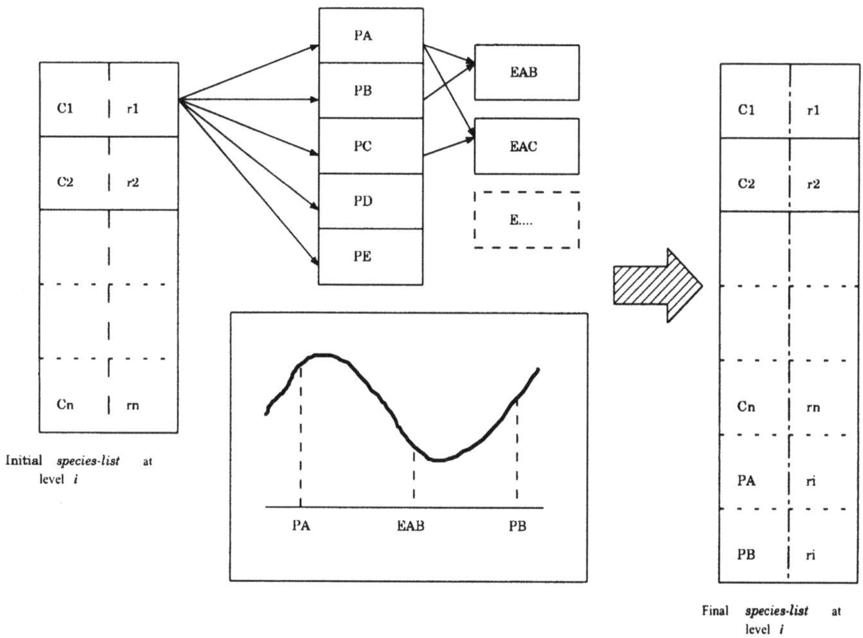

Figure 1. Species creation mechanism

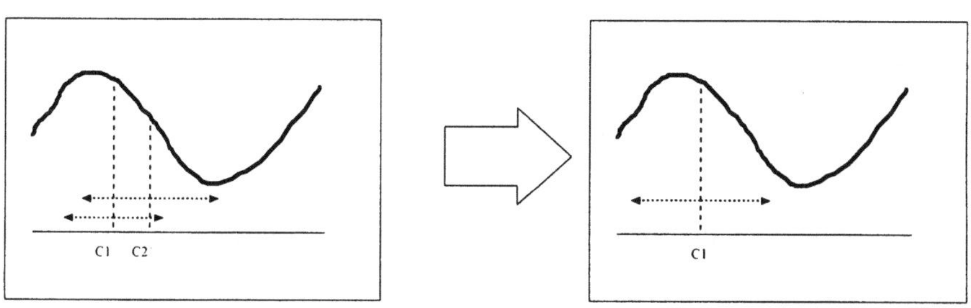

Figure 2. Fusion mechanism

`Shorten_species_list(max_spec_num)`. Delete species to reduce the list length to the given value. Higher level species are deleted first.

`Optimize_species(evals)`. Start the optimizer for every species with the given evaluation number (i.e. every single species in the actual list receives the given number of evaluations). See Section 2.1.

It is clear that if for some level i the species list is shorter than the allowed maximal length, `max_spec_num`, the overall number of function evaluations will be smaller than n_i (see Algorithm 1, `optimize_species`).

Finally, let us make a remark about a possible parallel implementation. The most time-consuming parts of the basic algorithm is the creation and optimization of the species. Note that these two steps can be done independently for every species, so each species can be assigned to a different processor. Notice also that a species is defined by its centre and its level, so the amount of information used in communications is really small.

2.3. Parameters of uego

The most important parameters are those that depend on the level (i): the radii (r_i) and two function evaluation numbers for species creation (new_i) and optimization (n_i) (see Algorithm 1). In this section a method is described that determines the value of these parameters, using a few easy-to-understand parameters provided by the user.

We will now make use of the notation introduced in Section 2.2. The user-given parameters are listed below. Short notations (in brackets) that will be used in equations in the subsequent sections are also given.

`evals` (N): The maximal number of function evaluations the user allows for the whole optimization process. Note that the actual number of function evaluations may be less than this value.

`levels` (l): The maximal level value (see Algorithm 1).

`threshold` (ν): The meaning of this parameter will be explained later.

`max_spec_num` (M): The maximal length of the species list.

`min_r` (r_l): The radius that is associated with the maximal level, i.e. `levels`.

The parameter setting algorithm to be described can use any four of the above five values while the remaining parameters are set automatically.

Speed of the optimizer. Before presenting the parameter setting method, the notion of the *speed* of the optimizer must be introduced.

As explained earlier, the optimizer cannot make a larger step in the search space than the current radius of the species. Furthermore, if the centre of a species is far from the local optima, then these steps will be larger while if the centre is already close to a local optimum then the steps will be very small. Given a certain number of evaluations, it is possible to measure the distance the given species moves during the optimization process assuming that the species is suboptimal. This distance can be approximated (as a function of the radius and evaluations) for certain optimizers using ideal landscapes (such as linear functions) with the help of mathematical models or experimental results. This naturally leads to a notion of speed that will characterize a given domain (assuming e.g. a linear landscape) and will depend on the species radius. Speed will be denoted by $v(r)$. As we will not give any actual approximations here, the reader should refer to [7].

The parameter-setting method is based on intuitive and reasonable *principles*. These principles are now described below.

Principle of equal chance. At a level, every species moves a certain distance from its original centre due to optimization. This principle ensures that every species will receive the number of evaluations that is enough to make it move at least a fixed distance, assuming that the speed of this motion is $v(r)$. A species will not necessarily move so far but the definition of speed is such that if the species is far from the local optima, then it will move approximately the given distance. This common distance is defined by $r_1\nu$. The meaning of `threshold` can now be given: it directly controls the distance a species is allowed to cover, so it actually controls the probability that a species will eventually represent a local optimum: the further a species can go, the higher the probability of reaching a local optimum is (and the more expensive the optimization is). Recall that r_1 is always the diameter of the search space. Now, the principle can be formalized:

$$\frac{v(r_i)n_i}{M} = r_1\nu \qquad (i = 2, \ldots, l) \tag{1}$$

Principle of exponential radius decreasing. This principle is quite straightforward; given the smallest radius and the largest one (r_l and r_1) the remaining radii are expressed by the exponential function:

$$r_i = r_1 \left(\frac{r_l}{r_1}\right)^{\frac{i-1}{l-1}} \qquad (i = 2, \ldots, l). \tag{2}$$

Principle of constant species creation chance. This principle ensures that even if the length of the species list is maximal, there is a chance of creating at least two more species for each old species. It also makes a strong simplification, namely, all the evaluations should be set to the same constant value.

$$new_i = 3M \qquad (i = 2, \ldots, l) \tag{3}$$

Decomposition of N. Let us define $new_1 = 0$ for the sake of simplicity since new_1 is never used by UEGO. The decomposition of N results in the trivial equation

$$\sum_{i=1}^{l}(n_i + new_i) = (l-1)3M + \sum_{i=1}^{l} n_i = N \tag{4}$$

making use of (3) in the process. One more simplification is possible; set $n_1 = 0$ whenever $l > 1$. Note that if $l = 1$ then UEGO is simply the optimizer.

Expressing n_i from (1) and substituting it for (4), we can write

$$(l-1)3M + \sum_{i=2}^{l} \frac{Mr_1\nu}{v(r_i)} = N \tag{5}$$

Using (2) as well, it is quite evident that the unknown parameters in (5) are just the user given parameters and due to the monotony of this equation in every variable, any of the parameters can be given using effective numerical methods, provided that the other parameters are known. Using the above principles, the remaining important parameters (n_i, new_i and r_i) can be calculated as well. Note, however, that some of the configurations set by the user may be unfeasible.

2.4. PREPARING EXPERIMENTS

The optimizer used by UEGO was the hill climber suggested in [10] (SASS) where the parameter ρ_{ub} (maximal step size) was set to the value of the species radius from which the optimizer is called and the accuracy of the search was set to $\min(\rho_{ub}/10^3, 10^{-5})$. The rest of parameters of the algorithm was set as the authors suggested in [10]. No fine tuning of the parameters of the optimizer was carried out.

In [6] and [8] several experiments with test functions were made, and those experiments showed that UEGO was a reliable global optimization method for finding not only the global maximum/maxima, but also several local optima. They also allowed confirmation that both the clustering technique and the level-based "cooling" technique showed some advantages over their predecessors. Comparisons to other algorithms showed that UEGO was competitive, at least on the examined test problems.

	Function	Domain	N. Optima
F1	$\sum_{i=1}^{10} \frac{1}{(k_i \cdot (x-a_i))^2 + c_i}$	$[0, 10]$	8
F2	$\sum_{i=1}^{5} i \cdot \sin((i+1)x + i)$	$[-10, 10]$	> 20
F3	$-\sin(x_1^2 + 2x_2^2) \cdot e^{-x_1^2 - x_2^2}$	$[-3, 3]^2$	> 30
F4	$-\sin^2(\pi y_1) - \sum_{i=1}^{2} (y_2 - 1)^2 (1 + \sin^2(\pi y_{i+1}))$ $-(y_3 - 1)^2$ where $y_i = 1 + (x_i - 1)/4$	$[-10, 10]^4$	> 1000
F5	$-\sin^2(\pi y_1) - \sum_{i=1}^{3} (y_i - 1)^2 (1 + 10\sin^2(3\pi y_{i+1}))$ $-(y_4 - 1)^2$ where $y_i = 1 + (x_i - 1)/4$	$[-10, 10]^4$	> 100
F6	$2.5 \cdot \prod_{i=1}^{n} \sin(x_i - \pi/2) + \prod_{i=1}^{n} \sin(5(x_i - \pi/2))$	$[0, \pi]^n$	> 100
F7	$0.5 \cdot \sum_{i=1}^{n} (x_i - 2)^2$	$[-1, 1]^{20}$	> 200
F8	$0.5 \cdot \sum_{i=1}^{n} (x_i + 0.5)^2$	$[-1, 1]^{20}$	> 200
F9	$10 \cdot \sum_{i=1}^{n} x_i^2$	$[-1, 1]^{20}$	> 200
F10	$0.5 \cdot \sum_{i=1}^{n} (x_i - 8)^2$	$[-1, 1]^{20}$	> 200
F11	$0.5 \cdot \sum_{i=1}^{n} i(x_i - 2)^2$	$[-1, 1]^{20}$	> 200

Table 1. Set of Test Functions

The goal of this work is to evaluate two different parallel implementations of UEGO but not the sequential version which is studied in [8]. *Table 1* shows the set of test functions (see [5,7,11,12]) used for evaluating both parallel versions of UEGO where the name, dimension and number of local optima of each function can be seen.

Due to the stochastic nature of UEGO, all the numerical results given in this work are average values of one hundred executions.

3. Parallel Strategy: Synchronous Global Model (PSUEGO)

Regarding the parallelization of UEGO, it seems that the most convenient model is the centralized one where a master-worker communication is required (see *Figure 3*). It is interesting to remark that the most time-consuming parts of the basic sequential algorithm is the creation and optimization of the species, therefore, these two steps can be carried out independently for every species, so each species can be assigned to a different processor.

3.1. DESCRIPTION OF PSUEGO STRATEGY

The first parallel strategy is a synchronous algorithm where the slave processors carry out the creation and optimization procedures (see *Figure 4*), while the master processor manages the global species list (fuse, delete, etc.).

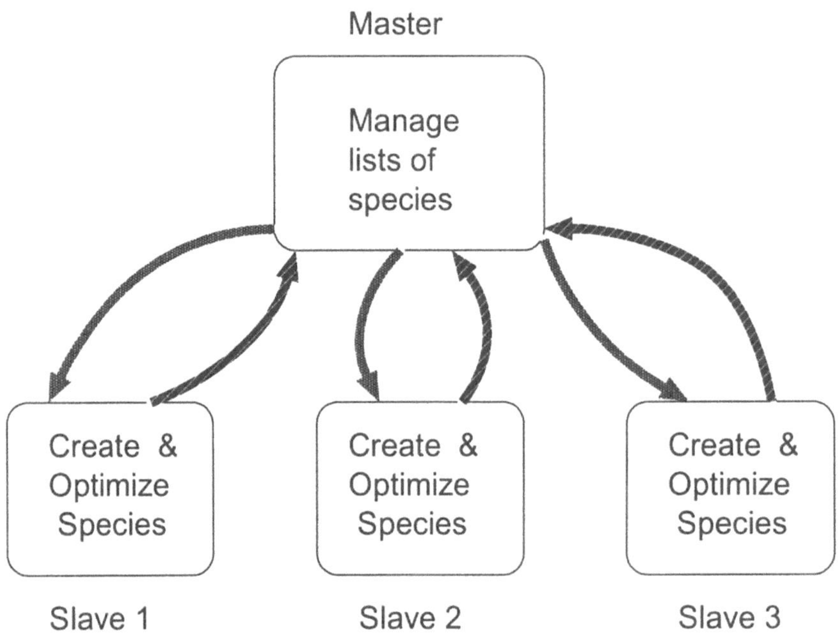

Figure 3. Parallel Strategy: Global Model

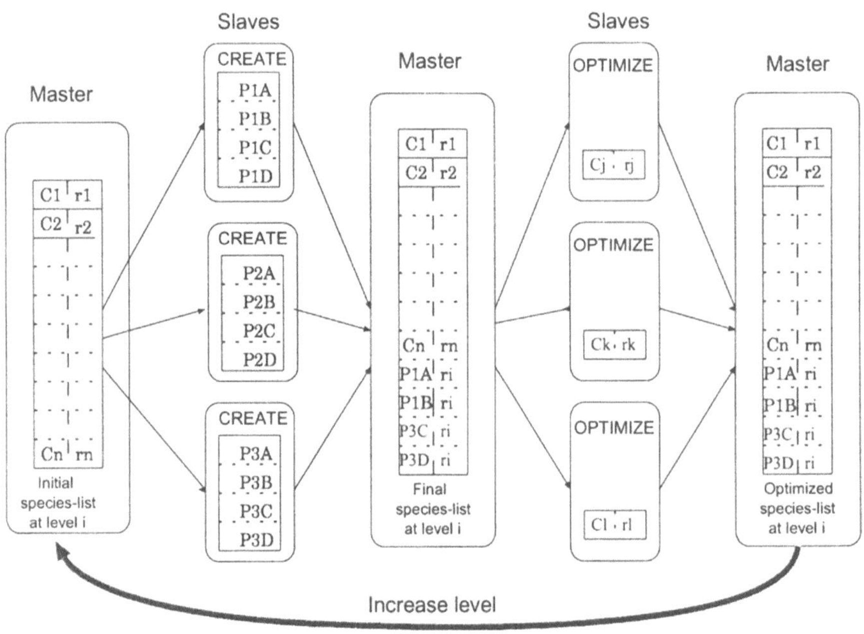

Figure 4. PSUEGO strategy

In order to run the "**optimize**" and "**create_species**" procedures, the slave processors need to receive from the master processor only the two parameters of a species (i.e. its centre and its radius), so the amount of information involved in the communication procedure is quite small.

The **optimize** and **create_species** procedures do not need any additional information; each procedure depends on a species only and does not depend on other parameters or species. For this reason, these procedures can be run independently on several slave processors at the same time.

Algorithms 2 and 3 show a description of the master and the slave processor algorithms.

ALGORITHM 2. *Description of* MASTER_PSUEGO *strategy*

Begin MASTER_PSUEGO
  init_species_list
  SEND *species to be optimized to slaves*
  RECEIVE *optimized species from slaves*
  for $i = 2$ *to levels*
    while *there are any species not sent*
      SEND *species (to create more new ones) to slaves*
      RECEIVE *lists of created species from slaves*
    end *while*
    join new lists to the old list of species
    fuse_species
    shorten_species_list
    while *there are any species not sent*
      SEND *species to be optimized to slaves*
      RECEIVE *optimized species from slaves*
    end *while*
    fuse_species
  end *for*
End MASTER_PSUEGO

The master processor, (see Algorithm 2), initializes the species list. In this parallel version, the **init_species_list** procedure does not create a single individual. Instead, it creates NP individuals randomly where NP is the number of slave processors. These NP individuals will be the centres of the first NP species, which will be included in the species list. Once the species list has been initialized, the master processor sends one species to each slave processor, so all species can be optimized simultaneously. Later, the master processor receives the optimized species from the slave processors and starts the iterative optimization process. The value of the level increases in each iteration (as was previously mentioned, **levels** is a user parameter that indicates the maximum number of levels).

ALGORITHM 3. *Description of* SLAVE_PSUEGO *strategy*

Begin SLAVE_PSUEGO
  RECEIVE *species from master*
  optimize_species
  SEND *optimized species to master*
  for $i = 2$ *to levels*
    while *there is any species not sent in master*
      RECEIVE *species from master*
      create_species
      SEND *lists of created species to master*
    end *while*
    while *there is any species not sent in master*
      RECEIVE *species from master*
      optimize_species
      SEND *optimized species to master*
    end *while*
  end *for*
End SLAVE_PSUEGO

At every iteration the master processor sends species from the species list to the slave processors. Then, when a slave processor generates a new species list, it sends this sublist back to the master processor. The master processor receives the sublist from a slave processor and sends another species (if there is a species from the species list that has not been sent yet) to the slave processor. Once the master processor has received all the sublists from the slave processors, it joins (**join** procedure) them to the previous species list and executes the **fuse_species** and **shorten_species** procedures. Then, the master processor distributes the species among the slave processors in the same way as in the previous case. When the master processor receives all the optimized species, it executes the **fuse_species** procedure and finally the level is increased starting a new iteration.

3.2. Evaluation and results of PSUEGO

For analyzing the performance of this synchronous parallel version of UEGO (PSUEGO), the set of test functions described in Table 1 was used. All experiments were run 100 times, and results are the average value of 100 executions. Experiments were done on a Cray T3E using up to 33 processors.

Results have been obtained by using some of the tools available at Cray T3E (PAT program). These numerical results showed that in this parallel strategy the computational load was not well balanced; some processors worked more than others. In addition, due to the synchronization procedures, processors spent a large amount of time waiting for information, so they were quite frequently idle. *Table 2* shows the percentage of idle time for every test function for several values of the number of processors. Results show that this percentage ranges from 16% to 50%.

As a consequence of the unbalanced workload and the synchronization procedures, PSUEGO does not exhibit a very good performance. This fact is shown in *Figure 5*, where the values of *speedup* versus the number of processors (NP) were depicted. The *speedup* for NP processors is defined by: $speedup = \frac{t_1}{t_{NP}}$ where t_1 is the execution time when the algorithm is run, using a single processor (sequential case), and t_{NP} is the execution time, using NP processors.

Function	% Idle time						
	NP=2	NP=3	NP=4	NP=5	NP=9	NP=17	NP=33
F1	49	34	26	20	18	23	36
F2	49	34	25	20	20	28	26
F3	49	36	27	21	18	21	38
F4	49	34	26	21	19	23	36
F5	49	35	28	21	18	21	36
F6	49	38	25	16	19	28	38
F7	49	36	27	21	18	17	35
F8	49	34	26	20	20	29	36
F9	49	33	26	20	19	16	35
F10	50	34	28	21	18	26	36
F11	49	34	26	21	18	21	33

Table 2. Percentage of idle time for PSUEGO for several processors

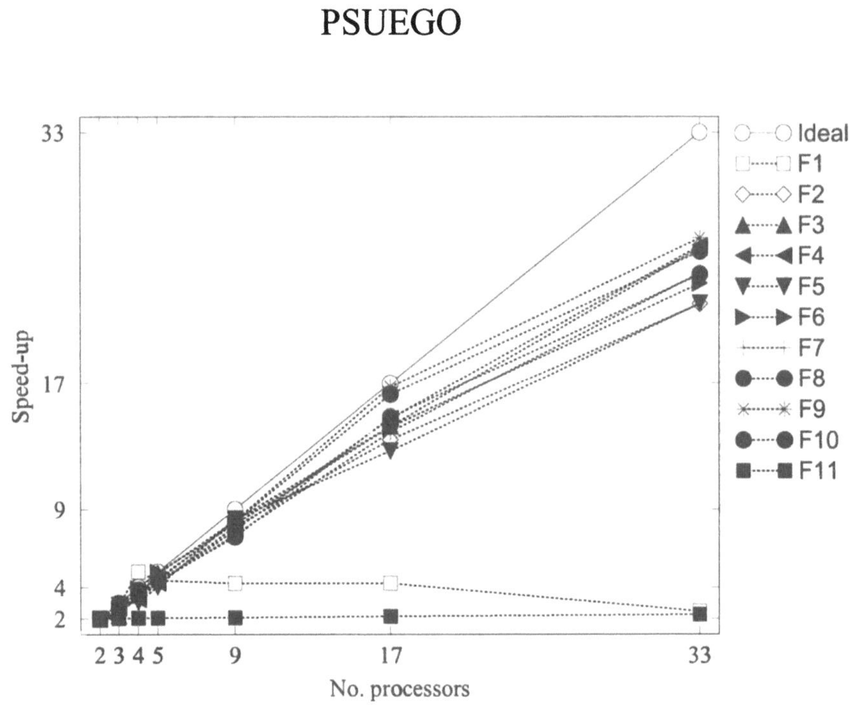

Figure 5. Speedup of PSUEGO for the set of test functions described in Table 1

4. Parallel Strategy: Asynchronous Global Model (PAUEGO)

The second parallel strategy (PAUEGO) is intended to solve some of the previous problems. In this new implementation the load has been balanced forcing the master processor to optimize and create species while the slave processors are working. Another change is the reduction in the number of synchronization points. Now the master processor can start to carry out the synchronous operations over the species before receiving all the information, so the idle time for slave processors can be considerably reduced.

4.1 DESCRIPTION OF PAUEGO STRATEGY

Algorithms 4 and 5 describe the master and slave strategies for PAUEGO algorithm. The first difference from PSUEGO is in the initialization phase. Now, every slave processor has a species initialization procedure, i.e. every slave processor chooses two or more individuals as centres of new species to optimize. When a slave processor has optimized all its species, it creates new species and the new species sublist is sent to the master processor. On the other hand, the master processor only initializes a single individual as the centre of a single species (it does not choose NP individuals like in PSUEGO), optimizes it and creates new species. Once this new sublist has been created, the master processor is prepared to receive any information (sublists of species) from any slave processor. If the master processor does not receive any sublist, it begins to fuse species saved in the current list of species. This fusion process stops when the master processor receives a sublist from a slave processor, and goes on when no more information is received. Once the master processor has received all sublists created by the slave processors and has applied the *fuse* procedure, the iterative loop starts.

ALGORITHM 4. *Description of* MASTER_PAUEGO *strategy*

Begin MASTER_PAUEGO
  init_species_list and optimize_species
  create_species
  RECEIVE *a new species's sublist from each slave*
  while *there are any species to be received*
    if *any slave processor has sent a sublist*
      RECEIVE *the sublist from the slave*
    else *fuse_species*
  end *while*
  for i = *2 to levels*
    while *there are any species not sent*
      SEND *species to slave*
      if *any slave processor has sent a sublist*
        RECEIVE *the sublist from the slave*
      else *fuse_species*
    end *while*
    shorten_species_list
    while *there are any species not sent*
      SEND *species to be optimized to slave*
      while *any optimized species has been sent*
        optimize a not yet optimized species
        fuse_species
      end *while*
      RECEIVE *optimized species from slave*
    end *while*
  end *for*
End MASTER_PAUEGO

The iterative part of PAUEGO for slave processors does not have any modification compared to the iterative part of PSUEGO. However, for

PAUEGO, the master processor always checks the arrival of information (a newly created sublist of species or an optimized species) from slave processors and it sends a new species on information arrival. In addition, the master processor contributes to the optimization process by creating, fusing and optimizing species. The processes executed by the master processor are always interrupted when new information arrives, and they are resumed later.

ALGORITHM 5. *Description of* SLAVE_PAUEGO *strategy*

Begin SLAVE_PAUEGO
  init_species_list and optimize_species
  while *there is any species to send to master*
    create_species
    SEND *lists of created species to master*
  end *while*
  for *i = 2 to levels*
    while *there is any species not sent in master*
      RECEIVE *species from master*
      create_species
      SEND *lists of created species to master*
    end *while*
    while *there is any species not sent in master*
      RECEIVE *species from master processor*
      optimize_species
      SEND *optimized species to master*
    end *while*
  end *for*
End SLAVE_PAUEGO

4.2. Evaluation and results of PAUEGO

For the analysis of the performance of PAUEGO, a set of experiments using the same set of test functions (see Table 1), was performed.

From the experiments we can conclude that the percentage of idle time for PAUEGO is much smaller than for PSUEGO as it can be seen in *Table 3*. For the asynchronous version, the percentage of idle time is no greater than 4% (which only appears for some test functions using up to 33 processors), while the same bound was 50% in the synchronous case.

Function	% idle time						
	NP=2	NP=3	NP=4	NP=5	NP=9	NP=17	NP=33
F1	0	1	1	2	2	3	4
F2	0	0	1	1	1	1	2
F3	0	0	1	2	2	2	4
F4	0	0	0	1	1	2	2
F5	0	1	1	0	1	1	2
F6	1	1	1	2	2	3	4
F7	0	1	1	0	1	1	2
F8	0	0	1	1	0	1	2
F9	0	0	1	1	0	1	1
F10	0	0	1	1	2	2	3
F11	0	0	1	0	1	2	2

Table 3. Percentage of idle time for PAUEGO for several processors

Another conclusion based on the results is that the computational load is much more balanced: every processor executes a similar number of operations. *Table 4* shows an example of the results achieved for test function F10 using up to 9 processors. Similar results were obtained for all test functions and number of processors.

In Table 4, first column (**PE**) indicates the processor identifier number where the identifier 0 corresponds to the master processor. The **cycles** column shows the number of cycles consumed by each processor during the execution. The **operations** column shows the number of operations made by each processor. It can be seen that the differences in the number of operations are quite small. The **oper/sec** column indicates the number of operation per second, and it is directly related to the workload balance. Data in this column show that approximately all the processors execute the same number of operations. The **dcache** column shows the cache misses, and the **miss/sec** column the average value of cache misses per second. It can be seen from here that the master processor has much fewer misses than the slave processors.

PE	cycles	operations	ops/sec	dcache	misses/sec
0	42368.98	2645.99	28.11	0.83	0.01
1	42227.79	2659.16	28.34	14.59	0.16
2	38129.47	2393.44	28.25	15.85	0.19
3	42306.36	2645.33	28.14	13.53	0.14
4	38244.14	2379.04	28.00	13.50	0.16
5	42542.94	2622.72	27.74	10.79	0.11
6	38443.14	2367.07	27.71	10.74	0.13
7	38223.73	2360.04	27.79	12.09	0.14
8	38218.41	2360.70	27.80	12.05	0.14

Table 4. Results from an execution of PAUEGO for F10 test function using up 9 processors

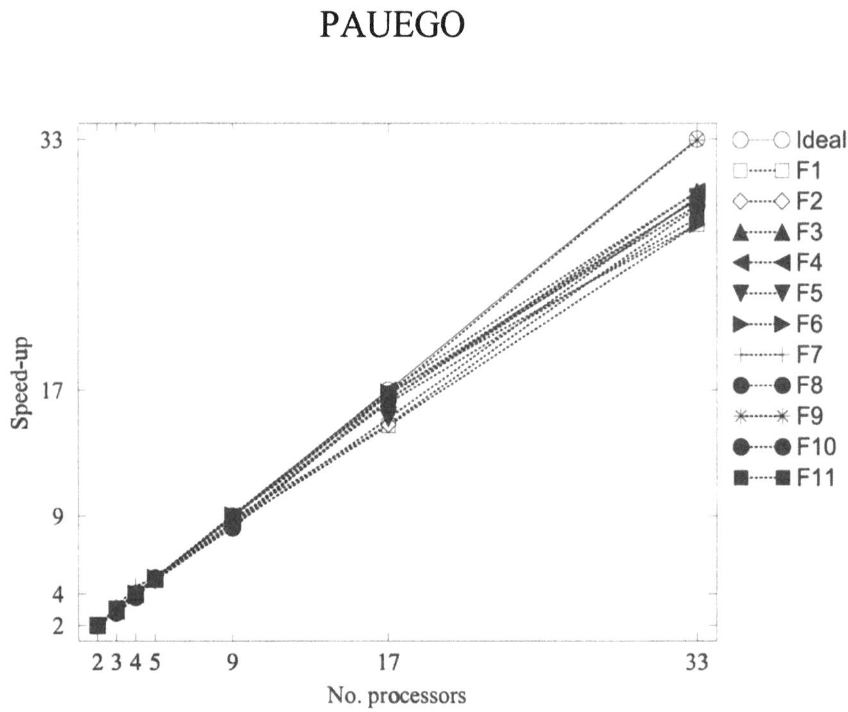

Figure 6. Speedup of PAUEGO for the set of test functions described in Table 1

Figure 6 shows the *speedup* for PAUEGO versus the number of processors. It can be seen that the *speedup* is closer to the ideal case, mainly for a small number of processors. Applying PAUEGO to a problem more expensive from computation point of view, the *speedup* would be better due to the overlapping of communication and computation times.

5. Summary

In this paper, a new general technique (UEGO) for accelerating existing search methods has been described and two parallel implementations have been evaluated.

The first parallel strategy is a synchronous algorithm where the slave processors carry out the creation and optimization procedures while the master processor manages the global species list (fuse, delete, etc.). The main drawback of this strategy is that, due to points of synchronization, the slave processors are idle for a considerable amount of time, so the parallel program is inefficient. Another problem is that the load is not balanced, the master processor spends a long time waiting for information from the slave processors.

The second parallel strategy is intended to solve some of the previous problems. In this new implementation the load has been balanced, forcing the master to optimize and create species, while the slave processors are working. Another change was the reduction in the number of points of synchronization. Now the master processor can start to carry out the synchronous operations over the species before receiving all the information from the slaves processors, so the idle time for slaves was considerably reduced.

References

1. D. Beasley, D. R. Bull, and R. R. Martin. A sequential niche technique for multimodal function optimization. *Evolutionary Computation*, 1(2):101–125, 1993.
2. K. Deb. Genetic algorithms in multimodal function optimization. TCGA report no. 89002, The University of Alabama, Dept. of Engineering mechanics, 1989.
3. K. Deb and D. E. Goldberg. An investigation of niche and species formation in genetic function optimization. In J. D. Schaffer, editor, *The Proceedings of the Third International Conference on Genetic Algorithms*. Morgan Kaufmann, 1989.
4. A. Ferreira and P. M. Pardalos. *Solving Combinatorial Optimization Problems in Parallel: Methods and Techniques*, volume 1054 of *Lecture Notes in Computational Science*. Springer-Verlag, 1996.
5. C. A. Floudas and P. M. Pardalos (eds.). *A Collection of Test Problems for Constrained Global Optimization Algorithms. Lecture Notes in Computer Science.* Springer-Verlag, Berlin, 1978.
6. M. Jelasity. UEGO, an abstract niching technique for global optimization. In Agoston E. Eiben, Thomas Bäck, Marc Schoenauer, and Hans-Paul Schwefel, editors, *Parallel Problem Solving from Nature - PPSN V*, volume 1498 of *Lecture Notes in Computational Science*, pages 378–387. Springer-Verlag, 1998.
7. M. Jelasity and J. Dombi. GAS, a concept on modeling species in genetic algorithms. *Artificial Intelligence*, 99(1):1–19, 1998.
8. M. Jelasity, P. M. Ortigosa, and I. García. UEGO, an abstract clustering technique for multimodal global optimization. *Journal of Heuristics*, To appear.
9. A. Migdalas, P. M. Pardalos, and S. Storoy. *Parallel Computing in Optimization.* Kluwer Academic Publishers, 1998.
10. F. J. Solis and R. J.-B. Wets. Minimization by random search techniques. *Mathematics of Operations Research*, 6(1):19–30, 1981.
11. G. W. Walster, E. R. Hansen, and S. Sengupta. Test results for a global optimization algorithm. In P. T. Boggs, Richard H. Byrd, and Robert B. Schnabel, editors, *Numerical Optimization*, pages 272–287. SIAM, 1984.
12. Z. B. Zabinsky, D. L. Craesser, M. E. Tuttle, and G. L. Kimm. Global optimization of composite laminates using improving hit-and-run. In C. Floudas and P. Pardalos, editors, *Recent Advances in Global Optimization*, pages 343–368. Princeton University Press, Princeton, 1992.

Convexification of Functions by Nonlinear Coordinate Transformations

T. Rapcsák
Computer and Automation Institute,
Hungarian Academy of Sciences

Abstract. The convexification of continuous and smooth functions are related to the choice of Riemannian metrics originated from regular nonlinear coordinate transformations.

1. Coordinate representations in R^n

In smooth optimization, the main theoretical results do not depend on nonlinear coordinate transformations [8]. Thus, theoretical results and numerical representations can be separated, which is important to construct efficient methods [4,6]. In differential geometry, the terms "curvilinear coordinate system", "nonlinear coordinate representation" and "nonlinear coordinate transformation" are synonyms, and deep mathematical results are known to characterize subclasses (subgroups) of transformations preserving geometric properties. In optimization theory, we will rather use "nonlinear coordinate representation", because we do not deal with the group theoretical description of the subclass introduced here, but with its applications.

The n-dimensional Euclidean space R^n is the product of the ordered n-tuples of real numbers and it forms an n-dimensional differentiable manifold.

DEFINITION 1.1 *A curvilinear coordinate system in a domain* $A \subseteq R^n$ *is a system of smooth functions* $\mathbf{u}$ *which is a one-to-one map of the domain* A *onto the domain* $\mathbf{u}(A)$ *so that the determinant of the Jacobian matrix, say* $\det(J\mathbf{u}) = \det(\frac{\partial \mathbf{u}}{\partial \mathbf{x}})$, *is not zero at every point of the domain* A.

If the set of smooth functions $\mathbf{u}$ has the property that the determinant of the Jacobian matrix is not zero in a domain of R^n, then for each point of this domain, there exists an open neighbourhood where $\mathbf{u}$ defines a local coordinate system. In this case, the inverse mapping $\mathbf{u}^{-1}$ also defines local curvilinear coordinates in a domain of R^n.

[1] This research was supported in part by the Hungarian National Research Foundation, Grant No. OTKA-T029572.

F. Giannessi et al. (eds.), Optimization Theory, 179–189.

In optimization theory, mainly linear transformations were used to improve the structure. By introducing curvilinear coordinate systems in R^n, they may be considered regular nonlinear coordinate transformations. It is known that all the regular nonlinear coordinate transformations of a given domain form a group under the operation of the composition of functions. Thus, if $\mathbf{u}$ and $\mathbf{z}$ are transformations, then the set of composite functions $\mathbf{z}(\mathbf{u})$ is a transformation as well and the group axioms hold. In differential geometry, special subgroups of transformations preserving geometric quantities are stressed. Here, the aim is to revisit the convexification of functions by using regular nonlinear coordinate transformations.

From theoretical point of view, the inverse nonlinear coordinate transformations or parametrizations in the sense of differential geometry are interesting, because in this case, the function f can be expressed in any neighbourhood U in $\mathbf{u}(A)$ as $f(\mathbf{x}(\mathbf{u})), \quad \mathbf{u} \in U$. Note that this is the standard way to handle differentiable manifolds, which makes tensor characterizations in optimization possible.

From the point of view of numerical analysis, nonlinear coordinate transformations should be taken into account in the original domain as well. Our aim is to extend this idea by studying transformations so that a given function become a geodesic convex function with respect to the Riemannian metric generated by a suitable transformation. An advantage of these transformations is that a geodesic convex function fulfils the local-global property [3] which is important in optimization theory, but mainly related to generalized convex functions [1]. This approach provides tools for characterizing the convexification of functions by coordinate representations. In the first part, continuous functions are considered, while in the second part, smooth functions in the positive orthant.

2. Convexification of functions by domain transformations

The n-dimensional Euclidean space R^n forms an n-dimensional differentiable manifold, so it is possible to introduce a Riemannian metric different from the Euclidean one.

DEFINITION 2.1 *A Riemannian metric is given in a domain of* R^n *if, in any curvilinear coordinate system, a set of smooth functions* g_{ij}, $i,j = 1,\dots,n$, *a matrix function* G *is defined so that the following conditions hold:*

(1) $g_{ij} = g_{ji}$, $i,j = 1,\dots,n$, *(the matrix function* G *is symmetric),*

(2) G *is nonsingular and positive definite,*

(3) under curvilinear coordinate transformations the value of the quadratic forms defined by G does not change.

If indefiniteness is required instead of positive definiteness, then the Riemannian metric is indefinite. A differentiable manifold endowed with a Riemannian metric is called a Riemannian manifold.

In optimization theory, the manifold R^n is generally endowed with the Euclidean metric which is a special Riemannian one, i.e., $G(\mathbf{x}) = I$, $\mathbf{x} \in R^n$, where I is the identity matrix.

DEFINITION 2.2 *Let M be a Riemannian manifold. A curve of M is a geodesic if its tangent is parallel along the curve. A set $A \subseteq M$ is geodesic convex if any two points of A are joined by a geodesic belonging to A. A singleton is geodesic convex.*

DEFINITION 2.3 *Let M be a Riemannian manifold. If $A \subseteq M$ is a geodesic convex set, then a function f $A \to R$ is geodesic (strictly) convex if its restrictions to all geodesic arcs belonging to A are (strictly) convex in the arc-length parameter.*

Our approach introduced in [6,8] is that the regular nonlinear coordinate transformations in the Euclidean space R^n can be considered by introducing suitable Riemannian metrics depending on the transformations. Namely, in the case of a transformation $\mathbf{u}$ defined in a subset of R^n, the corresponding Riemannian metric is $J\mathbf{u}^T J\mathbf{u}$. It follows that the convexification of functions by domain transformations, i.e., a one-to-one transformation to the domain of a nonquasiconvex function so that its upper-level sets be transformations into convex sets is equivalent to the geodesic convexity of functions with respect to a coordinate representation of the Euclidean metric. Geodesic convexity is studied in [3,8] where it is pointed out that a convex function is geodesic convex with respect to the Euclidean metric. In any coordinate representation, the geodesic convexity property can be considered by the second Christoffel symbols.

DEFINITION 2.4 *If a set of functions satisfies the equalities*

$$\Gamma^{l_3}_{l_1 l_2} = \Gamma^{l_3}_{l_2 l_1} \qquad \text{for all} \quad l_1, l_2, l_3 = 1, \ldots, n,$$

in every system of local coordinates, then it is symmetric.

A fundamental theorem of Riemannian manifolds is as follows (see, e.g., [9]).

THEOREM 2.1 *Let G be a symmetric matrix function defining the metric on a Riemannian manifold M in any system of local coordinates. Then, there exists a unique symmetric system of functions such that*

$$\Gamma_{l_1 l_2}^{l_3} = \sum_{\alpha=1}^{n} 1/2\left(\frac{\partial g_{l_1\alpha}}{\partial x_{l_2}} + \frac{\partial g_{l_2\alpha}}{\partial x_{l_1}} - \frac{\partial g_{l_1 l_2}}{\partial x_\alpha}\right)(G^{-1})_{l_3\alpha} \tag{2.1}$$

$$\text{for all} \quad l_1, l_2, l_3 = 1, \dots, n.$$

The coefficient functions $\Gamma_{l_1 l_2}^{l_3}$, $l_1, l_2, l_3 = 1, \dots, n$, uniquely determined by the Riemannian metric are called the second Christoffel symbols. If the Riemannian metric is Euclidean, then $\Gamma_{l_1 l_2}^{l_3} = 0$, $l_1, l_2, l_3 = 1, \dots, n$. The transformation law of Christoffel symbols is given in the following definition:

DEFINITION 2.5 *Let the smooth functions $\Gamma_{l_1 l_2}^{l_3}$ and $\bar{\Gamma}_{l_1 l_2}^{l_3}$, $l_1, l_2, l_3 = 1, \dots, n$, be given in a regular coordinate representation $\mathbf{x}$ and $\mathbf{u}$, respectively. Then, they are transformed by the rule*

$$\Gamma_{l_1 l_2}^{l_3} = \bar{\Gamma}_{k_1 k_2}^{k_3} \frac{\partial x_{l_3}}{\partial u_{k_3}} \frac{\partial u_{k_1}}{\partial x_{l_1}} \frac{\partial u_{k_2}}{\partial x_{l_2}} + \frac{\partial^2 u_{l_3}}{\partial x_{l_1} \partial x_{l_2}} \frac{\partial x_{l_3}}{\partial u_{k_3}}, \tag{2.2}$$

$$\text{for all} \quad l_1, l_2, l_3, k_1, k_2, k_3 = 1, \dots, n,$$

where two coinciding indices mean summation.

The second Christoffel symbols are not tensors and this fact seems to be one reason that they are rarely used explicitly in different coordinate representations. In order to make their use easier, let us introduce the notations as follows. If

$$\mathbf{y} = (y_1, \dots, y_n)^T \in R^n, \qquad \mathbf{w} = (w_1, \dots, w_n)^T \in R^n,$$

then

$$\mathbf{y}^T H\mathbf{u}(\mathbf{x}) = H\mathbf{u}(\mathbf{x})\mathbf{y} = \sum_{i=1}^{n} y_i H u_i(\mathbf{x}),$$

$$\mathbf{w}^T H\mathbf{u}(\mathbf{x})\mathbf{w} = \begin{pmatrix} \mathbf{w}^T H u_1(\mathbf{x})\mathbf{w} \\ \vdots \\ \mathbf{w}^T H u_n(\mathbf{x})\mathbf{w} \end{pmatrix},$$

$$\lambda \mathbf{w}^T H\mathbf{u}(\mathbf{x})\mathbf{w} = \mathbf{w}^T \lambda H\mathbf{u}(\mathbf{x})\mathbf{w}, \qquad \lambda \in R,$$

$$J\mathbf{u}(\mathbf{x})H\mathbf{u}(\mathbf{x}) = \begin{pmatrix} \nabla u_1(\mathbf{x})H\mathbf{u}(\mathbf{x}) \\ \vdots \\ \nabla u_n(\mathbf{x})H\mathbf{u}(\mathbf{x}) \end{pmatrix}, \qquad \mathbf{x} \in U \subseteq R^n,$$

where U is a subset of R^n and Hu_i denotes the Hessian matrix of the function u_i for every index i.

The first part of the next statement is a basic fact in differential geometry but explicit formulas (2.3) do not figure in textbooks though they seem to be useful.

THEOREM 2.2 *Let* $\mathbf{u}$ *be a curvilinear coordinate system in an open domain* $\hat{A} \subseteq R^n$ *and* γ *a geodesic belonging to a set* $A \subseteq \hat{A}$. *Then,* $\mathbf{u}(\gamma)$ *is a geodesic in* $\mathbf{u}(A)$, *moreover,*

$$\begin{aligned} H\mathbf{u}(\mathbf{x}) &= -J\mathbf{u}(\mathbf{x})^T \Big[J\mathbf{u}(\mathbf{x}) H_{\mathbf{u}}\mathbf{x}(\mathbf{u}(\mathbf{x})) \Big] J\mathbf{u}(\mathbf{x}), \qquad \mathbf{x} \in A, \\ \bar{\Gamma}(\mathbf{u}(\mathbf{x})) &= J\mathbf{u}(\mathbf{x}) H_{\mathbf{u}}\mathbf{x}(\mathbf{u}(\mathbf{x})), \qquad \mathbf{x} \in A, \end{aligned} \tag{2.3}$$

where $\bar{\Gamma}(\mathbf{u}(\mathbf{x}))$, $\mathbf{x} \in A$, *are the second Christoffel symbols on* $\mathbf{u}(A)$ *originated from the transformation* $\mathbf{u}(\mathbf{x})$, $\mathbf{x} \in A$.

Proof. The geodesic with respect to the Euclidean metric between two arbitrary points $\mathbf{x}_1$ and $\mathbf{x}_2$ of A is given by the line segment

$$\gamma(t) = (1-t)\mathbf{x}_1 + t\mathbf{x}_2, \qquad t \in [0,1]. \tag{2.4}$$

In order to show that

$$\mathbf{u}(\gamma(t)) = \mathbf{u}\Big((1-t)\mathbf{x}_1 + t\mathbf{x}_2\Big), \qquad t \in [0,1], \tag{2.5}$$

is a geodesic in $\mathbf{u}(A)$, it has to satisfy the system of differential equations of geodesics considered in the following form:

$$\begin{gathered} (\mathbf{x}_2 - \mathbf{x}_1)^T H_{\mathbf{x}}\mathbf{u}(\gamma(t))(\mathbf{x}_2 - \mathbf{x}_1) = \frac{d^2\mathbf{u}(\gamma(t))}{dt^2} = \\ -\frac{d\mathbf{u}(\gamma(t))}{dt}^T \bar{\Gamma}(\mathbf{u}(\gamma(t))) \frac{d\mathbf{u}(\gamma(t))}{dt} = \\ -(\mathbf{x}_2 - \mathbf{x}_1)^T J_{\mathbf{x}}\mathbf{u}(\gamma(t))^T \bar{\Gamma}(\mathbf{u}(\gamma(t))) J_{\mathbf{x}}\mathbf{u}(\gamma(t))(\mathbf{x}_2 - \mathbf{x}_1), \\ t \in [0,1]. \end{gathered} \tag{2.6}$$

It follows that it is sufficient to prove that

$$H\mathbf{u}(\mathbf{x}) = -J\mathbf{u}(\mathbf{x})^T \bar{\Gamma}(\mathbf{u}(\mathbf{x})) J\mathbf{u}(\mathbf{x}), \qquad \mathbf{x} \in A. \tag{2.7}$$

By rewriting (2.2) in matrix form, we have that

$$\begin{gathered} \Gamma(\mathbf{x}) = J\mathbf{u}(\mathbf{x})^T [J_{\mathbf{u}}\mathbf{x}(\mathbf{u}(\mathbf{x}))\bar{\Gamma}(\mathbf{u}(\mathbf{x}))] J\mathbf{u}(\mathbf{x}) + J_{\mathbf{u}}\mathbf{x}(\mathbf{u}(\mathbf{x})) H\mathbf{u}(\mathbf{x}), \\ \mathbf{x} \in A. \end{gathered} \tag{2.8}$$

In R^n, the Riemannian metric is Euclidean, thus $\Gamma^{l_3}_{l_1 l_2} = 0, \;\; l_1, l_2, l_3 = 1, \ldots, n$, and

$$Ju(x)^T [J_u x(u(x)) \bar{\Gamma}(u(x))] Ju(x) = -J_u x(u(x)) H u(x), x \in A. \tag{2.9}$$

On the other hand,

$$J_u x(u(x)) Ju(x) = I, \qquad x \in A, \tag{2.10}$$

and by differentiating this formula with respect to $\mathbf{x}$, we obtain that

$$-J_u x(u(x)) H u(x) = Ju(x)^T H_u x(u(x)) Ju(x), \qquad x \in A, \tag{2.11}$$

which is equivalent to the first formula of (2.3). Based on (2.9) and (2.11),

$$\begin{aligned} \bar{\Gamma}(u(x)) &= ((Ju(x))^{-1})^T H u(x) (Ju(x))^{-1} = \\ &= Ju(x) H_u x(u(x)), \qquad x \in A, \end{aligned} \tag{2.12}$$

which proves the second formula of (2.3), furthermore, (2.9) and (2.12) the statement. □

COROLLARY 2.1 *Under the assumptions of Theorem 2.2, the geodesic convexity property of subsets in a Riemannian manifold is a tensor property.*

The next statement characterizes the convexification of non-necessarily smooth functions by domain transformations and it further developed the results of [2] and [5,8].

THEOREM 2.3 *Let $A \subseteq R^n$ be an open subset and $f : A \to R$ a continuous function. Then, there exists a curvilinear coordinate system $\mathbf{u}$ on A such that the function f is convex on the convex set $\mathbf{u}(A)$ iff f is geodesic convex on A with respect to the metric*

$$((Ju(x))^{-1})^T (Ju(x))^{-1}, \qquad x \in A. \tag{2.13}$$

Proof. I. Assume that there exists a curvilinear coordinate system $\mathbf{u}$ such that the function f is convex on the convex set $\mathbf{u}(A)$. Since the geodesic with respect to the Euclidean metric between two arbitrary points $\mathbf{u}(\mathbf{x}_1)$ and $\mathbf{u}(\mathbf{x}_2)$ of $\mathbf{u}(A)$ is given by the line segment

$$u(\gamma(t)) = (1-t)u(x_1) + t u(x_2), \qquad t \in [0,1], \tag{2.14}$$

and the parametrization given by (2.14) is equivalent to that of Definition 2.3, the convexity property of the function f on $\mathbf{u}(A)$ is equivalent to the geodesic convexity of the function f on $\mathbf{u}(A)$ with respect to the Euclidean metric.

By Corollary 9.3.1 in [8], the geodesic convexity of smooth functions defined on open geodesic convex sets is a tensor property, i.e., it does not depend on regular nonlinear coordinate transformations. By Corollary 2.1, the geodesic convexity of open subsets in a Riemannian manifold is a tensor property as well, and the parametrization given by (2.14) is equivalent to that of arc-length parameter, thus the geodesic convexity of functions defined on open geodesic convex sets is a tensor property.

By the tensor character of the geodesic convexity, every geodesic can be explicitly given in any regular coordinate representation. In our case, by Theorem 2.2, all the geodesics in $\mathbf{u}(A)$ can be transformed into a geodesic in A by using the inverse transformation and formula

$$\gamma(t) = \mathbf{x}\Big((1-t)\mathbf{u}(\mathbf{x}_1) + t\mathbf{u}(\mathbf{x}_2)\Big), \qquad t \in [0,1], \tag{2.15}$$

for every pair of points $\mathbf{x}_1, \mathbf{x}_2$, preserving the convexity property of the function f along $\gamma(t)$, $t \in [0,1]$. Since,

$$I = J\mathbf{u}(\mathbf{x})^T((J\mathbf{u}(\mathbf{x}))^{-1})^T(J\mathbf{u}(\mathbf{x}))^{-1}J\mathbf{u}(\mathbf{x}), \qquad \mathbf{x} \in \mathbf{A}, \tag{2.16}$$

and the introduction of the inverse transformation on $\mathbf{u}(A)$ is equivalent to the introduction of the Riemannian metric

$$\begin{gathered} J_{\mathbf{u}}\mathbf{x}(\mathbf{u}(\mathbf{x}))^T I J_{\mathbf{u}}\mathbf{x}(\mathbf{u}(\mathbf{x})) = ((J\mathbf{u}(\mathbf{x}))^{-1})^T(J\mathbf{u}(\mathbf{x}))^{-1}, \\ \mathbf{x} \in \mathbf{A}, \end{gathered} \tag{2.17}$$

on the set A, the statement is proved.

II. Assume that the function f is convex on the convex set $\mathbf{u}(A)$. Then, the function f is geodesic convex on $\mathbf{u}(A)$ with respect to the Euclidean metric, and by following the proof of the first part, the statement is obtained. □

We remark that the set A is not necessarily convex in Theorem 2.3, and the geodesic given by formula (2.15) is equivalent to the well-known $\mathbf{u}$-mean-value function denoted by

$$\gamma(t) = H(\mathbf{x}_1, \mathbf{x}_2; t) : [0,1] \to A\,. \tag{2.18}$$

THEOREM 2.4 *Let $A \subseteq R^n$ be an open subset, $f : A \to R$ a twice continuously differentiable function and $\mathbf{u}$ a curvilinear coordinate system. Then, the function f is geodesic (strictly) convex on A with respect to the Riemannian metric $(J\mathbf{u})^T J\mathbf{u}$ generated by the transformation $\mathbf{u}$ iff*

(1) the set A is geodesic convex with respect to $(J\mathbf{u})^T J\mathbf{u}$ and

(2) the matrices

$$Hf(\mathbf{x}) - \nabla f(\mathbf{x})(J\mathbf{u}(\mathbf{x}))^{-1}H\mathbf{u}(\mathbf{x}), \qquad \mathbf{x} \in A, \tag{2.19}$$

are positive (definite) semidefinite.

Proof. By Definition 2.3, a geodesic convex function with respect to a Riemannian metric is defined on a geodesic convex set with respect to the same Riemannian metric.

By Theorem 9.3.1 in [8, p.149], if $A \subseteq R^n$ is an open subset and $f : A \to R$ is a twice continuously differentiable function, then f is geodesic (strictly) convex w.r.t. a Riemannian metric iff the matrices

$$Hf(\mathbf{x}) - \nabla f(\mathbf{x})\Gamma(\mathbf{x}), \qquad \mathbf{x} \in A, \tag{2.20}$$

are positive (definite) semidefinite matrices where Γ denotes the second Christoffel symbols.

If we consider Definition 2.5, formula (2.8) and assume that $\bar{\Gamma}$ denotes the second Christoffel symbols related to the Euclidean metric, then it follows that

$$\Gamma(\mathbf{x}) = J_{\mathbf{u}}\mathbf{x}(\mathbf{u}(\mathbf{x}))H\mathbf{u}(\mathbf{x}), \qquad \mathbf{x} \in A, \tag{2.21}$$

where Γ denotes the second Christoffel symbols related to the metric $(J\mathbf{u})^T J\mathbf{u}$.

Since $J_{\mathbf{u}}\mathbf{x}(\mathbf{u}(\mathbf{x})) = (J\mathbf{u}(\mathbf{x}))^{-1}$, $\mathbf{x} \in A$, formulas (2.20) and (2.21) result in the statement. □

3. Convexification of smooth functions in the positive orthant

In optimization theory, the positive orthant seems to be important considering interior point methods and entropy optimizations. Geodesic convexity with respect to the affine and other Riemannian metrics was introduced and investigated in the positive orthant [7]. Here, this approach is extended.

THEOREM 3.1 *Let R^n_+ be the positive orthant, $f : R^n_+ \to R$, a twice continuously differentiable function and*

$$\mathbf{u}(\mathbf{x}) = (e^{-cx_1}, \dots, e^{-cx_n}), \qquad \mathbf{x} \in R^n_+, \quad c > 0 \quad \text{constant}, \tag{3.1}$$

a curvilinear coordinate system. Then, the function f is geodesic (strictly) convex on R^n_+ with respect to the Riemannian metric generated by the transformation $\mathbf{u}$ iff the matrices

$$Hf(\mathbf{x}) + c \,\text{diag}\left(\frac{\partial f(\mathbf{x})}{\partial x_1}, \dots, \frac{\partial f(\mathbf{x})}{\partial x_n}\right), \qquad \mathbf{x} \in R^n_+, \tag{3.2}$$

are positive (definite) semidefinite where the symbol diag *denotes a diagonal matrix with the components in the parentheses.*

Proof. By Theorem 2.4, the first step is to prove that the set R^n_+ is geodesic convex with respect to the metric $(J\mathbf{u})^T J\mathbf{u}$. By transformation law (2.8), the second Christoffel symbols can be given by formulas

$$\begin{aligned} \Gamma(\mathbf{x}) &= (J\mathbf{u}(\mathbf{x}))^{-1} H\mathbf{u}(\mathbf{x}), \\ (J\mathbf{u}(\mathbf{x}))^{-1} &= -\frac{1}{c}\operatorname{diag}\left(\frac{1}{e^{-cx_1}}, \dots, \frac{1}{e^{-cx_n}}\right), \\ Hu_i(\mathbf{x}) &= c^2 \operatorname{diag}(0, \dots, e^{-cx_i}, \dots, 0), \qquad i = 1, \dots, n, \\ &\mathbf{x} \in R^n_+, \end{aligned} \tag{3.3}$$

which result in

$$\Gamma^i(\mathbf{x}) = \operatorname{diag}(0, \dots, -c, \dots, 0), \qquad i = 1, \dots, n, \tag{3.4}$$

where in the i-th diagonal matrices only the i-th components are different from zero. Based on (3.4), the following system of differential equations has to be solved:

$$x_i''(s) = c(x_i'(s))^2, \qquad s \in [s_1, s_2] \subseteq R, \qquad i = 1, \dots, n, \tag{3.5}$$

whose solution is $\gamma(s) = (x_1(s), \dots, x_n(s))$,

$$x_i(s) = -\frac{1}{c}\log(a_i s + b_i), \qquad s \in [s_1, s_2] \subseteq R, \qquad i = 1, \dots, n, \tag{3.6}$$

where the values $a_i s + b_i$, $s \in [s_1, s_2]$, $i = 1, \dots, n$, must be positive. In order to show the geodesic convexity of the positive orthant, the connecting geodesic of every two points $\mathbf{x}, \mathbf{y} \in R^n_+$ is given by

$$x_i(t) = -\frac{1}{c}\log(te^{-cx_i} + (1-t)e^{-cy_i}), \qquad t \in [0, 1], \qquad i = 1, \dots, n, \tag{3.7}$$

where $x_i(0) = y_i$, $x_i(1) = x_i$ for every i and the coordinate functions are positive.

By Theorem 2.4, the second step is to verify the fulfilment of (2.19) in our case. From (2.19) and (3.4), the statement follows. □

Let $\mathbf{q}_1, \dots, \mathbf{q}_n$ denote the columns of a matrix Q.

COROLLARY 3.1 *If Q is a nonnegative and symmetric matrix and $\mathbf{p} \in R^n_+$, then the function*

$$\mathbf{x}^T Q\mathbf{x} + \mathbf{p}^T\mathbf{x}, \qquad \mathbf{x} \in R^n_+, \tag{3.8}$$

is geodesic strictly convex with respect to the Riemannian metric generated by the coordinate representation (3.1).

Proof. By Theorem 3.1, R^n_+ is geodesic convex with respect to the chosen Riemannian metric.

Since there exists a constant c such that the matrices

$$Q + c\,\mathrm{diag}(\mathbf{x}^T\mathbf{q}_1 + p_1, \dots, \mathbf{x}^T\mathbf{q}_n + p_n), \qquad \mathbf{x} \in R^n_+, \tag{3.9}$$

are positive definite, the statement is proved. □

Now, we consider the affine Riemannian metric studied in [7,8].

THEOREM 3.2 *Let R^n_+ be the positive orthant, $f : R^n_+ \to R$, a twice continuously differentiable function and*

$$\mathbf{u}(\mathbf{x}) = (\log(x_1), \dots, \log(x_n)), \qquad \mathbf{x} \in R^n_+, \tag{3.10}$$

a curvilinear coordinate system. Then, the function f is geodesic (strictly) convex on R^n_+ with respect to the affine Riemannian metric generated by the transformation $\mathbf{u}$ iff the matrices

$$Hf(\mathbf{x}) + \mathrm{diag}\left(\frac{\partial f(\mathbf{x})}{\partial x_1}\frac{1}{x_1}, \dots, \frac{\partial f(\mathbf{x})}{\partial x_n}\frac{1}{x_n}\right), \qquad \mathbf{x} \in R^n_+, \tag{3.11}$$

are positive (definite) semidefinite.

Proof. By Theorem 10.1.1 in [8], the positive orthant is geodesic convex with respect to the affine metric.

The Jacobian matrix of the coordinate representation, its inverse matrix and the Hessian matrices of the components of the coordinate representation are as follows:

$$\begin{aligned} J\mathbf{u}(\mathbf{x}) &= \mathrm{diag}\left(\frac{1}{x_1}, \dots, \frac{1}{x_n}\right), \\ (J\mathbf{u}(\mathbf{x}))^{-1} &= \mathrm{diag}\Big(x_1, \dots, x_n\Big), \\ Hu_i(\mathbf{x}) &= \mathrm{diag}(0, \dots, -\frac{1}{x_i^2}, \dots, 0), \end{aligned} \tag{3.12}$$

$$\mathbf{x} \in R^n_+,$$

thus the second Christoffels symbols are

$$\Gamma^1 = \begin{pmatrix} -1/x_1 & \cdots & 0 \\ \vdots & \ddots & \vdots \\ 0 & \cdots & 0 \end{pmatrix}, \dots, \Gamma^n = \begin{pmatrix} 0 & \cdots & 0 \\ \vdots & \ddots & \vdots \\ 0 & \cdots & -1/x_n \end{pmatrix}, \tag{3.13}$$

$$\mathbf{x} \in R^n_+,$$

from which the statement follows. □

By Theorem 10.1.4 in [8], the posynomials are geodesic convex with respect to the affine metric. Another example is as follows.

EXAMPLE 3.1 Let us consider the function

$$f(x,y) = xy, \qquad (x,y) \in R^2_+ ,$$

its gradients and Hessian matrices

$$\nabla f(x,y) = (y,x),$$

$$Hf(x,y) = \begin{pmatrix} 0 & 1 \\ 1 & 0 \end{pmatrix}, \qquad (x,y) \in R^2_+ .$$

By Theorem 3.2, the consequence of the positive semidefiniteness of the matrices

$$\begin{pmatrix} \frac{y}{x} & 1 \\ 1 & \frac{x}{y} \end{pmatrix}, \qquad (x,y) \in R^2_+ .$$

is the geodesic convexity of the function f with respect to the affine metric.

References

1. Avriel, M., Diewert, W. E., Schaible, S. and Zang, I., *Generalized concavity*, Plenum Press, 1988.
2. Ben-Tal, A., On generalized means and generalized convex functions, *Journal of Optimization Theory and Applications* 21 (1977) 1–13.
3. Rapcsák, T., Geodesic convexity in nonlinear optimization, *Journal of Optimization Theory and Applications* 69 (1991) 169–183.
4. Rapcsák, T. and Csendes, T., Nonlinear coordinate transformations for unconstrained optimization, II. Theoretical background, *Journal of Global Optimization* 3 (1993) 359–375.
5. Rapcsák, T., Geodesic convexity on R^n, in: *Generalized convexity*, Lecture Notes in Economics and Mathematical Systems 405, Komlósi, S., Rapcsák, T. and Schaible, S. (eds.), Springer-Verlag, Berlin, Heidelberg, New York (1994) 91–103.
6. Rapcsák, T. and Thang, T. T., On nonlinear coordinate representations of smooth optimization problems, *Journal of Optimization Theory and Applications* 86 (1995) 459–489.
7. Rapcsák, T., Geodesic convexity on R^n_+, *Optimization* 37 (1996) 341-355.
8. Rapcsák, T., *Smooth nonlinear optimization in* R^n, Kluwer Academic Publishers, 1997.
9. Spivak, M., *A comprehensive introduction to differential geometry I-V*, Publish or Perish, Berkeley, 1979.

Packing Equal Circles in a Square I. — Problem Setting and Bounds for Optimal Solutions

P. G. Szabó
Department of Applied Informatics,
József Attila University, Szeged, Hungary
e-mail: pszabo@inf.u-szeged.hu

T. Csendes
Department of Applied Informatics,
József Attila University, Szeged, Hungary
e-mail: csendes@inf.u-szeged.hu

L. G. Casado
Department of Computer Architecture and Electronics,
University of Almería, Spain
e-mail: leo@miro.ualm.es

I. García
Department of Computer Architecture and Electronics,
University of Almería, Spain
e-mail: inma@iron.ualm.es

Abstract. In the paper, a short review of the problem of finding the densest packing of n equal circles in a square is made. There will be new lower bounds for this problem defined on the basis of regular arrangements. Also, there will new upper bounds be established based on the computation of the areas of circle and minimum gap between circles and between circles and sides of the square. The paper also contains all the known exact values of optimal packings and the corresponding minimal polynomials.

Key words: Optimal Packing, Minimal Polynomials, Regular Patterns.

1. Introduction

The general problem of finding the densest packing of n equal objects in a bounded space is a classical one which arises in many scientific and engineering fields. For the two-dimensional case, this problem has been studied for several different shapes of the bounded space; e.g. packing n equal circles in a circle [7, 8, 9, 12, 15, 20, 28, 42, 43], or circles in an equilateral triangle [13, 26, 27, 29, 38] or circles in a square where circles are non-overlapping and the radius of equal circles should

[1] This work was supported by the SOCRATES-ERASMUS program (25/ERMOB/1998-99), by the Ministry of Education of Spain (CICYT TIC 96-1125-C03-03) and by the Consejería de Educación de la Junta de Andalucía (07/FSC/MDM), and by the Grants FKFP 0739/97, OTKA T 016413 and T 017241.

F. Giannessi et al. (eds.), Optimization Theory, 191–206.

be maximized. In the paper all the discussions will only concern the packing circles in a square problem.

The packing problem has a long history in the mathematical literature [3, 4, 5, 6], but the packing circles in a square problem is 40 years old only. In 1960, Leo Moser was the first to study it when he wrote the next conjecture [32]: eight points in or on a unit square determine a distance $\leq \frac{1}{2}\sec 15^o$. This problem for up to nine circles ($n = 2, \ldots, 9$) was solved in 1965 [44, 46], although the first proof for $n = 6$ was reported in 1970 [48] and for $n = 7$ in 1996 [34]. Between 1970 and 1990, at least ten papers have reported solutions for $n = 10$ [11, 16, 17, 18, 30, 31, 41, 45, 47, 49], but the optimal solution was given in 1990 [17]. R. Peikert *et al.* [40] proposed an elimination procedure and found optimal results for $n = 10, \ldots, 20$. Until recently the problem has been solved for $n \leq 27$ [35, 40] using computer-aided methods for optimality proofs [34, 40] and the $n = 36$ [19] case was proven by mathematical tools. Optimality proofs of packings obtained manually (without using a computer) exist up to 9 [25, 44, 46, 48] and for the $n = 14$ [51], 16 [50], 25 [52], and 36 [19] cases. Albeit Schaer and Meir in [44] claimed that for $n = 7$ they had found the optimality proof, but did not explain it further on, and until now it has not been published. A more detailed history of the packing equal circles in a square problem can be found in [53].

In addition to the above optimal solutions, there has been published good approximations to the problem solution for $n \leq 52$ and n =54, 55, 56, 60, 61, 62, 72, and 78. These numerical results were obtained by using several different strategies; for instance, using billiard simulation [14, 21, 22], minimization of the energy function [33], standard BFGS quasi-Newton algorithm [40], nonlinear programming solver (MINOS 5.3) [24] or the Cabri-Géomètre software in [31]. Although it is not known whether these solutions are optimal, in some cases these numerical results can help to find better solutions (if any) when they are used as lower bounds of the optimal (maximal) solutions. For instance, several strategies described in the packing literature, as that of R. Peikert *et al.* in [40], are based on the knowledge of a good lower bound of the solution. In this sense, lower bounds are useful for the problem solution.

The paper provides new results, lower and upper bounds on the radius of circles for up to one hundred circles. The paper is organized as follows: in Section 2 four equivalent definitions of the packing n circles in a square problem as well as notations and parameters used are given. Section 3 is devoted to set up several proposals for defining upper and lower bounds on the radius for the packing n circles in a unit square problem. Upper bounds are based on the computation of the circle areas as well as on the minimum gap areas between circles and between circles and the sides of the square. Our suggestion for lower bounds are based on specific regular patterns to arrange n equal circles in a unit square. In Section 4, for several specific values of n, theoretical solutions and exact results of the optimal values of the problem at hand are given.

2. Definitions and notation

The packing circles in a square problem can be described by the following equivalent problem settings:

1. Locate n points in a unit square, such that the minimum distance m_n between any two points is maximal.

2. Find the value of the maximum circle radius, r_n, such that n equal non-overlapping circles can be placed in a unit square.

3. Give the smallest square of side s_n, which contains n equal and non-overlapping circles where the radius of circles is 1.

4. Determine the smallest square of side σ_n that contains n points with mutual distance of at least 1.

From these statements it is easy to prove that for the defined parameters m_n, r_n, s_n and σ_n the following relations hold:

$$r_n = \frac{m_n}{2(m_n+1)}, \qquad m_n = \frac{2r_n}{1-2r_n}, \qquad s_n = \frac{1}{r_n}, \qquad \sigma_n = \frac{1}{m_n}. \tag{1}$$

The problem of maximizing the minimal pairwise distance of n points being in a unit square (Problem 1) can be formulated as the following continuous global optimization problem:

$$\begin{gathered} \mu(x,y) = \min_{1 \le i < j \le n} \sqrt{(x_i - x_j)^2 + (y_i - y_j)^2} \\ \max \mu(x,y) \\ x, y \in [0,1]^n, \; n > 1 \text{ integer}, \end{gathered} \tag{2}$$

where x_i, y_i are the coordinates of the i-th point. The goal is not only to obtain the maximum of the minimum distance between any two points $(\max \mu(x,y))$, but also to find the respective locations of the n points in the unit square (the coordinates x_i, y_i; $1 \le i \le n$).

DEFINITION 1. *Given an $\epsilon \geq 0$ (tolerance error), we say that* two circles i, j *(Problem 1) located at* (x_i, y_i) *and* (x_j, y_j) are in contact *if*

$$\sqrt{(x_i - x_j)^2 + (y_i - y_j)^2} \leq m_n + \epsilon .$$

Similarly to this, a circle i is in contact with one side of the square if $x_i \leq \epsilon$, $y_i \leq \epsilon$, $x_i \geq 1 - \epsilon$, *or* $y_i \geq 1 - \epsilon$.

The exact mathematical definition for the contact of two circles is that m_n is equal to $\sqrt{(x_i - x_j)^2 + (y_i - y_j)^2}$, and also when a point is on the side of the square. We have considered the above exact form of definition in the theoretical results as well as in Definition 1 in numerical results [1], since usually a finite arithmetic is adapted to computers. For the study we set $\epsilon = 10^{-10}$.

DEFINITION 2. *Let us suppose that there is a given solution of the problem (2). We say that a circle is* free *if its center can be moved towards a positive distance point without causing the others' overlapping.*

We can point out that when a packing contains one or more free circles, then the solution is obviously not unique, moreover, the locations of the center of any free circles form a non-empty interior and connected set. In the paper the number of contacts will be denoted by c_n, while the number of free circles by f_n. In all the figures a contact will be represented by a short line section and free circles will be indicated by dark shading.

DEFINITION 3. *The density d_n of a packing can be given by*

$$d_n = n\pi r_n^2 = \frac{n\pi m_n^2}{4(m_n + 1)^2} .$$

3. Lower and upper bounds of the maximized minimum distance between n points in a unit square

An approximate solution of (2) can be used as a lower bound m_n^l of the optimal solution m_n. Let us suppose that it is possible to determine a good value for the upper bound m_n^u of the packing problem, so an approximate solution m_n of (2) can be considered a good solution of the packing problem if it is close enough to m_n^u.

On the other hand, it has been proved in [6] that the densest packing of circles in the plane is a regular (hexagonal) arrangement (see

Figure 1) and the density of this packing is $\sqrt{3}\pi/6 \approx 0.9068996821$. In addition, it can also be proved that this value can provide an upper bound for the solution of (2), although upper bound values better than that can be proposed. A hexagonal pattern can give an asymptotic formula for m_n, where $m_n \approx \sqrt{\frac{2}{\sqrt{3}n}}$ [3].

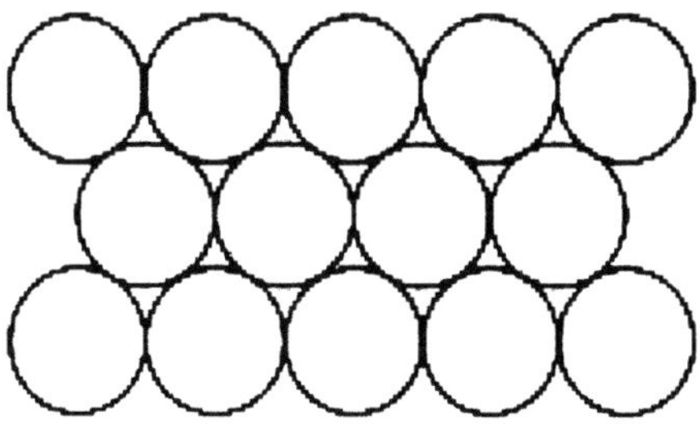

Figure 1. The arrangement of the densest packing in the plane

It is also known that for several values of n, the optimal solutions of the packing circles in a unit square problem are regular arrangements. Keeping these ideas in mind, it is easy to guess that for problem (2), other regular arrangements may also provide at least a good lower bound of the optimal solution. In the coming section improved lower and upper bounds are discussed.

3.1. Lower bounds of m_n by using patterns

THEOREM 4. *Consider the packing circles in a unit square problem according to Problem 1. The maximum of the minimal distance between n points in the unit square m_n is not smaller than*

$$m_n^l = \max(L_1(n), L_2(n), L_{3a}(n), L_{3b}(n), L_4(n), L_5(n))$$

with

$$L_1(n) = \frac{1}{\lceil\sqrt{n}\rceil - 1},$$

$$L_2(n) = \frac{1}{\lceil\sqrt{n+1}\rceil - 3 + \sqrt{2+\sqrt{3}}},$$

$$L_{3a}(n) = \frac{1}{\lceil\sqrt{n+2}\rceil - 2 + \frac{1}{2}\sqrt{3}},$$

$$L_{3b}(n) = \frac{1}{\lceil\sqrt{n+2}\rceil - 5 + 2\sqrt{2+\sqrt{3}}},$$

$$L_4(n) = \frac{k^2-k-\sqrt{2k}}{k^3-2k^2}, \text{ if } n = k(k+1), \text{ and } 0, \text{ otherwise },$$

$$L_5(n) = \sqrt{\frac{1}{p^2}+\frac{1}{q^2}}, \text{ if } n = \left\lceil\frac{(p+1)(q+1)}{2}\right\rceil, \; p^2 \le 3q^2, \text{ and } q^2 \le 3p^2 \;;$$

and 0, otherwise .

(Here $\lceil . \rceil$ denotes the smallest integer not smaller than the argument.)

Proof. The new lower bounds are based on the use of some patterns proposed in [14, 23, 33], also, on our own new patterns and at the assumption that $m_{n+1} \leq m_n$. Thus m^l_{n+1} or any values of the m_{n+1} of (2) are also lower bounds of m_n. In this way lower bounds for all the values of n $(2 \leq n \leq 100)$ can be obtained. The set of patterns we will take into account for defining lower bounds are (see them illustrated in *Figure 3*):

PAT1 For $n = k^2$ ($k \geq 2,\ k \in I\!N$, where $I\!N$ is the set of nonnegative integers) the packings have an obvious square grid pattern of $k \times k$ points [14], for which the minimum distance between the points is $L_1(n) = \frac{1}{k-1}$. This lower bound can be applied also for n values not conforming the equation $n = k^2$ with the $k = \lceil \sqrt{n} \rceil$ substitution.

PAT2 For $n = k^2 - 1 = (k-1)^2 + 2(k-1)$ ($k \geq 2,\ k \in I\!N$) the pattern can be considered as a square pattern such as PAT1 of $(k-1)^2$ circles into which one row and one column of shifted circles are inserted [14]. In this case $L_2(n) = \frac{1}{k-3+\sqrt{2+\sqrt{3}}}$, and $k = \lceil \sqrt{n+1} \rceil$, because $m_n(k - 3 + 2\cos(15^o)) = 1$.

PAT3 For $n = k^2 - 2 = (k-1)^2 + 2(k-2) + 1$ the pattern is similar to PAT2, but in this case there are two shifted columns and two shifted rows of fitting length [14]. In these cases
a) $L_{3a}(n) = \frac{1}{k-2+\frac{1}{2}\sqrt{3}}$, ($k \geq 3,\ k \in I\!N$),
where $k = \lceil \sqrt{n+2} \rceil$, since $m_n(k - 2 + \cos(30^o)) = 1$, and
b) $L_{3b}(n) = \frac{1}{k-5+2\sqrt{2+\sqrt{3}}}$, ($k \geq 5,\ k \in I\!N$),
where $k = \lceil \sqrt{n+2} \rceil$, since $m_n(k - 5 + 2\cos(15^o) + 2\cos(15^o)) = 1$.

PAT4 For $n = k(k+1)$ ($k \geq 2,\ k \in I\!N$) the pattern consists of $(k-1)$ alternating columns with k circles each [14]. In Figure 3 this pattern is demonstrated for $n = 20$. In this case

$$L_4(n) = \frac{k^2 - k - \sqrt{2k}}{k^3 - 2k^2},$$

as $m_n(k - 1 + \sin(\alpha)) = 1$, where $\cos(\alpha) = \frac{1}{m_n k}$.

PAT5 Focusing on the new pattern class, we can see that the centers of the circles are located in a grid which is built by composing two shifted rectangular grids. An example of this pattern is demonstrated in *Figure 2* and in *Figure 3*. The number of non-overlapping circles (n) which can be arranged in a square by using this pattern is a function of two integers $p, q \in \mathbb{N}$, $p \neq q$, subject to the following constraints:

$$p^2 \leq 3q^2\,,$$
$$q^2 \leq 3p^2\,.$$

For the sake of clarity, three cases will be distinguished:

- p and q are both even; then $p = 2p^*$; $q = 2q^*$, and $n = 2p^*q^* + p^* + q^* + 1$
- p is even and q is odd; then $p = 2p^*$; $q = 2q^* + 1$, and $n = (2p^* + 1)(q^* + 1)$
- both p and q are odd, then $p = 2p^* + 1$; $q = 2q^* + 1$, and $n = 2(p^* + 1)(q^* + 1)$

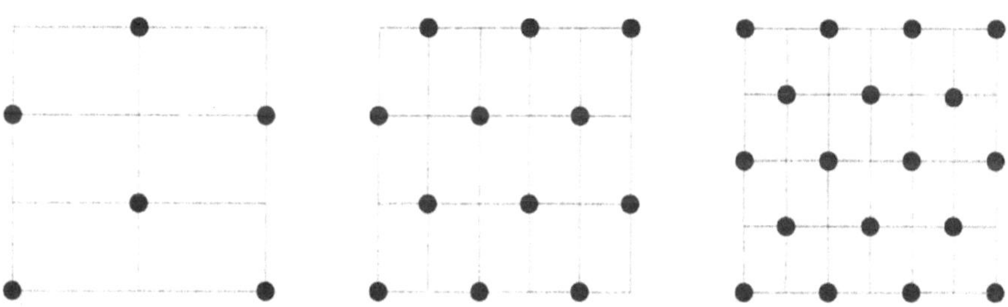

Figure 2. Examples (p, q, n) using PAT5, (2,3,6), (5,3,12) and (6,4,18)

and for all the cases $L_5(n) = \sqrt{\frac{1}{p^2} + \frac{1}{q^2}}$ holds for this lower bound.

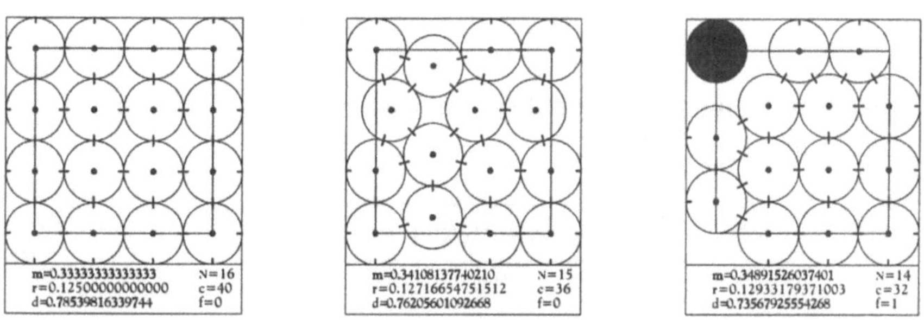

PAT1 **PAT2** **PAT3a**

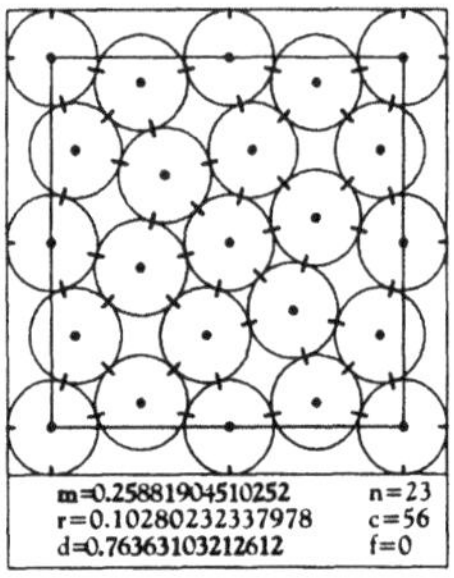

PAT3b

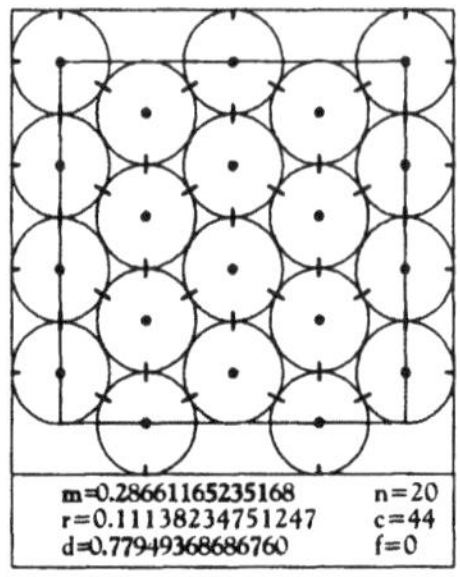

PAT4

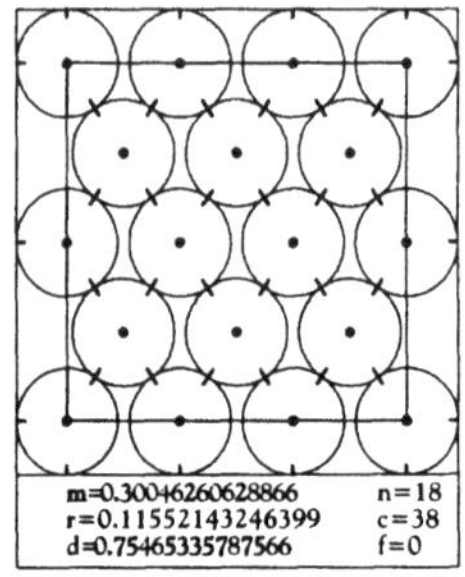

PAT5

Figure 3. Patterns for $n = 16$ (PAT1), $n = 15$ (PAT2), $n = 14$ (PAT3a), $n = 23$ (PAT3b), $n = 20$ (PAT4) and $n = 18$ (PAT5). These examples are optimal packings.

The maximum of the derived lower bounds provides the statement of the theorem with the suggested substitutions, wherever necessary. □

Further on, *Table 1* will summarize the results obtained by applying the discussed patterns. With patterns PAT1 to PAT5 only 48 out of 99 values of n can directly be modelled. In columns headers n, pattern and m_n^l denote the number of circles, the pattern used and the value of the lower bound of m_n, respectively. If there are two different patterns for the same value of n, then the better one has been chosen. In column n, boldface means that this packing is known as the optimal solution for the given packing circles in a square problem. All the values of m_n^l in Table 1 are those calculated by the respective formula of the specific pattern.

In [36], using a similar pattern as in PAT5, it was proved that if $N_p(\sigma)$ is the maximum number of points with mutual distance of at least 1, that can be placed into a square of side σ, then

$$\sigma^2 + \frac{1-\sqrt{3}}{2}\sigma \leq \frac{\sqrt{3}}{2} N_p(\sigma).$$

Resulting from this inequality we can provide the following lower bound of m_n:

$$\frac{1-\sqrt{3}+\sqrt{4-2\sqrt{3}+8\sqrt{3}n}}{2\sqrt{3}n} \leq m_n \tag{3}$$

After an analysis of (3) and m_n^l, it can be seen that m_n^l is always greater than the lower bound given by (3) for $2 \leq n \leq 100$.

n	pattern	m_n^l	n	pattern	m_n^l	n	pattern	m_n^l
2	PAT5	1.414213	21	PAT5	0.260341	52	PAT5	0.165386
3	PAT2	1.035276	**23**	PAT3b	0.258819	56	PAT4	0.156156
4	PAT1	1.000000	**24**	PAT2	0.254333	59	PAT5	0.150231
5	PAT5	0.707106	**25**	PAT1	0.250000	63	PAT5	0.146772
6	PAT5	0.600925	**27**	PAT5	0.235849	64	PAT1	0.142857
7	PAT3a	0.535898	30	PAT4	0.224502	65	PAT5	0.138888
8	PAT2	0.517638	32	PAT5	0.208333	72	PAT4	0.135416
9	PAT1	0.500000	34	PAT3b	0.205604	75	PAT5	0.132089
10	PAT5	0.416666	35	PAT2	0.202763	80	PAT5	0.129576
12	PAT5	0.388730	**36**	PAT1	0.200000	81	PAT1	0.125000
13	PAT5	0.353553	39	PAT5	0.194365	83	PAT5	0.122890
14	PAT3a	0.348915	42	PAT4	0.184277	88	PAT5	0.120185
15	PAT2	0.341081	44	PAT5	0.174379	90	PAT4	0.119501
16	PAT1	0.333333	47	PAT3b	0.170540	94	PAT4	0.117924
18	PAT5	0.300462	48	PAT5	0.169329	99	PAT4	0.116018
20	PAT4	0.286611	49	PAT1	0.166666	100	PAT1	0.111111

Table 1. Lower bounds of m_n obtained by using repeated patterns. The proven optimal solutions are denoted by boldface in columns n.

3.2. An upper bound of m_n

THEOREM 5. *Consider the packing circles in a unit square problem according to Problem 2. The maximum of the minimal distance between n points in the unit square m_n is not greater than*

$$m_n^u = \min(U_1(n), U_2(n))$$

with

$$U_1(n) = \frac{2}{\sqrt{n\pi + C_n(\sqrt{3} - \frac{\pi}{2}) + (4\lfloor\sqrt{n}\rfloor - 2)(2 - \frac{\pi}{2}) - 2}}, \text{ where}$$

$$C_n = \begin{cases} n-2 & \textit{if } 3 \le n \le 6, \\ n-1 & \textit{if } 7 \le n \le 9, \\ 3\left\lfloor\frac{n}{2}\right\rfloor - 5 + n \bmod 2 & \textit{otherwise.} \end{cases}$$

$$U_2(n) = \frac{1 + \sqrt{1 + (n-1)\frac{2}{\sqrt{3}}}}{n-1}.$$

(Here $\lfloor . \rfloor$ denotes the greatest integer not greater than the argument.)

Proof. Let us describe an analytical expression for the upper bounds of the packing problem. It is based on the area of n circles and on the area of their neighbourhood within the unit square:

- U_1: The area of n circles ($nr^2\pi$) plus the gaps' area between circles and between circles and the sides of the square must be less than 1. The convex hull of the circles' centers divides the square into two regions. Let us compute now the area of the minimal gaps in these regions.

  Circles-to-side case. When packing n circles in a square, it is evident that the minimum gap between circles and the sides of the square can be gained when $4\lfloor\sqrt{n}\rfloor - 2$ gaps (with the respective corrections at the corners) are in contact with the sides of the square of $r_n^2(2-\pi/2)$ areas per gaps (see *Figure 4*).

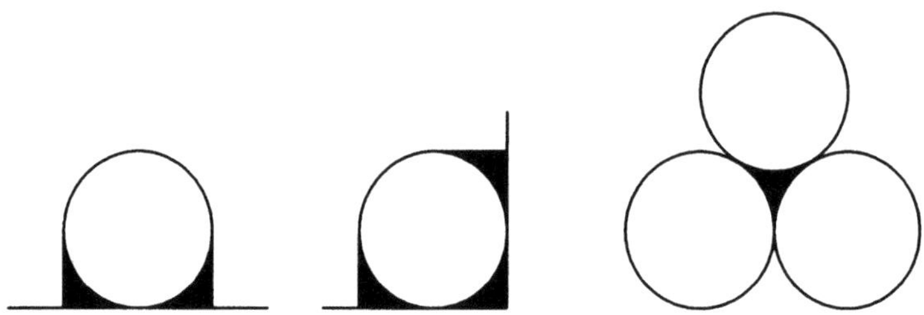

Figure 4. The minimal gaps between a circle and the sides of a square and among three circles

Circles-to-circles case. In a similar way, the minimal gap between circles occurs when there are three circles in the packing which have all the possible three contacts, as it is shown in Figure 4. For a packing of n circles the minimal number of circle-to-circle gaps is given by:

$$C_n = \begin{cases} n-2 & \text{if } 3 \le n \le 6, \\ n-1 & \text{if } 7 \le n \le 9, \\ 3\left\lfloor\frac{n}{2}\right\rfloor - 5 + n \bmod 2 & \text{otherwise.} \end{cases}$$

when the circles are in the densest (hexagonal) arrangement. The area of this minimum gap between three circles is $r_n^2(\sqrt{3}-\frac{\pi}{2})$, thus the sum of the area of the circles and gaps will be:

$$S = r_n^2\left(n\pi + C_n(\sqrt{3}-\frac{\pi}{2}) + (4\lfloor\sqrt{n}\rfloor - 2)(2-\frac{\pi}{2})\right).$$

Due to $S \le 1$,

$$r_n^u \le \left(n\pi + C_n(\sqrt{3}-\frac{\pi}{2}) + (4\lfloor\sqrt{n}\rfloor - 2)(2-\frac{\pi}{2})\right)^{-\frac{1}{2}},$$

$$m_n^u \leq U_1(n) = \frac{2}{\sqrt{n\pi + C_n(\sqrt{3} - \frac{\pi}{2}) + (4\lfloor\sqrt{n}\rfloor - 2)(2 - \frac{\pi}{2}) - 2}}.$$

- U_2: From Oler theorem [37] the following statement implies [10]: if X is a compact convex subset of the plane, then the number of points with a mutual distance of at least 1 can be at most

$$\frac{2}{\sqrt{3}}A(X) + \frac{1}{2}P(X) + 1, \qquad (4)$$

where $A(X)$ is the area and $P(X)$ is the perimeter of X. If X is a σ side of square, then $A(X) = \sigma^2$ and $P(X) = 4\sigma$. Based on (4) and (1), we can conclude that

$$m_n \leq U_2(n) = \frac{1 + \sqrt{1 + (n-1)\frac{2}{\sqrt{3}}}}{n-1}.$$

□

By using the values of the lower and upper bounds of m_n, described beforehand and our own numerical results, obtained by the algorithm described in [1], *Figure 5* has been built.

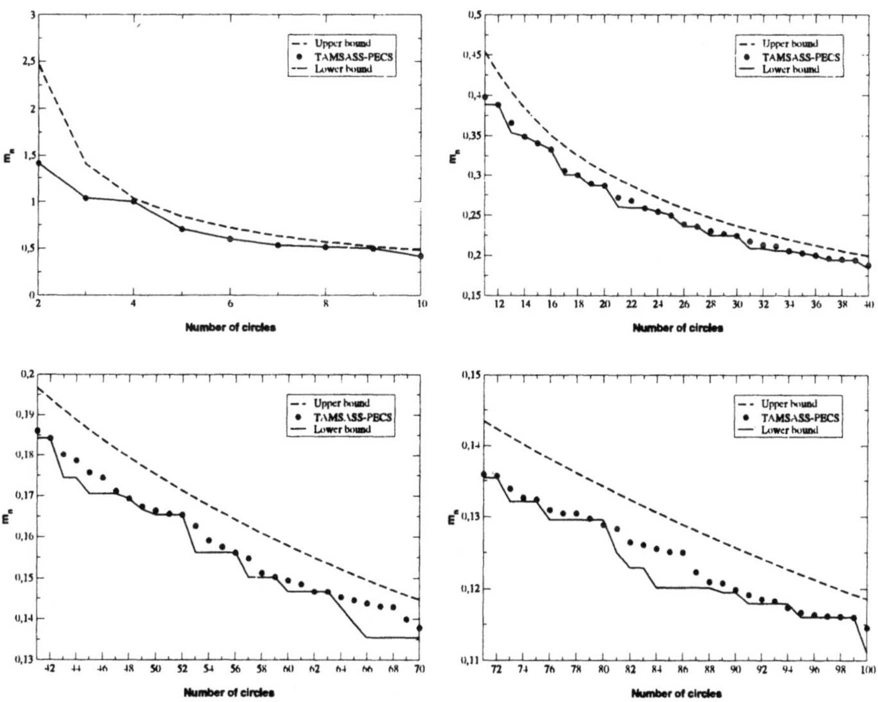

Figure 5. Lower (m_n^l) and upper bound (m_n^u) on the m_n values and the best m_n values found by our program [1] (denoted by circles) are displayed.

4. Minimal polynomials of optimal packings and exact results

Actually, we know the exact values for the solution of the packing circles in a unit square problem for several values of n. The set of known exact values is given in *Table 2.*

n	m_n	n	m_n
2	$\sqrt{2}$	15	$(1+\sqrt{2}-\sqrt{3})/2$
3	$\sqrt{6}-\sqrt{2}$	16	$1/3$
4	1	18	$\sqrt{13}/12$
5	$\sqrt{2}/2$	20	$(6-\sqrt{2})/16$
6	$\sqrt{13}/6$	23	$(\sqrt{6}-\sqrt{2})/4$
7	$2(2-\sqrt{3})$	24	$4+2\sqrt{3}-\sqrt{26+15\sqrt{3}}$
8	$(\sqrt{6}-\sqrt{2})/2$	25	$1/4$
9	$1/2$	27	$\sqrt{89}/40$
12	$\sqrt{34}/15$	36	$1/5$
14	$2(4-\sqrt{3})/13$		

Table 2. Exact values of m_n for some n

In case we are familiar with the optimal solution for the problem of packing n circles in a square, the definition of the corresponding minimal polynomial may be given. The procedure to obtain this minimal polynomial involves the generation of a system of equations representing the corresponding optimal packing. The variables of the system of equations are the coordinates of the centers of the circles and the maximum minimal distance, m_n. So the number of variables for the system is $2n+1$. For every two points (centers of the circles) of which distance is equal to m_n, an equation can be written. In addition, it is also well-known that every point located at the border of the square has got at least one coordinate which is equal to 0 or 1, so the corresponding equation(s) can be written. After reducing the system of equations, a symbolic algebra package e.g. MAPLE, MATHEMATICA, etc., can be used to calculate the Gröbner basis of the system [40] and the minimal polynomial for which the smallest positive root is m_n.

n	Minimal polynomials
2	$m^2 - 2$
3	$m^4 - 16m^2 + 16$
4	$m - 1$
5	$2m^2 - 1$
6	$36m^2 - 13$
7	$m^2 - 8m + 4$
8	$m^4 - 4m^2 + 1$
9	$2m - 1$
10	$1180129m^{18} - 11436428m^{17} + 98015844m^{16} - 462103584m^{15} +$ $+ 1145811528m^{14} - 1398966480m^{13} + 227573920m^{12} + 1526909568m^{11} -$ $- 1038261808m^{10} - 2960321792m^9 + 7803109440m^8 - 9722063488m^7 +$ $+ 7918461504m^6 - 4564076288m^5 + 1899131648m^4 - 563649536m^3 +$ $+ 114038784m^2 - 14172160m + 819200$
11	$m^8 + 8m^7 - 22m^6 + 20m^5 + 18m^4 - 24m^3 - 24m^2 + 32m - 8$
12	$225m^2 - 34$
13	$5322808420171924937409m^{40} + 5867739593380498861732332m^{39} +$ $+ 13024448845332271203266928m^{38} - 12988409567056909990170432m^{37} +$ $+ \ldots + 2960075719794736758784m^2 - 174103532094609162240m +$ $+ 4756927106410086400$
14	$13m^2 - 16m + 4$
15	$2m^4 - 4m^3 - 2m^2 + 4m - 1$
16	$3m - 1$
17	$m^8 - 4m^7 + 6m^6 - 14m^5 + 22m^4 - 20m^3 + 36m^2 - 26m + 5$
18	$-144m^2 + 13$
19	$242m^{10} - 1430m^9 - 8109m^8 + 58704m^7 - 78452m^6 - 2918m^5 + 43315m^4 +$ $+ 39812m^3 - 53516m^2 + 20592m - 2704$
20	$128m^2 - 96m + 17$
21	*
22	*
23	$16m^4 - 16m^2 + 1$
24	$m^4 - 16m^3 + 20m^2 - 8m + 1$
25	$4m - 1$
26	*
27	$1600m^2 - 89$
36	$5m - 1$

Table 3. The minimal polynomials of certain optimal solutions. Here * denotes cases with unknown minimal polynomials.

The minimal polynomial can be obtained for the most cases, but e.g. for $n = 13$ this method was unable [40] to provide the minimal polynomial. In *Table 3* the minimal polynomials for several values of n are shown [17, 39]. The degrees of these polynomials indicate how difficult it might be to find the relevant solution of the packing problem. When the degree of the minimal polynomial became less than five, we were able to calculate the exact value of m_n, which values can be found in Table 2. For $n = 10$, 11, 17 and 19, we were able to provide close bounds of m_n using a reliable algorithm for finding the first root of a function [2]. The mentioned algorithm is based on a branch and bounds technique with interval arithmetic. The numerical results for these four values of n are given in interval form in *Table 4*.

n	Lower and upper bounds for m_n
10	[0.421279541378085, 0.421279544064904]
11	[0.398207310236837, 0.398207310236850]
17	[0.306153985300327, 0.306153985300338]
19	[0.289541991994965, 0.289541991994996]

Table 4. Reliable numerical results for $n = 10, 11, 17, 19$.

5. Summary

First of all, in this series of papers a short introduction has been given to the problem of equal circles packing in a unit square. Then, in a nutshell the most important results have been summarized. Finally, new and improved lower and upper bounds on the maximal distance of the centers of the circles are demonstrated. These bounds are based on geometrical considerations and on an extended set of patterns of the related circles, moreover, they can be applied directly in some computational procedures for finding the best packings. Last but not least, they can also be well-utilised in the evaluation of the approximate solutions provided by our numerical algorithm discussed in the second paper of this series. The known exact optimal values are comprised in a table together with the related minimal polynomials for some cases.

References

1. L. G. Casado, I. García, P. G. Szabó and T. Csendes. Packing equal circles in a square II.
New results for up to 100 circles using TAMSASS-PECS algorithm. This volume. Available at http://www.inf.u-szeged.hu/ ~csendes/pack2.ps.gz.
2. L. G. Casado, I. García and Ya. D. Sergeyev. Interval Branch and Bound algorithm for finding the first-zero- crossing-point in one-dimensional functions. *Reliable Computing*, 6:179-191, 2000.
3. H. T. Croft, K. J. Falconer and R. K. Guy. *Unsolved problem in geometry.* Springer-Verlag, New York, 1991

4. DingZhu Du and P. M. Pardalos. Minimax and Applications *Kluwer Academic Publishers*, 1995
5. G. Fejes Tóth. *Handbook of Discrete and Computational Geometry.* CRC Press, Boca Raton, 1997.
6. L. Fejes Tóth. *Lagerungen in der Ebene, auf der Kugel und im Raum.* Springer-Verlag, Berlin, 1972.
7. F. Fodor. The densest packing of 12 congruent circles in a circle. *Beiträge zur Algebra und Geometrie*, (Submitted for publication), 1998
8. F. Fodor. The densest packing of 13 congruent circles in a circle. (Submitted for publication), 1999.
9. F. Fodor. The densest packing of 19 congruent circles in a circle. *Geometriae Dedicata*, 74:139-145, 1999.
10. J. H. Folkman and R. L. Graham. A packing inequality for compact convex subsets of the plane. *Canadian Mathematical Bulletin*, 12:745-752, 1969.
11. M. Goldberg. The packing of equal circles in a square. *Mathematics Magazine*, 43:24–30, 1970.
12. M. Goldberg. Packing of 14,16,17 and 20 circles in a circle. *Mathematics Magazine*, 44:134–139, 1971.
13. R. L. Graham and B. D. Lubachevsky. Dense packings of equal disks in an equilateral triangle: from 22 to 34 and beyond. *The Electronic J. of Combinatorics*, 2, 1995.
14. R. L. Graham and B. D. Lubachevsky. Repeated patterns of dense packings of equal disks in a square. *The Electronic J. of Combinatorics*, 3:1–16, 1996.
15. R. L. Graham, B. D. Lubachevsky, K. J. Nurmela, and P. R. J. Östergård. Dense packings of congruent circles in a circle. *Discrete Mathematics*, 181:139–154, 1998.
16. M. Grannell. An even better packing of ten equal circles in a square. *Manuscript.*, 1990.
17. C. de Groot, R. Peikert, and D. Würtz. The optimal packing of ten equal circles in a square. *IPS Research Report, ETH Zürich No. 90-12*, August 1990.
18. B. Grünbaum. An improved packing of ten circles in a square. *Manuscript.*, 1990.
19. K. Kirchner and G. Wengerodt. Die dichteste Packung von 36 Kreisen in einem Quadrat. *Beiträge zur Algebra und Geometrie*, 25:147–159, 1987.
20. S. Kravitz. Packing cylinders into cylindrical containers. *Mathematics Magazine*, 40:65–71, 1967.
21. B. D. Lubachevsky. How to simulate billiards and similar systems. *J. Computational Physics*, 94:255–283, 1991.
22. B. D. Lubachevsky and F. H. Stillinger. Geometric properties of random disk packings. *J. Statistical Physics*, 60:561–583, 1990.
23. B. D. Lubachevsky, R. L. Graham and F. H. Stillinger. Patterns and structures in disk packings. *Periodica Mathematica Hungarica*, 34:123–142, 1997
24. C. D. Maranas, C. A. Floudas, and P. M. Pardalos. New results in the packing of equal circles in a square. *Discrete Mathematics*, 142:287–293, 1995.
25. H. Melissen. Densest packing of six equal circles in a square. *Elemente der Mathematik*, 49:27–31, 1994.
26. J. B. M. Melissen and P. C. Schuur. Packing 16, 17 or 18 circles in an equilateral triangle *Discrete Mathematics*, 145:333–342, 1995.
27. J. B. M. Melissen. Densest packings for congruent circles in an equilateral triangle. *American Mathematical Monthly*, 100:916–925, 1993.
28. J. B. M. Melissen. Densest packing of eleven congruent circles in a circle. *Geometriae Dedicata*, 50:15–25, 1994.
29. J. B. M. Melissen. Optimal packings of eleven equal circles in an equilateral triangle. *Acta Mathematica Hungarica*, 65:389–393, 1994.

30. R. Milano. Configurations optimales de desques dans un polygone régulier. *Mémoire de Licence*, 1987. Université Libre de Bruxelles.
31. M. Mollard and C. Payan. Some progress in the packing of equal circles in a square. *Discrete Mathematics*, 84:303–307, 1990.
32. L. Moser. Problem 24 (corrected). *Canadian Mathematical Bulletin*, 3:78, 1960.
33. K. J. Nurmela and P. R. J. Östergård. Packing up to 50 equal circles in a square. *Discrete & Computational Geometry*, 18:111–120, 1997.
34. K. J. Nurmela and P. R. J. Östergård. Optimal packings of equal circles in a square. *Proc. of the Eighth Int. Conf. on Graph Theory, Combinatorics, Algorithms, and Applications*, (In press.)
35. K. J. Nurmela and P. R. J. Östergård. More optimal packings of equal circles in a square. *Discrete & Computational Geometry*. (To appear.)
36. K. J. Nurmela, P. R. J. Östergård and R. aus dem Spring. Asymptotic behaviour of optimal circle packings in a square. *Canadian Mathematical Bulletin*. (To appear.)
37. N. Oler. An inequality in the geometry of numbers. *Acta Math.*, 105:19–48, 1961
38. N. Oler A finite packing problem. *Canadian Mathematical Bulletin*, 4:153-155, 1961
39. R. Peikert. Dichteste Packungen von gleichen Kreisen in einem Quadrat. *Elemente der Mathematik*, 49:16–26, 1994.
40. R. Peikert, D. Würtz, M. Monagan, and C. de Groot. Packing circles in a square: a review and new results. In P. Kall (ed.), *System Modelling and Optimization*, volume 180 of *Lecture Notes in Control and Information Sciences*, 45–54. Springer-Verlag, Berlin, 1992.
41. J. Petris and N. Hungerbüler. *Manuscript*, 1990.
42. U. Pirl. Der Mindestabstand von n in der Einheitskreisscheibe gelegenen Punkten. *Math. Natchr.*, 40:111-124, 1969
43. G. E. Reis. Dense packings of equal circles within a circle. *Mathematics Magazine*, 48:33–37, 1975
44. J. Schaer. The densest packing of nine circles in a square. *Canadian Mathematical Bulletin*, 8:273–277, 1965.
45. J. Schaer. On the densest packing of ten equal circles in a square. *Mathematics Magazine*, 44:139–140, 1971.
46. J. Schaer and A. Meir. On a geometric extremum problem. *Canadian Mathematical Bulletin*, 8:21–27, 1965.
47. K. Schlüter. Kreispackung in Quadraten. *Elemente der Mathematik*, 34:12–14, 1979.
48. B. L. Schwartz. Separating points in a square. *J. Recreational. Math.*, 3:195–204, 1970.
49. G. Valette. A better packing of ten circles in a square. *Discrete Mathematics*, 76:57–59, 1989.
50. G. Wengerodt. Die dichteste Packung von 16 Kreisen in einem Quadrat. *Beiträge zur Algebra und Geometrie*, 16:173–190, 1983.
51. G. Wengerodt. Die dichteste Packung von 14 Kreisen in einem Quadrat. *Beiträge zur Algebra und Geometrie*, 25:25–46, 1987.
52. G. Wengerodt. Die dichteste Packung von 25 Kreisen in einem Quadrat. *Ann. Univ. Sci. Budapest Eötvös Sect. Math.*, 30:3–15, 1987.
53. D. Würtz, M. Monagan, and R. Peikert. The history of packing circles in a square. *Maple Technical Newsletter*, 0:35–42, 1994. Maple in Mathematics and the Sciences — A Special issue.

Packing Equal Circles in a Square II. — New Results for up to 100 Circles Using the TAMSASS-PECS Algorithm

L. G. Casado
Department of Computer Architecture and Electronics,
University of Almería, Spain
e-mail: leo@miro.ualm.es

I. García
Department of Computer Architecture and Electronics,
University of Almería, Spain
e-mail: inma@iron.ualm.es

P. G. Szabó
Department of Applied Informatics,
József Attila University, Szeged, Hungary
e-mail: pszabo@inf.u-szeged.hu

T. Csendes
Department of Applied Informatics,
József Attila University, Szeged, Hungary
e-mail: csendes@inf.u-szeged.hu

Abstract. In this work we propose a new stochastic optimization algorithm for solving the problem of optimal packing of n non-overlapping equal circles in a square. It will be shown that our procedure can find most of the optimal solutions for all the problems previously solved and reported in the literature. Results obtained by our algorithm for up to 100 circles are given in relevant numerical and graphical form. For n = 32, 37, 47, 62 and 72 the algorithm has obtained better solutions than those reported on in the literature on packing. In addition, forty new and unpublished packing results are reported on. The arrangements obtained were validated by interval arithmetic computations.

Key words: Optimal Packing, Equal Circles, Unit Square.

1. Introduction

The packing circles in a unit square problem has been already introduced in details in the first paper [16]. The version of the problem discussed in this paper is: locate n points in a unit square, such that the minimum distance m_n between any two points will be maximal. The problem of maximizing the minimal pairwise distance of n points,

[1] This work was supported by the SOCRATES-ERASMUS program (25/ER-MOB/ 1998-99), by the Ministry of Education of Spain (CICYT TIC 96-1125-C03-03) and by the Consejería de Educación de la Junta de Andalucía (07/FSC/MDM), and by the Grants FKFP 0739/97, OTKA T 016413 and T 017241.

F. Giannessi et al. (eds.), Optimization Theory, 207–224.

which are contained in a unit square, can be formulated as the following continuous global optimization problem:

$$\begin{gathered}\mu(x,y) = \min_{1\le i<j\le n} \sqrt{(x_i - x_j)^2 + (y_i - y_j)^2} \\ \max \mu(x,y) \\ x, y \in [0,1]^n,\ n > 1 \text{ integer}, \end{gathered} \tag{1}$$

where x_i, y_i are the coordinates of the i-th point.

The computational results for the packing circles in a unit square problem, reported in the literature, were obtained by using several different procedures. As an example, the method used by R.L. Graham and B.D. Lubachevsky [3] is based on billiards simulation, K.J. Nurmela and P.R.J. Östergård [11] built a minimization of energy function strategy, standard BFGS quasi-Newton algorithm were used by R. Peikert *et al.* [13], nonlinear programming solver (MINOS 5.3) by C.D. Maranas *et al.* in [6] or Cabri-Géomètre software by M. Mollard and C. Payan in [9].

Most of the algorithmic suggestions used for solving (1) are based on Global Optimization stochastic approaches, nevertheless, there have been several papers ([5], [13]) published with the description of deterministic approaches. Deterministic methods can ensure that their solutions are optimal solutions for (1), and by applying these kinds of method optimal solutions for $n \le 27$ have, actually, been found. The main problem of these deterministic techniques is that as the number of spread points increases, an explosion of the computational burden occurs. For instance, for $n = 7$ a nontrivial procedure involving a serious amount of computations must run 8 times, but for $n = 14$ this number is 9,808 and for $n = 23$ it is 288,873,270. Optimal solutions have been found only for up to $n = 27$ even with the best known deterministic algorithm. However, for these kinds of algorithms it is also necessary to a priori know an approximate value or a lower bound of the solution. Thus, stochastic algorithms are useful, at least for revealing a good lower bound.

The paper is a source of new results to the literature on packing for up to one hundred circles. Some of our results represent improvements compared to the best known results; some others confirm some results of the literature and the rest correspond to unpublished results. All our computational results were obtained by the stochastic algorithm called TAMSASS-PECS (Threshold Accepting Modified Single Stochastic Search for Packing Equal Circles in a Square) described in Section 2. Our numerical results are precise in all digits, and they were verified by interval arithmetic computations. The numerical results and their graphical presentation in Section 3 show to what extent TAMSASS-PECS is able to find optimal or good solutions.

2. The TAMSASS-PECS algorithm

The TAMSASS-PECS algorithm is based on the Threshold Accepting method [2] and on our modified version of SASS (Single Agent Stochastic Search) [4, 7, 8, 14, 15] (MSASS). The TAMSASS-PECS algorithm has specifically been built to solve the packing n equal circles in a square problem, but can be easily extended to pack equal circles in other regular shapes.

The Threshold Accepting framework and the SASS local search algorithm are used for finding a local maximum of a function defined over a search space. As for all stochastic local search algorithms, the probability of finding a global solution grows with the number of trial points. These methods are based on stepwise improvements of the current solution by movements in the solution space. The following notations and definitions, as in [10], will be used.

DEFINITION 1. *Let $S_F \subseteq S$ be the set of feasible solutions for a maximization problem and S the search space. We require that S be finite. The goal of the* maximization problem *is to:*

$$\text{maximize } f(s) \quad \text{subject to } s \in S,$$

where $f : S \to I\!R$ is the objective (cost) function and $I\!R$ is the set of real numbers.

DEFINITION 2. *A* move *is defined as :*

$$d : S(d) \to S,$$

where $S(d) \subseteq S$ is the domain of the move. The set of all the moves of the problem is D. We claim that the union of the domains for all move sequences in D is the solution set, thus there is no point in the search space that cannot be reached from any points by multiplied use of moves. We denote the application of a move d to a candidate solution $s \in S$ by $d(s)$.

DEFINITION 3. *A candidate solution s' is a* neighbour *of the candidate solution s, if*

$$s' = d(s) \text{ for some } d \in D.$$

The neighbourhood $N(s)$ of a solution $s \in S(d)$ is the union of all the neighbours of s:

$$N(s) = \bigcup_{d \in D} d(s).$$

DEFINITION 4. *A candidate solution s is a* global optimum *if there is no other candidate solution s' in the search space S such that $f(s') > f(s)$. A candidate solution s is a* (weak) local optimum *if there is no other candidate solution s' in the neighbourhood $N(s)$ of s such that $f(s') > f(s)$. In case all the neighbours of s have a value of the objective function smaller than $f(s)$, then s is a* strong local optimum.

Formally, the Threshold Accepting algorithm [2] is very similar to the simulated annealing algorithms. In Threshold Accepting we accept every move that leads to a new solution not much worse than the current solution. To be more specific if s is the current solution, the proposed next solution $s' \in N(s)$ is accepted as the next solution if

$$\Delta = f(s) - f(s') < T_h,$$

where $T_h > 0$ is the threshold level. During the optimization process the threshold level has been gradually decreased like the temperature in simulated annealing. In Threshold Accepting a move which would give a solution much worse than the current solution is rejected, unlike in simulated annealing. The convergence of the Threshold Accepting algorithm has been proved for some cases [1].

The Threshold Accepting method is described in Algorithm 1. The inner loop is performed a constant L times and the stopping criterion is fulfilled when the final value of the threshold level is reached. A new candidate solution can be taken also when Δ is equal to the threshold level, keeping the sideway moving with the same function value allowed. In [2] a more adaptive inner loop stopping criterion is worked with: it ends the inner loop after no improvements were made in the last steps.

ALGORITHM 1. The Threshold Accepting Algorithm

Select an initial solution s
while stopping criterion not satisfied **do**
    while inner loop criterion not satisfied **do**
        Select $s' \in N(s)$
        $\Delta = f(s) - f(s')$
        if $\Delta \leq T_h$
            $s = s'$
    Decrease threshold T_h
return the best found solution

ALGORITHM 2. The Single Agent Stochastic Search

proc $SASS(s_0, ex, ct, Scnt, Fcnt, \sigma_{sup}, \sigma_{inf}, NIter)$
  var $b_0 := 0; k := 1; scnt := 0; fcnt := 0; \sigma_0 := 1;$
    while $k < NIter$ **do**

$$\sigma_k := \begin{cases} ex \cdot \sigma_{k-1} & \text{if } scnt > Scnt \\ ct \cdot \sigma_{k-1} & \text{if } fcnt > Fcnt \\ \sigma_{\text{sup}} & \text{if } \sigma_{k-1} < \sigma_{\text{inf}} \\ \sigma_{k-1} & \text{otherwise} \end{cases}$$

      Generate a random vector ξ_k with $N(b_k, \sigma_k \mathbf{I})$ distribution;
      if $f(s' = s_k + \xi_k) > f(s_k)$
        then
          $s_{k+1} := s';$
          $b_{k+1} := 0.2b_k + 0.4\xi_k;$
          $scnt := scnt + 1;$
          $fcnt := 0;$
        else
          if $f(s' = s_k - \xi_k) > f(s_k)$
            then
              $s_{k+1} := s';$
              $b_{k+1} := b_k - 0.4\xi_k;$
              $scnt := scnt + 1;$
              $fcnt := 0;$
            else
              $s_{k+1} := s_k;$
              $b_{k+1} := 0.5b_k;$
              $fcnt := fcnt + 1;$
              $scnt := 0;$
      $k := k + 1;$

Random optimization is traditionally based on single agent stochastic search strategies [8]. S.S. Rao [14] and D.C. Karnop [4] used a uniform random variable as the move function, $d(s)$. J. Matyas utilized Gaussian perturbations for the move function with a bias term to direct the search, $\xi \sim N(b, \sigma\mathbf{I})$ [7]. F.J. Solis and J.B. Wets [15] enhanced this approach by evaluating the objective function at $s' = s - \xi s$ if the evaluation at $s' = s + \xi s$ does not improve the current value of the objective function and incorporated a variable perturbation variance. The bias term and this additional function evaluation serve as stochastic equivalents to incorporating momentum and gradient information. The SASS method is given in Algorithm 2.

The variance of the perturbation size, ξ, is controlled by the number of repeated successes and failures, *scnt* and *fcnt*, respectively, in selecting a neighbour that decreases the value of the objective function. Note that conditions in line 4 are mutually exclusive. The contraction (ct) and expansion (ex) constants as well as the upper and lower bounds on the standard deviation of the random perturbation (σ_{sup}, σ_{inf}) are set by the user. The standard deviation of the perturbation is increased when it falls below a given lower bound σ_{inf} (line 4). The contraction and expansion of σ are performed when the number of successes, $scnt$, or failures, $fcnt$, are greater than the user given constants, $Scnt$ and $Fcnt$, respectively. The values 0.4 and 0.2, which modify the bias value are retained from F.J. Solis and J.B. Wets's paper [15]. Other values used in [15] are: $ex = 2$, $ct = 0.5$, $Scnt = 5$, $Fcnt = 3$, $\sigma_{\text{sup}} = 1.0$ and $\sigma_{\text{inf}} = 10^{-5}$. The standard deviation σ specifies the size of the sphere that most likely contains the perturbation vector and the bias term b locates the center of the sphere based on directions of a past success. In line 13 a reverse strategy, seeking for a better solution in the opposite direction of the originally failed move, is performed. The stopping criterion is based on the number of iterations.

The TAMSASS-PECS algorithm is established on a Threshold Accepting method and on a modified version of SASS (MSASS), which was particularly designed for improving the current solution s of the packing equal circles in a square problem. The objective function returns the minimum distance from the center of a circle to the other ones, and the problem at hand consists in maximizing this minimum distance for all centers. Thus, the program iteratively does a maximization of the minimum distance between circle centers.

MSASS, which is in charge of perturbing the location of a point s_i, has been built on the parameters provided by TAMSASS-PECS. This perturbation is intended to increase the minimum value of the distances, $(d_{i,j})$, between s_i and any point s_j $(1 \leq j \neq i \leq n)$. It moves the point s_i to a new location s'_i and computes the value of the new minimum distance $d'_{i,j}$. Following the Threshold Accepting strategy, a move is accepted if

$$d_{i,j} - d'_{i,j} < d_{i,j} T_h \,,$$

where $d_{i,j} T_h$ is the threshold level and $T_h > 0$. New trial locations of point s'_i are restricted to the neighbourhood of the current location of the point s_i. This neighbourhood is determined by a normal distribution $N(s_i, \sigma \mathbf{I})$ around s_i. While the Threshold Accepting condition is not satisfied, new locations for the point s_i should be tested following the classical SASS algorithm, although the number of trials is limited by a given maximum value (see Algorithm 4).

ALGORITHM 3. TAMSASS-PECS: a threshold accepting algorithm adapted to the packing equal circles in a square problem

```
Select an initial solution s
Select an initial value for Th
Select an initial standard deviation σ
while σ > σ_final do
    while All centers are not visited do
        s = MSASS(s, σ, Th, NextCenter(s))
        Decrease Th
        Decrease standard deviation σ
return the best found solution
```

The TAMSASS-PECS algorithm starts with a pseudorandom initial solution (the location of the n points s_i, $1 \leq i \leq n$, which are generated by dividing the square in $t = k \times k$ non-overlapping tiles, where $k = \lceil \sqrt{n} \rceil$, and $\lceil . \rceil$ indicates the smallest integer not less than the argument. Notice that the number of points n is less than or equal to the number of tiles t. The first point is located randomly at the center of the first or second tile. The following points are located at the center of a tile which is separated from the previous by one free tile in a row order. The remaining points are randomly allocated at the free tiles, (one point in one tile). Nevertheless, based on our experience if the threshold level is not very small, then the initial solution is not crucial because large random movements are allowed.

The initial value for the threshold level is $T_h = 0.02$ and the standard deviation σ is equal to the common diameter of the tiles. The TAMSASS-PECS algorithm tries to improve the current solution by an iterative procedure. At every step, the MSASS subroutine is executed for all the n points using the same values for both the standard deviation σ and the threshold level T_h. The criterion to stop this iterative procedure is based on the value of the standard deviation, which is decreased by a factor of 0.99 at every iteration. The threshold level is also decreased by the same factor.

At every iterative step, MSASS is performed for all the points n in an increasing order, which is determined by the value of the minimum distance from any point to the rest of points ($d_{i,j}$). The value of the minimum distance is updated after every execution of MSASS. The description of the TAMSASS-PECS method is given in Algorithms 3 and 4.

ALGORITHM 4. The Modified Single Agent Stochastic Search subroutine

```
proc MSASS(s, σ0, Th, i)
   var scnt := 0; fcnt := 0; Fcnt := 3; ct := 0.5; σ := σ0;
      while fcnt < 4 · Fcnt ∧ scnt = 0 do
          σ := { ct · σ   if fcnt > Fcnt
               { σ        otherwise
          Generate a random vector ξ with N(0, σI) distribution;
          s'(i) := s(i) + ξ;
          if f(s(i)) − f(s') ≤ f(s(i))Th
             then
                s(i) := s'(i);
                scnt := scnt + 1;
             else
                s'(i) := s(i) − ξ;
                if f(s(i)) − f(s') ≤ f(s(i))Th
                  then
                     s(i) := s'(i);
                     scnt := scnt + 1;
                  else
                     fcnt := fcnt + 1;
      return s
end
```

3. Computational results

In this section results obtained by carrying out the algorithm suggested in Section 2 will be compared to the best results found previously reported on in the literature. The programs were coded in C and run under Linux on a Pentium II PC with 266 Mhz. Since the underlying problems are very hard, sometimes the solution of a problem required multiple runs, which took CPU time between a few seconds to a few hours. The numerical results in details together with graphical representations are available at `http://www.inf.u-szeged.hu/ ~pszabo`.

Tables 1-2 show some results of the problem for $2 \leq n \leq 100$. In the second and third columns we can see the best results reported on in the literature with their references for every value of n. The rest of the columns, with headers m_n, c_n, f_n and ρ_n, refer to the numerical results of our algorithm, where:

m_n: the solution of the packing problem found by TAMSASS-PECS,

c_n: the number of contacts between circles and between circles and the sides of the square,

f_n: the number of free circles, and

ρ_n: the density values of the packing.

The definitions of c_n, f_n and ρ_n can be found in the previous paper [16]. The last column, (*), indicates whether the results obtained by our algorithm are better or worse than those previously published by other authors or by any chance equal to them. The meaning of symbols used in this column is as follows

+: our result improves all the previously published results,

−: an earlier result is better than our one,

=: the TAMSASS-PECS algorithm has obtained the same result as the best reported one,

N: it is a new result, previously unpublished.

Let us conclude the results of Table 1 and 2:

- In 20 out of 59 cases we have found the same results as reported on in the literature. Most of them are optimal solutions of the packing problem.
- In 34 cases our results have been worse than those of other authors. Nevertheless, for several cases our algorithm has obtained similar final geometrical arrangements, and the difference is only in the accuracy of the calculated numerical values.
- For $n = 32, 37, 47, 62$, and 72 our results improve all of those previously published.
- 40 new results have been generated. In all these cases the value of m_n is better than or equal to the new lower bound described in [16].

The accuracy of our results is 10^{-14}, and the feasibility of these solutions was verified by interval arithmetic.

On the one hand, there are no better packings in the very close neighbourhood of the given results according to these tests. However, these interval verification tests cannot prove global optimality. At the end of this section packing diagrams of the solutions of problem (1) obtained by the TAMSASS-PECS algorithm are shown (see Figures 1–5). Geometrical arrangements have been drawn for $n = 2$ to $n = 100$ and for $n = 150$ just as an example that TAMSASS-PECS is able to give a solution which is better than the one obtained by the lower bounds described in [16].

4. Summary

There has been a threshold accepting algorithm introduced, which is based on a modified single agent stochastic search procedure for the solution of the equal circles packing in the unit square problem. As a result of extensive computational efforts, we have found improved packings for 5 specific problems, while in 27 cases the available known packings were confirmed. In addition, forty new and unpublished packing results are reported on. Since the underlying algorithm does not make much use of the special structure of the present problem, we expect that it can be well utilised in other packing problems, too.

References

1. I. Althöfer and K. U. Koschnick. On the convergence of threshold accepting. *Appl. Math. Optim.*, 24:183–195, 1991.
2. G. Dueck and T. Scheuer. Threshold accepting: a general purpose optimization algorithm appearing superior to simulated annealing. *J. Comput. Phys.*, 90:161–175, 1990.
3. R. L. Graham and B. D. Lubachevsky. Repeated patterns of dense packings of equal disks in a square. *The Electronic Journal of Combinatorics*, 3:1–16, 1996.
4. D. C. Karnop. Random search techniques for optimization problems. *Automatica*, 1:111–121, 1963.
5. M. Locatelli and U. Raber. A Deterministic Global Optimization Approach for Solving the Problem of Packing Equal Circles in a Square. *International Workshop on Global Optimization (GO.99)*, Firenze, Italy, 1999
6. C. D. Maranas, C. A. Floudas, and P. M. Pardalos. New results in the packing of equal circles in a square. *Discrete Mathematics*, 142:287–293, 1995.
7. J. Matyas. Random optimization. *Automatization and Remote Control*, 26:244–251, 1965.
8. J. R. McDonnell and D. Waagen. Evolving recurrent perceptrons for time-series modeling. *IEEE Trans. on Neural Networks*, 5(1):24–38, 1994.
9. M. Mollard and C. Payan. Some progress in the packing of equal circles in a square. *Discrete Mathematics*, 84:303–307, 1990.
10. K. J. Nurmela. Constructing combinatorial designs by local search. Series A: Research Reports 27, Digital Systems Laboratory, Helsinki University of Technology, 1993.
11. K. J. Nurmela and P. R. J. Östergård. Packing up to 50 equal circles in a square. *Discrete & Computational Geometry*, 18:111–120, 1997.
12. R. Peikert. Dichteste Packungen von gleichen Kreisen in einem Quadrat. *Elemente der Mathematik*, 49:16–26, 1994.
13. R. Peikert, D. Würtz, M. Monagan, and C. de Groot. Packing circles in a square: a review and new results. In P. Kall (ed.), *System Modelling and Optimization*, volume 180 of *Lecture Notes in Control and Information Sciences*, 45–54. Springer-Verlag, Berlin, 1992.

14. S. S. Rao. *Optimization Theory and Applications.* John Willey and Sons, New York, 1978.
15. F. J. Solis and J. B. Wets. Minimization by random search techniques. *Math. of Operations Research*, 6(1):19–50, 1981.
16. P. G. Szabó, T. Csendes, L. G. Casado and I. García. Packing Equal Circles in a Square I. Problem Setting and Bounds for Optimal Solutions. This volume. Available at http://www.inf.u-szeged.hu/ ~csendes/pack1.ps.gz

n	The best result in the literature	Ref.	Our Results m_n	c_n	f_n	d_n	*
2	1.41421356237309	[12]	1.41421356237309	5	0	0.53901208445264	=
3	1.03527618041008	[12]	1.03527618041008	7	0	0.60964480874135	=
4	1.00000000000000	[12]	1.00000000000000	12	0	0.78539816339744	=
5	0.70710678118654	[12]	0.70710678118654	12	0	0.67376510556580	=
6	0.60092521257733	[12]	0.60092521257733	13	0	0.66395690946413	=
7	0.53589838486224	[12]	0.53589838486224	14	1	0.66931082684079	=
8	0.51763809020504	[12]	0.51763809020504	20	0	0.73096382525390	=
9	0.50000000000000	[12]	0.50000000000000	24	0	0.78539816339744	=
10	0.42127954398390	[3]	0.42127954398390	21	0	0.69003578526417	=
11	0.39820731023684	[12]	0.39820731023684	20	2	0.70074157775610	=
12	0.38873012632302	[12]	0.38873012632301	25	0	0.73846822388404	-
13	0.36609600769643	[12]	0.36609600769623	25	1	0.73326469490355	-
14	0.34891526037401	[12]	0.34891526037401	32	1	0.73567925554268	=
15	0.34108137740210	[12]	0.34108137740210	36	0	0.76205601092668	=
16	0.33333333333333	[12]	0.33333333333333	40	0	0.78539816339744	=
17	0.30615398530033	[12]	0.30615398530033	34	1	0.73355026330232	=
18	0.30046260628866	[12]	0.30046260628866	38	0	0.75465335787566	=
19	0.28954199199498	[12]	0.28954199199356	37	2	0.75230789673638	-
20	0.28661165235168	[12]	0.28661165235168	44	0	0.77949368686760	=
21	0.27181225535931	[3]	0.27181168718966	40	2	0.75335522651101	-
22	0.26795840155072	[3]	0.26795835833157	42	1	0.77167991577529	-
23	0.25881904510252	[3]	0.25881904510252	56	0	0.76363103212612	=
24	0.25433309503024	[3]	0.25433309503024	56	0	0.77496325975782	=
25	0.25000000000000	[3]	0.25000000000000	60	0	0.78539816339744	=
26	0.23873475724122	[3]	0.23873475724121	56	2	0.75846909048393	-
27	0.23584952830141	[3]	0.23584952830050	55	0	0.77231145646250	-
28	0.23053549364267	[3]	0.23053540627071	56	1	0.77185363595323	-
29	0.22688290074421	[3]	0.22688290074420	65	1	0.77890624177970	-
30	0.22450296453109	[3]	0.22450296453108	65	0	0.79201902646073	-
31	0.21754729161912	[3]	0.21754726920445	54	4	0.77729734717346	-
32	0.21308235294443	[3]	0.21317456258979	63	3	0.77600412447400	+
33	0.21132838414326	[3]	0.21132833175032	64	1	0.78885198102597	-
34	0.20560464675957	[3]	0.20560464675956	80	0	0.77664906433227	-
35	0.20276360086323	[3]	0.20276360086322	80	0	0.78122721299871	-
36	0.20000000000000	[3]	0.20000000000000	84	0	0.78539816339744	=
37	0.19623810145141	[3]	0.19642918368533	73	2	0.78330271742989	+
38	0.19534230412691	[3]	0.19534202418353	74	0	0.79704064569288	-
39	0.19436506316151	[3]	0.19436506314440	80	0	0.81117902717206	-
40	0.18817552201832	[3]	0.18817551865770	85	2	0.78797949519086	-
41	0.18609951184812	[3]	0.18609948922999	97	1	0.79272373684650	-
42	0.18427707211710	[3]	0.18427707211709	90	0	0.79868427865342	-
43	0.18019113545743	[3]	0.18019112440214	83	1	0.78726408457444	-
44	0.17863924567120	[3]	0.17863916198187	79	4	0.79384217677293	-
45	0.17571631417559	[3]	0.17571631417546	94	3	0.78944426849399	-
46	0.17445936087241	[3]	0.17445933585066	90	1	0.79718693728810	-
47	0.17126830721141	[3]	0.17127055746101	93	2	0.78929383201248	+
48	0.16938210954876	[3]	0.16938050793544	97	1	0.79094499029400	-
49	0.16738607686833	[3]	0.16738607686832	120	1	0.79121698952690	-
50	0.16645462588286	[3]	0.16645461022968	102	0	0.79967929970534	-

Table 1. Best results of the packing circles in a unit square problem for $2 \leq n \leq 50$.

n	The best result in the literature	Ref.	Our Results m_n	c_n	f_n	d_n	*
51	0.16561837431260	[3]	0.16561390218764	90	2	0.80861948195609	–
52	0.16538623796964	[3]	0.16538621483322	103	0	0.82253064459417	–
53	–	–	0.16264607633837	93	4	0.81462525898420	N
54	0.15913951630719	[3]	0.15913951414578	109	2	0.79940760435030	–
55	0.15755574752972	[3]	0.15755573439859	105	2	0.80027118200754	–
56	0.15615650046215	[3]	0.15615650046214	119	0	0.80235172950297	–
57	–	–	0.15474723074928	107	3	0.80396408840952	N
58	–	–	0.15130206629471	109	6	0.78673600284850	N
59	–	–	0.15029765718520	118	3	0.79108990822535	N
60	0.14950565404867	[3]	0.14945461308474	125	3	0.79666571193460	–
61	0.14854412669518	[3]	0.14854408720568	119	1	0.80137375361858	–
62	0.14569394327531	[3]	0.14671467098564	114	5	0.79710938267796	+
63	–	–	0.14667828199655	118	2	0.80961563357266	N
64	–	–	0.14532088220301	132	3	0.80922920583758	N
65	–	–	0.14456113665810	112	5	0.81438235586667	N
66	–	–	0.14380319561470	120	2	0.81934777745264	N
67	–	–	0.14304397643538	118	5	0.82409628091674	N
68	–	–	0.14288350731175	115	6	0.83475507341762	N
69	–	–	0.13993726983993	117	3	0.81666484211077	N
70	–	–	0.13787246738894	114	5	0.80715295095497	N
71	-	–	0.13593878298879	133	3	0.79859232191938	N
72	0.13549029317569	[3]	0.13569567607132	129	7	0.80729163319538	+
73	–	–	0.13389894938934	135	5	0.79949982777920	N
74	–	–	0.13265445602583	141	6	0.79720572964583	N
75	–	–	0.13241101073542	108	11	0.80536208356566	N
76	–	–	0.13094762506744	127	9	0.80022799448052	N
77	–	–	0.13047875975659	120	6	0.80562963362431	N
78	0.13046077259640	[3]	0.13045661660359	157	3	0.81584734690826	–
79	–	–	0.12976111574313	131	8	0.81852673455691	N
80	–	–	0.12886897358290	147	5	0.81882205826177	N
81	–	–	0.12833078028621	170	3	0.82293152331831	N
82	–	–	0.12646140351791	145	4	0.81168420885740	N
83	–	–	0.12609559154647	167	3	0.81736730901968	N
84	–	–	0.12555071002762	152	5	0.82087566500715	N
85	–	–	0.12511540460076	146	4	0.82553641420847	N
86	–	–	0.12500381019147	123	7	0.83392471939050	N
87	–	–	0.12230707667418	158	7	0.81150070836153	N
88	–	–	0.12099893725078	162	10	0.80523985787757	N
89	–	–	0.12080111590653	140	14	0.81201615746621	N
90	–	–	0.11988928132785	160	7	0.81010801424488	N
91	–	–	0.11916044953221	172	8	0.81023468623028	N
92	–	–	0.11854559378259	175	4	0.81159835604241	N
93	–	–	0.11827868195489	167	6	0.81711971813354	N
94	–	–	0.11733029717908	158	10	0.81409469757729	N
95	–	–	0.11668436071663	174	4	0.81466286707370	N
96	–	–	0.11633897630581	192	5	0.81887839722130	N
97	–	–	0.11618529717218	179	4	0.82545113090290	N
98	–	–	0.11610339843903	170	7	0.83290785939723	N
99	–	–	0.11601228667056	193	2	0.84022403037368	N
100	–	–	0.11456309641300	184	6	0.82979353948491	N

Table 2. Best results of the packing circles in a unit square problem for $51 \leq n \leq 100$.

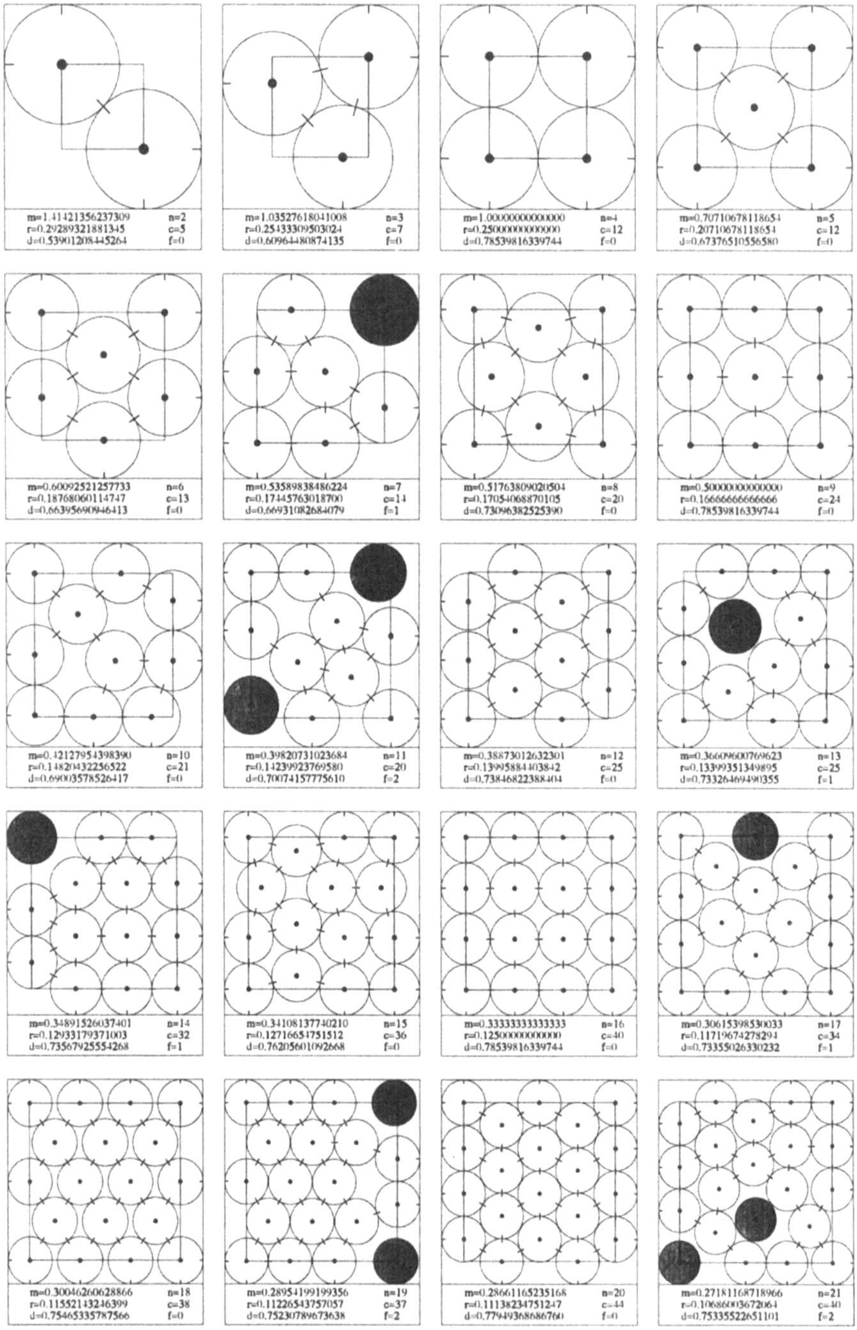

Figure 1. The best found packings for 2 – 21 circles (n is the number of circles, c the number of contacts, f the number of free circles, m the maxmin distance, r the radius of circles, and d is the density of the packing). The free circles are denoted by dark shading.

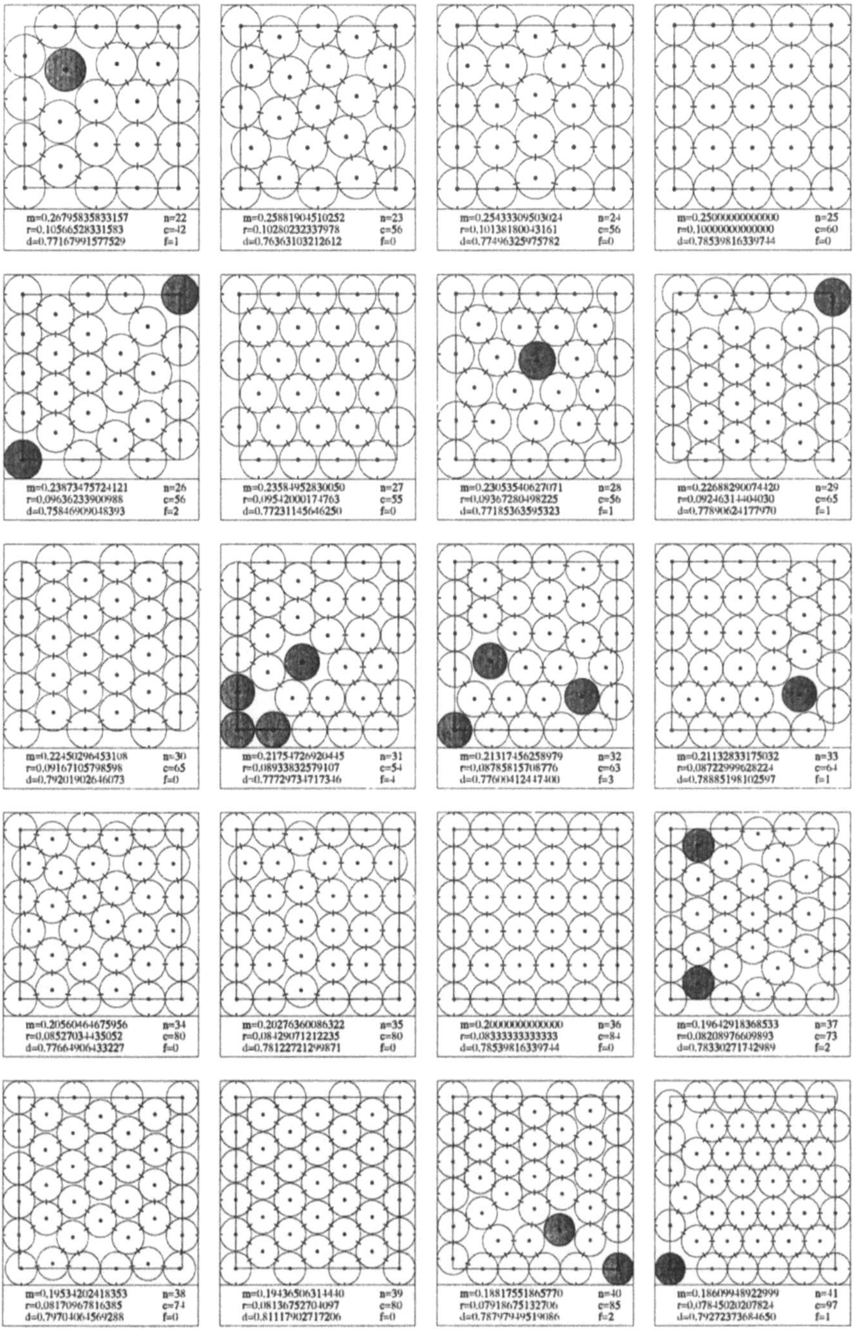

Figure 2. The best found packings for 22 - 41 circles (n is the number of circles, c the number of contacts, f the number of free circles, m the maxmin distance, r the radius of circles, and d is the density of the packing). The free circles are denoted by dark shading.

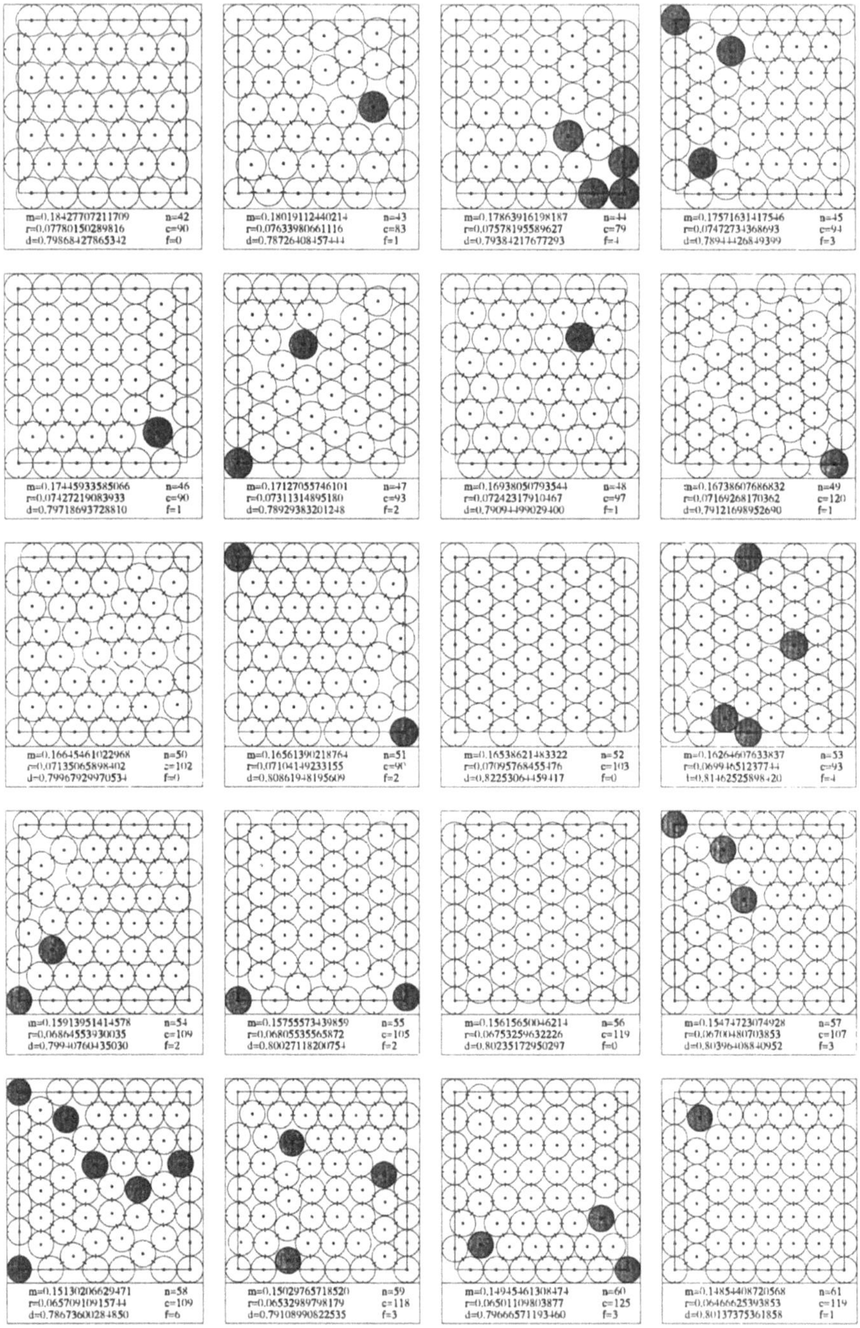

Figure 3. The best found packings for 42 – 61 circles (n is the number of circles, c the number of contacts, f the number of free circles, m the maxmin distance, r the radius of circles, and d is the density of the packing). The free circles are denoted by dark shading.

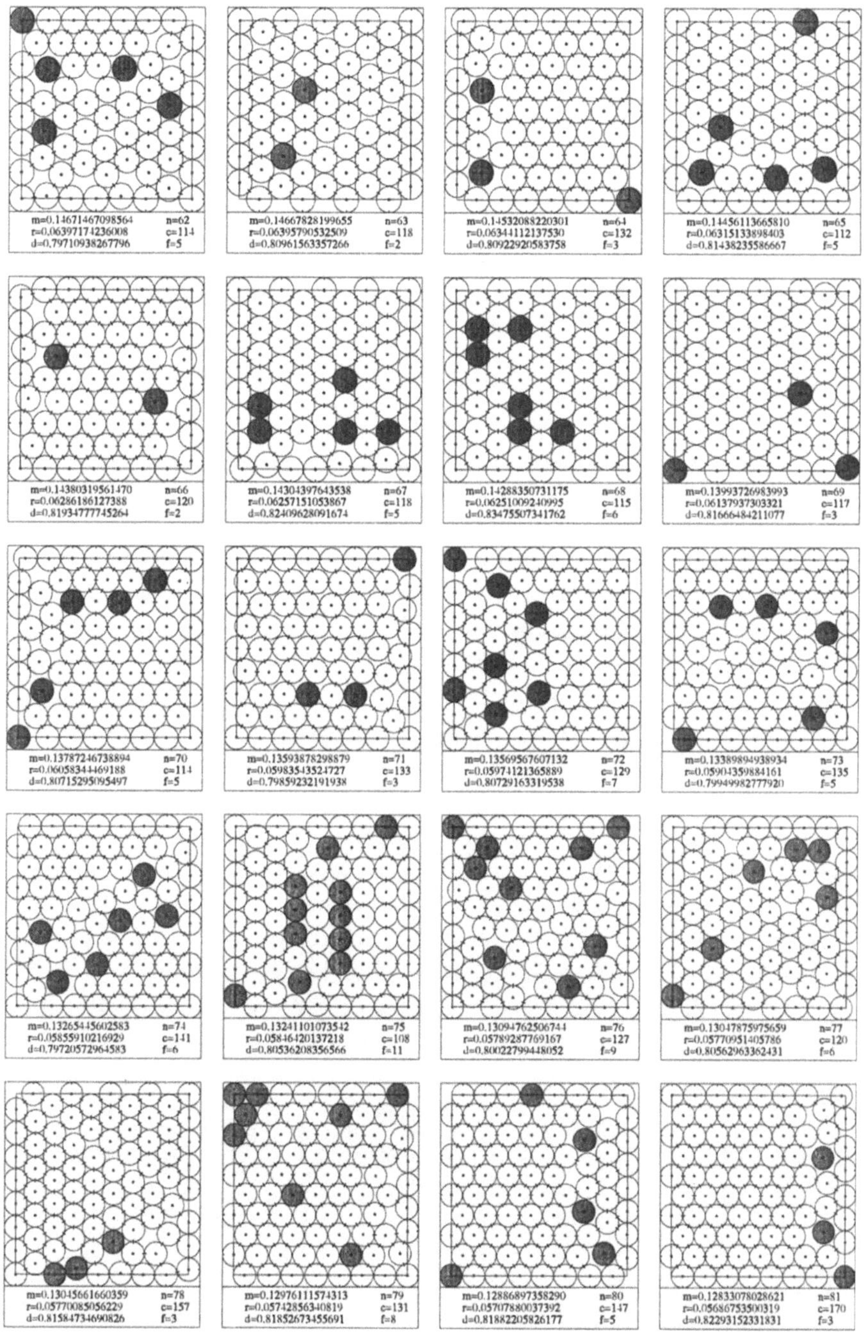

Figure 4. The best found packings for 62 – 81 circles (n is the number of circles, c the number of contacts, f the number of free circles, m the maxmin distance, r the radius of circles, and d is the density of the packing). The free circles are denoted by dark shading.

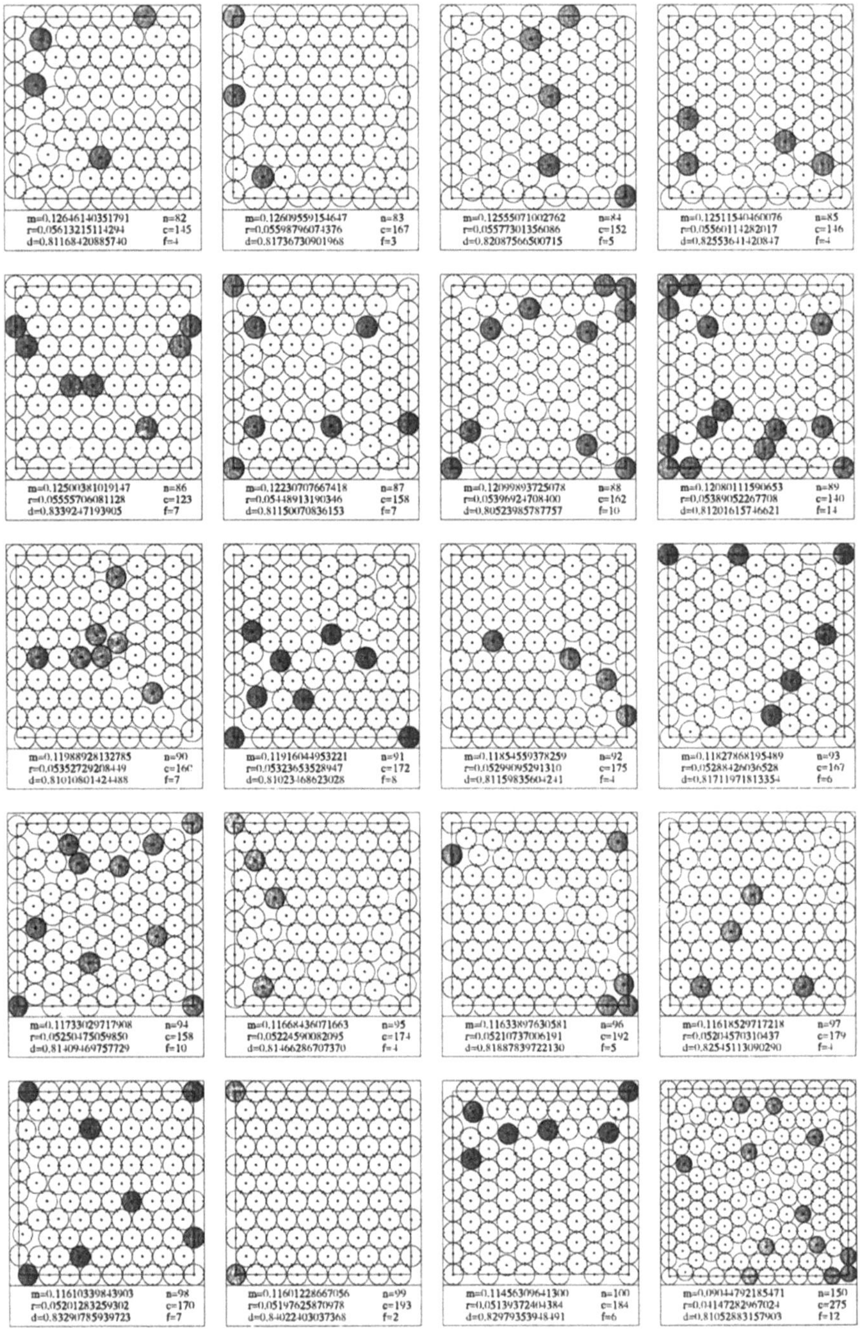

Figure 5. The best found packings for 82 – 100, and 150 circles (n is the number of circles, c the number of contacts, f the number of free circles, m the maxmin distance, r the radius of circles, and d is the density of the packing). The free circles are denoted by dark shading.

On Duality for Quasi-convex Supremization and Reverse Convex Infimization

I. Singer
Institute of Mathematics,
P. O. Box 1-764, RO-70700 Bucharest, Romania

Abstract. Using the tools of abstract convex analysis, in the first part of the paper we study the connections between the primal problem of quasi-convex supremization and two kinds of dual problems to it, and in the second part we study the connections between the primal problem of reverse convex infimization and two kinds of dual problems to it. To this end, in the general case we replace the duality (between families of subsets), used in previous papers, by a family of dualities depending on the primal constraint set. For the study of some particular cases, we introduce various families of dualities, which may have interest also for other applications.

Key words: Quasi-convex supremization, reverse convex infimization, dual problem, abstract convex analysis, family of dualities

Mathematics subject classification 1991: Primary: 49N15, 90C26; Secondary: 41A65

1. Introduction

Let X be a set, C a subset of X, which we shall assume non-empty throughout this paper, without any special mention, and $f : X \to \overline{R} = [-\infty, +\infty]$ a function, and let us consider the primal supremization problem

$$\alpha = \sup f(C). \tag{1.1}$$

In the paper [5] we have introduced and studied dual problems to (1.1) of the form

$$\beta_\Gamma = \sup_{w \in W} \inf f(\Gamma_{C,w}), \tag{1.2}$$

where W is a set and $\Gamma_{C,w} \subseteq X$ $(w \in W)$ is a family of subsets of X. On the other hand, in the recent paper [8] we have studied, suggested by a result of Volle ([10], corollary 5), dual problems to (1.1) of the form

$$\widetilde{\beta}_\Delta = \sup_{w \in W \setminus \Delta(C)} \inf f(X \setminus \Delta'(\{w\})), \tag{1.3}$$

[1] Research supported in part by the National Agency for Science, Technology and Innovation, grant nr. 4073/1998.

F. Giannessi et al. (eds.), Optimization Theory, 225–254.

where W is a set, $\Delta : 2^X \to 2^W$ is a duality (2^X denotes the family of all subsets of X), i.e., a mapping satisfying

$$\Delta(G) = \cap_{g \in G} \Delta(\{g\}) \qquad (G \subseteq X), \tag{1.4}$$

and $\Delta' : 2^W \to 2^X$ is the duality (called the "dual" of Δ) defined by

$$\Delta'(P) = \{x \in X \mid P \subseteq \Delta(\{x\})\} \qquad (P \subseteq W). \tag{1.5}$$

An essential difference between (1.2) and (1.3) is that in (1.2) the objective function $\lambda(w) := \inf f(\Gamma_{C,w})$ $(w \in W)$ depends on C, but the constraint set W does not depend on C, while in (1.3) the objective function $\lambda(w) := \inf f(X \setminus \Delta'(\{w\}))$ $(w \in W)$ does not depend on C, but the constraint set $W \setminus \Delta(C)$ depends on C. Among other results, in [8] we have shown (see [8], remark 3.2) that

$$\sup f(C) \geq \widetilde{\beta}_\Delta \tag{1.6}$$

and we have given conditions for the equality $\sup f(C) = \widetilde{\beta}_\Delta$. In [8] we have applied the general results on (1.3) mainly to an arbitrary set W of functions $w : X \to \overline{R}$ and the dualities $\Delta_1, \Delta_2 : 2^X \to 2^{W \times R}$ defined by

$$\Delta_1(G) = \{(w,d) \in W \times R \mid \sup w(G) \leq d\} \qquad (G \subseteq X), \tag{1.7}$$

$$\Delta_2(G) = \{(w,d) \in W \times R \mid w(g) < d \ (g \in G)\} \qquad (G \subseteq X), \tag{1.8}$$

and to the dualities $\Delta_3, \Delta_4 : 2^X \to 2^W$ defined by

$$\Delta_3(G) = \{w \in W \mid \sup w(G) \leq 1\} \qquad (G \subseteq X), \tag{1.9}$$

$$\Delta_4(G) = \{w \in W \mid w(g) < 1 \ (g \in G)\} \qquad (G \subseteq X), \tag{1.10}$$

which do not depend on the constraint set C of problem (1.1).

Formula (1.2) suggests to consider, instead of a single duality $\Delta : 2^X \to 2^W$, a family of dualities $\Delta_C : 2^X \to 2^W$ depending on $C \subseteq X$, and then to consider the family of subsets of X defined by

$$\Gamma_{C,w} = \Gamma^{\Delta_C}_{C,w} = X \setminus \Delta'_C(\{w\}) \qquad (w \in W), \tag{1.11}$$

and the dual problem (1.2) for this family, that is,

$$\beta_{\Delta_C} = \beta_{\Gamma^{\Delta_C}} = \sup_{w \in W} \inf f(X \setminus \Delta'_C(\{w\})). \tag{1.12}$$

Let us note that for a family of dualities $\Delta_C : 2^X \to 2^W$ and a fixed set $G \subseteq X$, the mapping $C \to \Delta_C(G)$ of 2^X into 2^W is not assumed to be a duality (i.e., we need not have $\Delta_C(G) = \cap_{c\in C}\Delta_{\{c\}}(G)$), as shown, for example, by the families of dualities $\Delta_C^i : 2^X \to 2^W$ $(i = 1,2,3,4)$ which will be defined in (3.3), (3.21), (3.33) and (3.47) below.

In the first part of this paper we shall study the connections between the primal problem (1.1) and the dual problems (1.12) and (1.3) with $\Delta = \Delta_C$, that is,

$$\widetilde{\beta}_{\Delta_C} := \sup_{w\in W\setminus\Delta_C(C)} \inf f(X\setminus\Delta_C'(\{w\})). \tag{1.13}$$

Note that in (1.13) both the constraint set and the objective function depend on C. Also, clearly,

$$\beta_{\Delta_C} \geq \widetilde{\beta}_{\Delta_C}. \tag{1.14}$$

Now let X be a set, C a subset of X, $f : X \to \overline{R}$ a function, and let us consider the primal infimization problem

$$\alpha_r = \inf f(X\setminus C). \tag{1.15}$$

In the papers [6] and [7] we have introduced and studied dual problems to (1.15) of the form

$$\beta_\Gamma^r = \inf_{w\in W} \inf f(\Gamma_{C,w}), \tag{1.16}$$

where W is a set and $\Gamma_{C,w} \subseteq X$ $(w \in W)$ is a family of subsets of X. On the other hand, in [8] we have studied, suggested by a result of Volle ([10], corollary 6), dual problems to (1.15) of the form

$$\widetilde{\beta}_\Delta^r = \inf_{w\in\Delta(C)} \inf f(X\setminus\Delta'(\{w\})), \tag{1.17}$$

where W is a set, $\Delta : 2^X \to 2^W$ is a duality and $\Delta' : 2^W \to 2^X$ is the dual (1.5) of Δ. An essential difference between (1.16) and (1.17) is that in (1.16) the objective function $\lambda(w) := \inf f(\Gamma_{C,w})$ $(w \in W)$ depends on C, but the constraint set W does not depend on C, while in (1.17) the objective function $\lambda(w) := \inf f(X\setminus\Delta'(\{w\}))$ $(w \in W)$ does not depend on C, but the constraint set $\Delta(C)$ depends on C. Among other results, in [8] we have shown (see [8], remark 5.2(b)) that

$$\inf f(X\setminus C) \leq \widetilde{\beta}_\Delta^r \tag{1.18}$$

and we have given conditions for the equality $\inf f(X\setminus C) = \widetilde{\beta}_\Delta^r$. In [8] we have applied the general results on (1.17) mainly to the dualities $\Delta_1, \Delta_2 : 2^X \to 2^{W\times R}$ defined by (1.7), (1.8), and to the dualities $\Delta_3, \Delta_4 : 2^X \to 2^W$ defined by (1.9), (1.10).

Formula (1.16) suggests to consider, instead of a single duality $\Delta : 2^X \to 2^W$, a family of dualities $\Delta_C : 2^X \to 2^W$ depending on $C \subseteq X$, and then to consider the family of subsets of X defined by (1.11) and the dual problem (1.16) for this family, that is,

$$\beta^r_{\Delta_C} = \beta^r_{\Gamma^{\Delta_C}} = \inf_{w\in W} \inf f(X \setminus \Delta'_C(\{w\})). \tag{1.19}$$

Let us note that in (1.12) and (1.19) one takes $\sup_{w\in W}$ and $\inf_{w\in W}$ respectively, of the same dual objective function

$$\lambda(w) := \inf f(X \setminus \Delta'_C(\{w\})) \qquad (w \in W).$$

In the second part of this paper we shall study the connections between the primal problem (1.15) and the dual problems (1.19) and (1.17) with $\Delta = \Delta_C$, that is,

$$\widetilde{\beta}^r_{\Delta_C} := \inf_{w\in \Delta_C(C)} \inf f(X \setminus \Delta'_C(\{w\})). \tag{1.20}$$

Note that in (1.20) both the constraint set and the objective function depend on C. Also, clearly,

$$\beta^r_{\Delta_C} \leq \widetilde{\beta}^r_{\Delta_C}. \tag{1.21}$$

Let us first observe the following simple lemma, which will allow us to express the dual problems (1.2) and (1.16) in the forms (1.12) and (1.19) respectively, and to express the results of [5], [6] and [7] on families of subsets $\Gamma_{C,w} \subseteq X$ ($C \subseteq X, w \in W$) in a more concise form, with the aid of families of dualities $\Delta_C : 2^X \to 2^W$ ($C \subseteq X$):

LEMMA 1. *Let X and W be two sets, C a subset of X, and $\Gamma_{C,w} \subseteq X$ ($w \in W$) a family of subsets of X. Then there exists a unique duality $\Delta_C : 2^X \to 2^W$ satisfying* (1.11), *namely the mapping defined by*

$$\Delta_C(G) := \{w \in W | G \subseteq X \setminus \Gamma_{C,w}\} = \{w \in W | G \cap \Gamma_{C,w} = \emptyset\} \quad (G \subseteq X\}. \tag{1.22}$$

Proof. By (1.22), for any set $G \subseteq X$ we have

$$\Delta_C(G) = \{w \in W | G \subseteq X \setminus \Gamma_{C,w}\} = \{w \in W | g \in X \setminus \Gamma_{C,w} \ (g \in G)\} = \\ = \cap_{g\in G} \Delta_C(\{g\}),$$

so the mapping Δ_C defined by (1.22) is a duality. Also, by (1.5) and (1.22), for each $w \in W$ we have

$$X \setminus \Delta'_C(\{w\}) = \{x \in X | w \notin \Delta_C(\{x\})\} = \{x \in X | x \notin X \setminus \Gamma_{C,w}\} = \Gamma_{C,w},$$

which proves (1.11). Finally, by (1.5) there holds

$$\Delta_C(\{x\}) = \{w \in W \mid x \in \Delta'_C(\{w\})\} \qquad (x \in X), \tag{1.23}$$

and hence, if Δ_C is a duality satisfying (1.11), then, by (1.4), (1.23) and (1.11), we obtain

$$\Delta_C(G) = \cap_{g \in G} \Delta_C(\{g\}) = \{w \in W \mid G \subseteq \Delta'_C(\{w\})\} = \\ = \{w \in W \mid G \subseteq X \setminus \Gamma_{C,w}\}.$$

□

Remark 1. By the above, formulas (1.22) and (1.11) yield a one-to-one correspondence between families of subsets $\Gamma_{C,w} \subseteq X$ ($C \subseteq X, w \in W$) and families of dualities $\Delta_C : 2^X \to 2^W$ ($C \subseteq X$). Moreover, given a set $C \subseteq X$ and a family of subsets $\Gamma_{C,w} \subseteq X$ ($w \in W$), if we associate to them the duality $\Delta_C : 2^X \to 2^W$ defined by (1.22), then the dual problems (1.3) and (1.17) for $\Delta = \Delta_C$ can be expressed, respectively, in the forms

$$\widetilde{\beta}_{\Delta_C} = \widetilde{\beta}_{\Gamma^{\Delta_C}} = \sup_{\substack{w \in W \\ C \cap \Gamma_{C,w} \neq \emptyset}} \inf f(\Gamma_{C,w}), \tag{1.24}$$

$$\widetilde{\beta}^r_{\Delta_C} = \widetilde{\beta}^r_{\Gamma^{\Delta_C}} = \inf_{\substack{w \in W \\ C \cap \Gamma_{C,w} = \emptyset}} \inf f(\Gamma_{C,w}). \tag{1.25}$$

Hence, the problem of the connections between the primal problem (1.1) and the dual problems (1.12) and (1.13) is equivalent to that of the connections between the primal problem (1.1) and the dual problems (1.2) and (1.24); also, the problem of the connections between the primal problem (1.15) and the dual problems (1.19) and (1.20) is equivalent to that of the connections between the primal problem (1.15) and the dual problems (1.16) and (1.25). In the present paper, from these two equivalent languages we shall choose that of the families of dualities, since the use of families of dualities $\Delta_C : 2^X \to 2^W$ ($C \subseteq X$) instead of families of subsets $\Gamma_{C,w} \subseteq X$ ($C \subseteq X, w \in W$) will not only allow us to study the relations between the primal problems and the dual problems mentioned above, but will also lead to generalizations and improvements of some of the results of [5], [6] and [7]. Also, due to the property (1.4), the use of families of dualities $\Delta_C : 2^X \to 2^W$ ($C \subseteq X$) will permit us to express the results in a more concise form. In the sequel, for simplicity, we shall use the following

Convention. Instead of "family of subsets $\Gamma_{C,w} \subseteq X$ $(C \subseteq X, w \in W)$" and "family of dualities $\Delta_C : 2^X \to 2^W$ $(C \subseteq X)$" we shall write: family of subsets $\Gamma_{C,w} \subseteq X$ $(w \in W)$ and, respectively, family of dualities $\Delta_C : 2^X \to 2^W$ (i.e., their dependence on C will be shown only by the subscript C).

In Section 2 we shall give some results on duality for the primal supremization problem (1.1), using arbitrary families of dualities $\Delta_C : 2^X \to 2^W$. In Section 3 we shall study the families of dualities $\Delta_C : 2^X \to 2^W$ associated (by lemma 1) to the families of subsets $\Gamma_{C,w} \subseteq X$ $(w \in W)$ used in [5] and to an additional family $\Gamma^4_{C,w}$ used in [6] (it will turn out that these are new dualities, not mentioned in [9], which may have some interest also for other applications), and we shall apply to the dual problems to (1.1) defined with the aid of these families the results of Section 2. In Section 4 we shall study duality for the primal infimization problem (1.15), using arbitrary families of dualities $\Delta_C : 2^X \to 2^W$. Finally, in Section 5 we shall study the families of dualities $\Delta_C : 2^X \to 2^W$ associated (by lemma 1) to the families of subsets $\Gamma_{C,w} \subseteq X$ $(w \in W)$ used in [6] and [7] (it will turn out that they coincide with those introduced in Section 3) and we shall apply to the dual problems to (1.15) defined with the aid of these families the results of Section 4.

Let us recall now some notations, concepts and results of abstract convex analysis, which will be used in the sequel. For a function $f : X \to Y$ and a set $C \subseteq X$, we shall denote, as usual, $f(C) := \{f(c) | c \in C\}$ (thus, for example, $\sup w(C) = \sup_{c\in C} w(c)$ in formula (1.26) below); this should lead to no confusion with the fact that for a *duality* $\Delta : 2^X \to 2^W$ and a set $C \subseteq X, \Delta(C)$ is the set $\cap_{c\in C}\Delta(\{c\})$ (by (1.4)), since Δ is defined only at subsets of X, not at elements $c \in X$.

Let X be a set, let $W \subseteq \overline{R}^X$, where $\overline{R}^X$ denotes the set of all functions $f : X \to \overline{R}$, and let $\Delta : 2^X \to 2^W$ be a duality, with dual $\Delta' : 2^W \to 2^X$. A set $C \subseteq X$ is said to be

(a) *W-convex*, in symbols, $C \in \mathcal{K}(W)$, if for each $x \in X \backslash C$ there exists $w \in W$ such that

$$\sup w(C) < w(x), \tag{1.26}$$

with the convention $\sup \emptyset = -\infty$, where $\emptyset$ denotes the empty set;

(b) *W-evenly convex*, in symbols, $C \in \mathcal{EK}(W)$, if for each $x \in X \backslash C$ there exists $w \in W$ such that

$$w(c) < w(x) \qquad (c \in C); \tag{1.27}$$

(c) *W-evenly coaffine*, in symbols, $C \in \mathcal{ECA}(W)$, if for each $x \in X \backslash C$ there exists $w \in W$ such that

$$w(x) \notin w(C) \tag{1.28}$$

(i.e., such that $w(x) \neq w(c)$ $(c \in C)$);

(d) *$\Delta'\Delta$-convex*, in symbols, $C \in \mathcal{C}(\Delta'\Delta)$, if for each $x \in X \backslash C$ there exists $w \in W$ such that

$$C \subseteq \Delta'(\{w\}),\ x \in X \setminus \Delta'(\{w\}). \tag{1.29}$$

Clearly,

$$\mathcal{K}(W) \subseteq \mathcal{EK}(W) \subseteq \mathcal{ECA}(W). \tag{1.30}$$

For a set $C \subseteq X$, and for its *W-convex hull* $\mathrm{co}_W C$ (i.e., the intersection of all W-convex sets containing C), we have

$$\mathcal{K}(W) = \{C \in 2^X \,|\, C = \mathrm{co}_W C\}, \tag{1.31}$$

and similar statements hold for $\mathcal{EK}(W)$ and the *W-evenly convex hull* $\mathrm{eco}_W C$ of C, for $\mathcal{ECA}(W)$ and the *W-evenly coaffine hull* $\mathrm{eca}_W C$ of C, and, respectively, for $\mathcal{C}(\Delta'\Delta)$ and the *$\Delta'\Delta$-convex hull* $\Delta'\Delta(C)$ of C. In the sequel we shall freely use some well-known elementary properties of dualities, for example, that every duality $\Delta : 2^X \to 2^W$ is antitone, i.e., $G_1 \subseteq G_2 (\subseteq X)$ implies $\Delta(G_2) \subseteq \Delta(G_1)$, and that $\Delta'\Delta : 2^X \to 2^X$ and $\Delta\Delta' : 2^W \to 2^W$ are hull operators, i.e., isotone, contractive and idempotent (see e.g. [9]).

A function $f : X \to \overline{R}$ is called

(a) *W-quasi-convex*, in symbols, $f \in Q(W)$, if all its (lower) level sets

$$S_d(f) := \{x \in X \,|\, f(x) \leq d\} \qquad (d \in R) \tag{1.32}$$

are W-convex, i.e., if

$$S_d(f) \in \mathcal{K}(W) \qquad (d \in R); \tag{1.33}$$

(b) *W-evenly quasi-convex*, in symbols, $f \in EQ(W)$, if

$$S_d(f) \in \mathcal{EK}(W) \qquad (d \in R); \tag{1.34}$$

(c) *W-evenly quasi-coaffine*, in symbols, $f \in EQA(W)$, if

$$S_d(f) \in \mathcal{ECA}(W) \qquad (d \in R); \tag{1.35}$$

(d) *$\Delta'\Delta$-quasi-convex*, in symbols, $f \in Q(\Delta'\Delta)$, if

$$S_d(f) \in \mathcal{C}(\Delta'\Delta) \qquad (d \in R). \tag{1.36}$$

By (1.30), we have

$$Q(W) \subseteq EQ(W) \subseteq EQA(W). \tag{1.37}$$

In the particular case when X is a locally convex space and $W = X^*$, the conjugate space of X, or $W = X^* \backslash \{0\}$, the W-convex, W-evenly convex and W-evenly coaffine sets and hulls of sets are just the usual closed convex, evenly convex and evenly coaffine sets and hulls of sets, respectively; also, the W-quasi-convex, W-evenly quasi-convex and W-evenly quasi-coaffine functions are just the usual lower semi-continuous quasi-convex, evenly quasi-convex and evenly quasi-coaffine functions respectively.

For a function $f : X \to \overline{R}$ we shall also use its strict level sets

$$A_d(f) := \{x \in X | \, f(x) < d\} \qquad (d \in R). \tag{1.38}$$

For any subset C of a locally convex space X, we shall denote by $\overline{C}$ and int C the closure and, respectively, the interior of the set C.

2. Duality for primal supremization, using arbitrary families of dualities $\Delta_C : 2^X \to 2^W$

LEMMA 2. *Let X and W be two sets, $\Delta : 2^X \to 2^W$ a duality and $x_0 \in X$. The following statements are equivalent:*

1°. *We have*

$$\Delta(\{x_0\}) = \emptyset. \tag{2.1}$$

2°. *We have*

$$x_0 \notin \cup_{w \in W} \Delta'(\{w\}). \tag{2.2}$$

Proof. If (2.2) does not hold, i.e., if there exists $w_0 \in W$ such that $x_0 \in \Delta'(\{w_0\})$, then $\Delta(\{x_0\}) \supseteq \Delta\Delta'(\{w_0\}) \ni w_0$, so (2.1) does not hold. Conversely, if we do not have (2.1), i.e., if there exists $w_0 \in \Delta(\{x_0\})$, then $\Delta'(\{w_0\}) \supseteq \Delta'\Delta(\{x_0\}) \ni x_0$, so (2.2) does not hold. □

THEOREM 1. *Let X and W be two sets, $f : X \to \overline{R}$ a function, and $\Delta_C : 2^X \to 2^W$ a family of dualities such that for any $C \subseteq X$ we have*

$$\Delta_{\{c\}}(\{c\}) = \emptyset \qquad (c \in C), \tag{2.3}$$

$$\inf f(X \setminus \Delta'_C(\{w\})) \leq \sup_{c \in C} \inf f(X \setminus \Delta'_{\{c\}}(\{w\})) \qquad (w \in W). \tag{2.4}$$

Then, given $C \subseteq X$, there holds

$$\sup f(C) \geq \beta_{\Delta_C} \geq \widetilde{\beta}_{\Delta_C}. \tag{2.5}$$

Moreover, if we have (2.3), (2.4) *and* $f \in Q(\Delta'_C \Delta_C)$, *then*

$$\sup f(C) = \beta_{\Delta_C} = \widetilde{\beta}_{\Delta_C}. \tag{2.6}$$

Proof. By (2.3) and lemma 2, we have $c \in X \setminus \Delta'_{\{c\}}(\{w\})$ $(c \in C, w \in W)$, whence $\inf f(X \setminus \Delta'_{\{c\}}(\{w\})) \leq f(c)$ $(c \in C, w \in W)$. Therefore, by (2.4),

$$\inf f(X \setminus \Delta'_C(\{w\})) \leq \sup_{c \in C} \inf f(X \setminus \Delta'_{\{c\}}(\{w\})) \leq \sup f(C) \quad (w \in W),$$

whence, by (1.12) and (1.14), we obtain (2.5). Furthermore, if also $f \in Q(\Delta'_C \Delta_C)$, then, by [10], corollary 5, we have $\sup f(C) = \widetilde{\beta}_{\Delta_C}$, whence, by (2.5), we obtain (2.6). □

THEOREM 2. *Let X and W be two sets, C a subset of X, $f : X \to \overline{R}$ a function, and $\Delta_C : 2^X \to 2^W$ a duality. The following statements are equivalent, where $\alpha = \alpha_C = \sup f(C)$:*

1°. *We have*

$$\Delta_C(S_d(f)) \neq \emptyset \qquad (d < \alpha). \tag{2.7}$$

2°. *We have*

$$\Delta_C(A_d(f)) \neq \emptyset \qquad (d < \alpha). \tag{2.8}$$

3°. *We have*

$$A_\alpha(f) \subseteq \cup_{w \in W} \Delta'_C(\{w\}). \tag{2.9}$$

4°. *We have*

$$\widetilde{\beta}_{\Delta_C} \leq \sup f(C) \leq \beta_{\Delta_C}. \tag{2.10}$$

Proof. 1° ⇒ 2°. If 1° holds, then, since $A_d(f) \subseteq S_d(f)$, we have $\Delta_C(A_d(f)) \supseteq \Delta_C(S_d(f)) \neq \emptyset$ $(d < \alpha)$.

2° ⇒ 3°. If $w_d \in \Delta_C(A_d(f))$ $(d < \alpha)$, then

$$A_d(f) \subseteq \Delta'_C \Delta_C(A_d(f)) \subseteq \Delta'_C(\{w_d\}) \subseteq \cup_{w \in W} \Delta'_C(\{w\}) \qquad (d < \alpha),$$

whence $A_\alpha(f) = \cup_{d<\alpha} A_d(f) \subseteq \cup_{w \in W} \Delta'_C(\{w\})$.

3° ⇒ 1°. If 3° holds, then for each $d < \alpha$ there exists $w_d \in W$ such that $S_d(f) \subseteq A_\alpha(f) \subseteq \Delta'_C(\{w_d\})$, whence $\Delta_C(S_d(f)) \supseteq \Delta_C \Delta'_C(\{w_d\}) \ni w_d$.

3° ⇔ 4°. We have 3° if and only if for each $d < \alpha$ there exists $w_d \in W$ such that $A_d(f) \subseteq \Delta'_C(\{w_d\})$, i.e., such that $A_d(f) \cap (X \setminus \Delta'_C(\{w_d\})) = \emptyset$, which, by [5], proposition 2.2, is equivalent to $\sup f(C) \leq \beta_{\Delta_C}$. Finally, by (1.6), the first inequality in (2.10) always holds. □

Remark 2. Theorem 2 gives both an upper and a lower estimate for $\sup f(C)$. In general, we may have one of the weak duality equalities $\sup f(C) = \beta_{\Delta_C}$ or $\sup f(C) = \tilde{\beta}_{\Delta_C}$ without having the other one.

COROLLARY 1. *If we have (2.9) and $\tilde{\beta}_{\Delta_C} = \beta_{\Delta_C}$, then there holds (2.6).*

Remark 3. Since

$$\beta_{\Delta_C} = \sup_{w\in W} \inf f(X \setminus \Delta'_C(\{w\})) =$$
$$= \max\{\tilde{\beta}_{\Delta_C}, \sup_{w\in\Delta_C(C)} \inf f(X\setminus\Delta'_C(\{w\}))\},$$

condition $\tilde{\beta}_{\Delta_C} = \beta_{\Delta_C}$ is equivalent to

$$\tilde{\beta}_{\Delta_C} = \sup_{w\in W\setminus\Delta_C(C)} \inf f(X\setminus\Delta'_C(\{w\})) \geq \sup_{w\in\Delta_C(C)} \inf f(X\setminus\Delta'_C(\{w\})). \tag{2.11}$$

COROLLARY 2. *If for each $d < \sup f(C)$ we have $S_d(f) \in \mathcal{C}(\Delta'_C\Delta_C)$ (in particular, if $f \in Q(\Delta'_C\Delta_C)$), then there holds (2.10).*

Proof. The assumption $S_d(f) \in \mathcal{C}(\Delta'_C\Delta_C)$ $(d < \sup f(C))$ means that for each $d < \sup f(C)$ and $x \in X\setminus S_d(f)$ there exists $w = w_{d,x} \in W$ such that

$$S_d(f) \subseteq \Delta'_C(\{w_{d,x}\}),\ x \in X\setminus\Delta'_C(\{w_{d,x}\}). \tag{2.12}$$

Hence, for each $d < \sup f(C)$ we have $A_d(f) \subseteq S_d(f) \subseteq \cup_{w\in W}\Delta'_C(\{w\})$, whence (2.9), and thus, by theorem 2, implication $3° \Rightarrow 4°$, there follows (2.10). □

Combining theorems 1 and 2, we obtain

COROLLARY 3. *If we have (2.3), (2.4) and (2.9), then*

$$\tilde{\beta}_{\Delta_C} \leq \sup f(C) = \beta_{\Delta_C}. \tag{2.13}$$

3. Duality for primal supremization, using special families of dualities $\Delta_C : 2^X \to 2^W$

In this Section we shall assume, without any special mention, that X is a set and

$$W \subseteq R^X, \tag{3.1}$$

where R^X denotes the set of all functions $f : X \to R = (-\infty, +\infty)$, and, in some cases, even that X is a locally convex space and $W \subseteq X^*$, the conjugate space of X, or $W \subseteq X^* \setminus \{0\}$. For such X and W, and for $C \subseteq X$, we shall study the dualities $\Delta_C : 2^X \to 2^W$ associated (by lemma 1) to the families of sets $\Gamma_{C,w} \subseteq X$ $(w \in W)$ used in [5] (it will turn out that these are new dualities, not mentioned in [9], which may have some interest also for other applications), and we shall apply to the dual problems to (1.1) defined with the aid of these families the results of Section 2.

(1) Let

$$\Gamma^1_{C,w} = \{x \in X |\, w(x) \geq \sup w(C)\} \qquad (w \in W). \tag{3.2}$$

Then the associated duality is (by (1.22))

$$\Delta^1_C(G) = \{w \in W |\, w(g) < \sup w(C) \ (g \in G)\} \qquad (G \subseteq X). \tag{3.3}$$

Note that, for any $w \in W$ we have $\sup w(C) > -\infty$, since $C \neq \emptyset$. On the other hand, if $\sup w(C) = +\infty$, then, by (3.1), we have $\Gamma^1_{C,w} = \emptyset$ and $\Delta^1_C(G) = W$ for all $G \subseteq X$. Similar remarks are valid also for $\Gamma^2_{C,w}, \Delta^2_C$ and $\Gamma^3_{C,w}, \Delta^3_C$ below.

LEMMA 3. *(a) For any set $C \subseteq X$ the duality $\Delta_C = \Delta^1_C$ satisfies (1.11), (2.3) and*

$$W \setminus \Delta^1_C(C) = \{w \in W |\exists c \in C, w(c) = \sup w(C)\}. \tag{3.4}$$

(b) We have $f \in Q((\Delta^1_C)'\Delta^1_C)$ if and only if for each $d \in R$ and $x \in X \backslash S_d(f)$ there exists $w = w_{d,x} \in W$ such that

$$w(y) < \sup w(C) \leq w(x) \qquad (y \in S_d(f)). \tag{3.5}$$

Consequently,

$$Q((\Delta^1_C)'\Delta^1_C) \subseteq EQ(W). \tag{3.6}$$

Proof. (a) Formula (1.11) for Δ^1_C follows from lemma 1 applied to $\Gamma^1_{C,w}$ of (3.2). Also, by (3.3), there holds $\Delta^1_{\{c\}}(\{c\}) = \{w \in W |\, w(c) < w(c)\} = \emptyset$ $(c \in C)$. Finally, by (3.3), we have $W \setminus \Delta^1_C(C) = \{w \in W |\exists c \in C, w(c) \geq \sup w(C)\}$, whence (3.4).

(b) We have $f \in Q((\Delta^1_C)'\Delta^1_C)$ if and only if for each $d \in R$ and $x \in X \backslash S_d(f)$ there exists $w = w_{d,x} \in W$ satisfying (2.12) for $\Delta_C = \Delta^1_C$, that is, (3.5). Hence, by the definition of $EQ(W)$, we obtain (3.6). □

THEOREM 3. *Let X and W be two sets, $W \subseteq R^X, f : X \to \overline{R}$ and C a subset of X.*

(a) If

$$\inf_{\substack{x \in X \\ w(x) \geq \sup w(C)}} f(x) \leq \sup_{c \in C} \inf_{\substack{x \in X \\ w(x) \geq w(c)}} f(x) \qquad (w \in W), \tag{3.7}$$

then

$$\sup f(C) \geq \sup_{w \in W} \inf_{\substack{x \in X \\ w(x) \geq \sup w(C)}} f(x). \tag{3.8}$$

(b) There holds

$$\sup_{\substack{w \in W \\ \exists c \in C,\, w(c) = \sup w(C)}} \inf_{\substack{x \in X \\ w(x) \geq \sup w(C)}} f(x) \leq \sup f(C) \leq \sup_{w \in W} \inf_{\substack{x \in X \\ w(x) \geq \sup w(C)}} f(x) \tag{3.9}$$

if and only if for each $d < \sup f(C)$ there exists $w = w_d \in W$ such that

$$w(y) < \sup w(C) \qquad (y \in S_d(f)). \tag{3.10}$$

(c) If we have (3.7) and if for each $d < \sup f(C)$ there exists $w = w_d \in W$ satisfying (3.10) (by lemma 3(b), the latter condition is satisfied, e.g., when $f \in Q((\Delta_C^1)'\Delta_C^1)))$, then

$$\sup f(C) = \sup_{w \in W} \inf_{\substack{x \in X \\ w(x) \geq \sup w(C)}} f(x). \tag{3.11}$$

Proof. (a) This follows from theorem 1 for $\Delta_C = \Delta_C^1$ of (3.3), since

$$X \setminus (\Delta_C^1)'(\{w\}) = \{x \in X \,|\, w(x) \geq \sup w(C)\} \qquad (w \in W), \tag{3.12}$$

$$X \setminus (\Delta_{\{c\}}^1)'(\{w\}) = \{x \in X \,|\, w(x) \geq w(c)\} \qquad (c \in C, w \in W). \tag{3.13}$$

(b) This follows from theorem 2, equivalence $4° \Leftrightarrow 1°$, applied to Δ_C^1, using also (3.4), since formulas (3.9) and (3.10) mean, respectively, (2.10) for $\Delta_C = \Delta_C^1$ and $w \in \Delta_C^1(S_d(f))$.

(c) This follows by combining parts (a) and (b). □

COROLLARY 4. *Let X be a set, $W \subseteq R^X$, $C \subseteq X$, and $f : X \to \overline{R}$ a function such that for each $d < \sup f(C)$ the level set $S_d(f)$ is W-evenly convex (e.g., let $f \in EQ(W)$). Then we have (3.9).*

Proof. For each $d < \alpha = \sup f(C)$ there exists $c_d \in C$ such that $f(c_d) > d$, that is, $c_d \in X \setminus S_d(f)$. Hence, since $S_d(f)$ is W-evenly convex, there exists $w_d \in W$ such that

$$w_d(y) < w_d(c_d) \qquad (y \in S_d(f)). \tag{3.14}$$

Then, by (3.14) and $c_d \in C$, we have

$$w_d(y) < w_d(c_d) \leq \sup w_d(C) \qquad (y \in S_d(f)).$$

Consequently, by theorem 3(b), we obtain (3.9). □

Remark 4. Corollary 4 is an improvement of [5], proposition 3.2, according to which, if for each $d < \sup f(C)$ the level set $S_d(f)$ is W -convex (e.g., if $f \in Q(W)$), then we have (3.9).

For a set X and functions $f, w : X \to \overline{R}$, let

$$\varphi_{f,w}(d) = \inf_{\substack{x \in X \\ w(x) \geq d}} f(x) \qquad (d \in R). \tag{3.15}$$

We recall (see [5] and the references therein) that f is called regular with respect to w, if

$$\varphi_{f,w}(d) = \sup_{\substack{t \in R \\ t < d}} \varphi_{f,w}(t) \qquad (d \in R). \tag{3.16}$$

For example, if $f : R^n \to R$ and f is convex, then f is regular with respect to all $w \in (R^n)^*$ (see [5], remark 3.2). Using theorem 3(a), we can obtain again

COROLLARY 5. ([5], proposition 3.1). *Let X be a set, $W \subseteq R^X$, C a subset of X such that*

$$\sup w(C) \in R \qquad (w \in W), \tag{3.17}$$

and $f : X \to \overline{R}$ a function which is regular with respect to all $w \in W$. Then we have (3.8).

Proof. Let $w \in W$. Then, by our assumption of regularity, we have (3.16) for $d = \sup w(C) \in R$, i.e.,

$$\inf_{\substack{x \in X \\ w(x) \geq \sup w(C)}} f(x) = \sup_{\substack{t \in R \\ t < \sup w(C)}} \inf_{\substack{x \in X \\ w(x) \geq t}} f(x). \tag{3.18}$$

But, for each $t \in R$ with $t < \sup w(C)$ there exists $c = c_t \in C$ such that $t < w(c)$, whence $\{x \in X | w(x) \geq t\} \supseteq \{x \in X | w(x) \geq w(c)\}$. Consequently,

$$\sup_{\substack{t \in R \\ t < \sup w(C)}} \inf_{\substack{x \in X \\ w(x) \geq t}} f(x) \leq \sup_{c \in C} \inf_{\substack{x \in X \\ w(x) \geq w(c)}} f(x), \tag{3.19}$$

which, together with (3.18), yields (3.7). Hence, by theorem 3(a), we obtain (3.8). □

COROLLARY 6. *Let X be a set, $W \subseteq R^X$, C a subset of X satisfying (3.17), and $f : X \to \overline{R}$ a function which is regular with respect to all $w \in W$ and such that for each $d < \sup f(C)$ the level set $S_d(f)$ is W-evenly convex (the latter condition is satisfied, e.g., when $f \in EQ(W)$). Then we have (3.11).*

Remark 5. Corollary 6 is an improvement of [5], theorem 3.1, in which the stronger assumption of W-convexity of the level sets $S_d(f)$ $(d < \sup f(C))$ has been made.

(2) Let

$$\Gamma^2_{C,w} = \{x \in X \,|\, w(x) > \sup w(C)\} \qquad (w \in W). \tag{3.20}$$

Then the associated duality is (by (1.22))

$$\Delta^2_C(G) = \{w \in W \,|\, \sup w(G) \leq \sup w(C)\} \qquad (G \subseteq X). \tag{3.21}$$

LEMMA 4. *(a) For any set $C \subseteq X$ the duality $\Delta_C = \Delta^2_C$ satisfies (1.11) and*

$$\Delta^2_{\{c\}}(\{c\}) = W \qquad (c \in C), \tag{3.22}$$

$$\Delta^2_C(C) = W, \tag{3.23}$$

whence $\widetilde{\beta}_{\Delta^2_C} = -\infty$.

(b) We have $f \in Q((\Delta^2_C)'\Delta^2_C)$ if and only if for each $d \in R$ and $x \in X \backslash S_d(f)$ there exists $w = w_{d,x} \in W$ such that

$$\sup w(S_d(f)) \leq \sup w(C) < w(x). \tag{3.24}$$

Consequently,

$$Q((\Delta^2_C)'\Delta^2_C) \subseteq Q(W). \tag{3.25}$$

(c) There holds

$$\beta_{\Delta^1_C} \leq \beta_{\Delta^2_C}. \tag{3.26}$$

(d) If X is a locally convex space, $W \subseteq X^* \backslash \{0\}$ and $f : X \to \overline{R}$ is upper semi-continuous, then

$$\beta_{\Delta^1_C} = \beta_{\Delta^2_C}. \tag{3.27}$$

Proof. (a) Formula (1.11) for Δ^2_C follows from lemma 1 applied to $\Gamma^2_{C,w}$ of (3.20). Also, by (3.21), we have $\Delta^2_{\{c\}}(\{c\}) = \{w \in W \,|\, w(c) \leq w(c)\} = W$ $(c \in C)$ and $\Delta^2_C(C) = \{w \in W \,|\, \sup w(C) \leq \sup w(C)\} = W$. Hence, by (1.3) for $\Delta = \Delta^2_C$ and (3.23), we obtain $\widetilde{\beta}_{\Delta^2_C} = \sup \emptyset = -\infty$.

(b) We have $f \in Q((\Delta_C^2)'\Delta_C^2)$ if and only if for each $d \in R$ and $x \in X \backslash S_d(f)$ there exists $w = w_{d,x} \in W$ satisfying (2.12) for $\Delta_C = \Delta_C^2$, that is, (3.24). Hence, by the definition of $Q(W)$, we obtain (3.25).

(c) By the definitions, we have

$$\beta_{\Delta_C^1} = \sup_{w \in W} \inf_{\substack{x \in X \\ w(x) \geq \sup w(C)}} f(x) \leq \sup_{w \in W} \inf_{\substack{x \in X \\ w(x) > \sup w(C)}} f(x) = \beta_{\Delta_C^2}.$$

(d) If X is a locally convex space and $W \subseteq X^* \backslash \{0\}$, then

$$\{x \in X | w(x) \geq \sup w(C)\} = \overline{\{x \in X | w(x) > \sup w(C)\}} \qquad (w \in W),$$

and hence, if $f : X \to \overline{R}$ is upper semi-continuous, then (see e.g. [1], section 6.2, exercise 5)

$$\inf_{\substack{x \in X \\ w(x) \geq \sup w(C)}} f(x) = \inf_{\substack{x \in X \\ w(x) > \sup w(C)}} f(x) \qquad (w \in W),$$

which yields (3.27). □

Let us observe now that, by (3.26), any condition ensuring (3.9) ensures also

$$\sup f(C) \leq \sup_{w \in W} \inf_{\substack{x \in X \\ w(x) > \sup w(C)}} f(x) \tag{3.28}$$

Hence, for example, from corollary 4 there follows

COROLLARY 7. *Let X be a set, $W \subseteq R^X$, $C \subseteq X$, and $f : X \to \overline{R}$ a function such that for each $d < \sup f(C)$ the level set $S_d(f)$ is W-evenly convex (e.g., let $f \in EQ(W)$). Then we have (3.28).*

THEOREM 4. *There holds (3.28) if and only if for each $d < \sup f(C)$ there exists $w = w_d \in W$ such that*

$$\sup w(S_d(f)) \leq \sup w(C). \tag{3.29}$$

Proof. This follows from theorem 2, equivalence 4° ⇔ 1°, applied to Δ_C^2, since formulas (3.28) and (3.29) mean, respectively, (2.10) for $\Delta_C = \Delta_C^2$ and $w \in \Delta_C^2(S_d(f))$. □

Remark 6. (a) Condition (3.29) is also satisfied, in particular, if

$$S_d(f) \subseteq C \qquad (d < \sup f(C)). \tag{3.30}$$

(b) If $f \in Q((\Delta_C^2)'\Delta_C^2)$, then, by lemma 4(b), the above condition involving (3.29) is satisfied, and hence, by theorem 4, we have (3.28). However, this follows also from corollary 2 applied to Δ_C^2, or from corollary 7 above, since by (3.25) and (1.37) we have $Q((\Delta_C^2)'\Delta_C^2) \subseteq Q(W) \subseteq EQ(W)$.

COROLLARY 8. *Let X be a locally convex space, $W \subseteq X^* \setminus \{0\}, C$ a subset of X satisfying (3.17), and $f : X \to \overline{R}$ an upper semi-continuous function which is regular with respect to all $w \in W$, and such that for each $d < \sup f(C)$ the level set $S_d(f)$ is W-evenly convex (the latter condition is satisfied e.g. when f is evenly quasi-convex). Then*

$$\sup f(C) = \sup_{w \in W} \inf_{\substack{x \in X \\ w(x) > \sup w(C)}} f(x). \tag{3.31}$$

Proof. By corollary 5, we have (3.8), whence, by lemma 4(d), we obtain the inequality $\geq$ in (3.31). On the other hand, by corollary 7 we have the inequality $\leq$ in (3.31), and hence the equality. □

Remark 7. Corollary 8 is an improvement of [5], corollary 3.3, in which the stronger assumption of W-convexity of the level sets $S_d(f)$ ($d < \sup f(C)$) has been made.

(3) Let

$$\Gamma^3_{C,w} = \{x \in X \,|\, w(x) = \sup w(C)\} \qquad (w \in W). \tag{3.32}$$

Then the associated duality is (by (1.22))

$$\Delta^3_C(G) = \{w \in W |\, \sup w(C) \notin w(G)\} \qquad (G \subseteq X). \tag{3.33}$$

LEMMA 5. *(a) For any set $C \subseteq X$ the duality $\Delta_C = \Delta^3_C$ satisfies (1.11) and*

$$\Delta^3_{\{c\}}(\{c\}) = \emptyset \qquad (c \in C), \tag{3.34}$$

$$\Delta^1_C(G) \subseteq \Delta^3_C(G) \qquad (G \subseteq X). \tag{3.35}$$

$$\Delta^3_C(C) = \Delta^1_C(C). \tag{3.36}$$

(b) There holds

$$\beta_{\Delta^1_C} \leq \beta_{\Delta^3_C}. \tag{3.37}$$

Proof. (a) Formula (1.11) for Δ^3_C follows from lemma 1 applied to $\Gamma^3_{C,w}$ of (3.32). Also, by (3.33), we have

$$\Delta^3_{\{c\}}(\{c\}) = \{w \in W |\, \sup w(\{c\}) \notin w(\{c\})\} = \emptyset \qquad (c \in C)\,.$$

Finally, (3.35) and (3.36) follow from (3.33), (3.3), and the fact that $w(c) \leq \sup w(C)$ $(c \in C)$.

(b) By the definitions, we have

$$\beta_{\Delta^1_C} = \sup_{w \in W} \inf_{\substack{x \in X \\ w(x) \geq \sup w(C)}} f(x) \leq \sup_{w \in W} \inf_{\substack{x \in X \\ w(x) = \sup w(C)}} f(x) = \beta_{\Delta^3_C}\,.$$

□

Remark 8. We have $f \in Q((\Delta_C^3)'\Delta_C^3)$ if and only if for each $d \in R$ and $x \in X \backslash S_d(f)$ there exists $w = w_{d,x} \in W$ satisfying (2.12) for $\Delta_C = \Delta_C^3$, that is,

$$w(x) = \sup w(C) \notin w(S_d(f)). \tag{3.38}$$

Hence, by the definition of $EQA(W)$, we obtain

$$Q((\Delta_C^3)'\Delta_C^3) \subseteq EQA(W). \tag{3.39}$$

THEOREM 5. *If $C \subseteq X$ and $f : X \to \overline{R}$ satisfy*

$$\inf_{\substack{x \in X \\ w(x) = \sup w(C)}} f(x) \leq \sup_{c \in C} \inf_{\substack{x \in X \\ w(x) = w(c)}} f(x) \qquad (w \in W), \tag{3.40}$$

then

$$\sup f(C) \geq \sup_{w \in W} \inf_{\substack{x \in X \\ w(x) = \sup w(C)}} f(x) \geq \sup_{\substack{w \in W \\ \exists c \in C,\, w(c) = \sup w(C)}} \inf_{\substack{x \in X \\ w(x) = \sup w(C)}} f(x). \tag{3.41}$$

Proof. Formula (3.41) follows from lemma 5(a) and theorem 1 applied to $\Delta = \Delta_C^3$ of (3.33), since

$$X \backslash (\Delta_C^3)'(\{w\}) = \{x \in X |\, w(x) = \sup w(C)\} \qquad (w \in W), \tag{3.42}$$

$$X \backslash (\Delta_{\{c\}}^3)'(\{w\}) = \{x \in X |\, w(x) = w(c)\} \qquad (c \in C, w \in W). \tag{3.43}$$

□

Using theorem 5, one can obtain again

COROLLARY 9. ([3], part of theorem 2.1). *Let X be a locally convex space, $W \subseteq X^* \backslash \{0\}, C$ a bounded subset of X, and $f : X \to R$ a finite convex function. Then we have (3.41).*

Proof. Let $w \in W$. Then, by our assumptions and [3], lemma 2.1, the function $\psi_{f,w} : R \to \overline{R}$ defined by

$$\psi_{f,w}(d) = \inf_{\substack{x \in X \\ w(x) = d}} f(x) \qquad (d \in R) \tag{3.44}$$

is finite, convex and continuous on R. Hence, if $c_n \in C$ $(n = 1, 2, ...)$ is any sequence such that $\lim_{n \to \infty} w(c_n) = \sup w(C)$ ($\in R$, since C is bounded), then

$$\psi_{f,w}(\sup w(C)) = \lim_{n \to \infty} \psi_{f,w}(w(c_n)) \leq \sup_{c \in C} \psi_{f,w}(w(c)),$$

that is, we have (3.40). Consequently, by theorem 5, we obtain (3.41). □

Remark 9. Under the same assumptions we also have, by [3], theorem 2.1, the opposite inequality to the first part of (3.41) (and, as shown by the separation argument of that proof, for the opposite inequality it is sufficient to assume that f is quasi-convex and lower semi-continuous), and hence the equality

$$\sup f(C) = \sup_{w\in W} \inf_{\substack{x\in X \\ w(x)=\sup w(C)}} f(x). \tag{3.45}$$

(4) Let

$$\Gamma^4_{C,w} = \{x \in X \mid w(x) \notin w(C)\} \qquad (w \in W). \tag{3.46}$$

Then the associated duality is (by (1.22))

$$\Delta^4_C(G) = \{w \in W \mid w(G) \subseteq w(C)\} \qquad (G \subseteq X). \tag{3.47}$$

Some properties of this duality are given in

LEMMA 6. *(a) For any set $C \subseteq X$ the duality $\Delta_C = \Delta^4_C$ satisfies (1.11) and*

$$\Delta^4_{\{c\}}(\{c\}) = W \qquad (c \in C), \tag{3.48}$$

$$\Delta^4_C(C) = W, \tag{3.49}$$

whence $\widetilde{\beta}_{\Delta^4_C} = -\infty$.

(b) We have $f \in Q((\Delta^4_C)'\Delta^4_C)$ if and only if for each $d \in R$ and $x \in X\backslash S_d(f)$ there exists $w = w_{d,x} \in W$ such that

$$w(S_d(f)) \subseteq w(C),\ w(x) \notin w(C). \tag{3.50}$$

Consequently,

$$Q((\Delta^4_C)'\Delta^4_C) \subseteq EQA(W). \tag{3.51}$$

(c) There holds

$$\beta_{\Delta^4_C} \le \beta_{\Delta^2_C}. \tag{3.52}$$

Proof. (a) Formula (1.11) for Δ^4_C follows from lemma 1 applied to $\Gamma^4_{C,w}$ of (3.46). Also, by (3.47), we have $\Delta^4_{\{c\}}(\{c\}) = \{w \in W \mid w(c) \in w(\{c\})\} = W$ $(c \in C)$ and $\Delta^4_C(C) = \{w \in W \mid w(C) \subseteq w(C)\} = W$. Hence, by (1.3) for $\Delta = \Delta^4_C$ and (3.49), we obtain $\widetilde{\beta}_{\Delta^4_C} = \sup \emptyset = -\infty$.

(b) We have $f \in Q((\Delta^4_C)'\Delta^4_C)$ if and only if for each $d \in R$ and $x \in X\backslash S_d(f)$ there exists $w = w_{d,x} \in W$ satisfying (2.12) for $\Delta_C = \Delta^4_C$, that is, (3.50). Hence, by the definition of $EQA(W)$, we obtain (3.51).

(c) By the definitions, we have

$$\beta_{\Delta^4_C} = \sup_{w\in W} \inf_{\substack{x\in X \\ w(x)\notin w(C)}} f(x) \le \sup_{w\in W} \inf_{\substack{x\in X \\ w(x)>\sup w(C)}} f(x) = \beta_{\Delta^2_C}.$$

□

THEOREM 6. *Let X and W be two sets, $W \subseteq R^X, f : X \to \overline{R}$ and C a subset of X.*

(a) If

$$\inf_{\substack{x\in X\\ w(x)\notin w(C)}} f(x) \leq \sup_{c\in C} \inf_{\substack{x\in X\\ w(x)\neq w(c)}} f(x) \qquad (w \in W), \tag{3.53}$$

then

$$\sup f(C) \geq \sup_{w\in W} \inf_{\substack{x\in X\\ w(x)\notin w(C)}} f(x). \tag{3.54}$$

(b) There holds

$$\sup f(C) \leq \sup_{w\in W} \inf_{\substack{x\in X\\ w(x)\notin w(C)}} f(x) \tag{3.55}$$

if and only if for each $d < \sup f(C)$ there exists $w = w_d \in W$ such that

$$w(S_d(f)) \subseteq w(C). \tag{3.56}$$

(c) If we have (3.53) and if for each $d < \sup f(C)$ there exists $w = w_d \in W$ satisfying (3.56) (by lemma 6(b), the latter condition is satisfied, e.g., when $f \in Q((\Delta_C^4)'\Delta_C^4)))$, then

$$\sup f(C) = \sup_{w\in W} \inf_{\substack{x\in X\\ w(x)\notin w(C)}} f(x). \tag{3.57}$$

Proof. (a) This follows from theorem 1 for $\Delta_C = \Delta_C^4$ of (3.47), since

$$X \setminus (\Delta_C^4)'(\{w\}) = \{x \in X \mid w(x) \notin w(C)\} \qquad (w \in W), \tag{3.58}$$

$$X \setminus (\Delta_{\{c\}}^4)'(\{w\}) = \{x \in X \mid w(x) \neq w(c)\} \qquad (c \in C, w \in W). \tag{3.59}$$

(b) This follows from theorem 2, equivalence 4° $\Leftrightarrow$ 1°, applied to Δ_C^4, since formulas (3.55) and (3.56) mean, respectively, (2.10) for $\Delta_C = \Delta_C^4$ and $w \in \Delta_C^4(S_d(f))$.

(c) This follows by combining parts (a) and (b). □

4. Duality for primal infimization, using arbitrary families of dualities $\Delta_C : 2^X \to 2^W$

In this Section X, C, f and W will have the same meaning as in Section 1, without any special mention.

LEMMA 7 ([2], lemma 2.1). *For any family of sets* $\{A_i\}_{i\in I}$ *and any function* $f : \cup_{i\in I} A_i \to \overline{R}$, *we have*

$$\inf_{i\in I} \inf f(A_i) = \inf f(\cup_{i\in I} A_i). \tag{4.1}$$

LEMMA 8. *For any duality* $\Delta : 2^X \to 2^W$ *and any set* $P \subseteq W$, *we have*

$$\cup_{w\in P} [X \setminus \Delta'(\{w\})] = X \setminus \Delta'(P). \tag{4.2}$$

Consequently, for any set $C \subseteq X$,

$$\beta^r_{\Delta_C} = \inf f(X \setminus \Delta'_C(W)), \tag{4.3}$$

$$\widetilde{\beta}^r_{\Delta_C} = \inf f(X \setminus \Delta'_C \Delta_C(C)). \tag{4.4}$$

Proof. By (1.4) (applied to Δ'),

$$\cup_{w\in P}[X \setminus \Delta'(\{w\})] = X \setminus \cap_{w\in P} \Delta'(\{w\}) = X \setminus \Delta'(P),$$

which proves (4.2). Hence, by (1.19), (4.1) applied to

$$\{A_i\}_{i\in I} = \{X \setminus \Delta'_C(\{w\})\}_{w\in W},$$

and (4.2), we obtain

$$\beta^r_{\Delta_C} = \inf_{w\in W} \inf f(X \setminus \Delta'_C(\{w\})) = \inf f(\cup_{w\in W}[X \setminus \Delta'_C(\{w\})]) \\ = \inf f(X \setminus \Delta'_C(W)),$$

which proves (4.3). Similarly, by (1.17), (4.1) applied to

$$\{A_i\}_{i\in I} = \{X \setminus \Delta'_C(\{w\})\}_{w\in \Delta_C(C)},$$

and (4.2), we obtain (4.4). □

Remark 10. (a) For $P = W$, formula (4.2) is known (see e.g. [8], formula (3.12)).

(b) Formula (4.4) yields immediately [10], corollary 6, according to which, if $C \in \mathcal{C}(\Delta'_C \Delta_C)$, then we have $\inf f(X \setminus C) = \widetilde{\beta}^r_{\Delta_C}$.

THEOREM 7. *If* $\Delta_C : 2^X \to 2^W$ *is a duality such that*

$$C \subseteq \Delta'_C(W), \tag{4.5}$$

then

$$\inf f(X \setminus C) \le \beta^r_{\Delta_C} \le \widetilde{\beta}^r_{\Delta_C}. \tag{4.6}$$

Moreover, if we have (4.5) and $C \in \mathcal{C}(\Delta'_C \Delta_C)$, *then*

$$\inf f(X \setminus C) = \beta^r_{\Delta_C} = \widetilde{\beta}^r_{\Delta_C}. \tag{4.7}$$

Proof. By (4.5) we have $X \setminus C \supseteq X \setminus \Delta'_C(W)$, whence, by (4.3) and (1.21), we obtain (4.6). Hence, if (4.5) holds and $C \in \mathcal{C}(\Delta'_C\Delta_C)$, then, by remark 10(b) and (4.6), we obtain (4.7). □

Remark 11. (a) If $\Delta_C : 2^X \to 2^W$ is a duality such that

$$\Delta'_C(W) = \emptyset, \tag{4.8}$$

then (4.5) implies that $C = \emptyset$, for which (4.6) is trivial.

(b) It is well known and immediate (see e.g. [9], p. 194, remark 6.3(a)) that we have (4.8) if and only if $\emptyset \in \mathcal{C}(\Delta'_C\Delta_C)$ (i.e., for each $x \in X$ there exists $w \in W$ such that $x \notin \Delta'_C(\{w\})$), or, equivalently, $\Delta_C(\{x\}) \neq W$ $(x \in X)$.

THEOREM 8. *If X is a topological space and $\Delta_C : 2^X \to 2^W$ is a duality such that*

$$\operatorname{int} C \subseteq \Delta'_C(W), \tag{4.9}$$

and if f is upper semi-continuous, then we have (4.6). Moreover, if also $C \in \mathcal{C}(\Delta'_C\Delta_C)$, then we have (4.7).

Proof. By (4.9), we have $\overline{X \setminus C} = X \setminus \operatorname{int} C \supseteq X \setminus \Delta'_C(W)$, whence, by (4.3), (1.21) and the well-known equality which holds for any upper semi-continuous function f (see e.g. [1], section 6.2, exercise 5)

$$\inf f(\overline{X \setminus C}) = \inf f(X \setminus C), \tag{4.10}$$

we obtain (4.6). Furthermore, if also $C \in \mathcal{C}(\Delta'_C\Delta_C)$, then, as in the above proof of theorem 7, we obtain (4.7). □

Remark 12. If we have (4.8), then condition (4.9) implies that $\operatorname{int} C = \emptyset$.

LEMMA 9. *If $\Delta_C : 2^X \to 2^W$ is a duality satisfying $C \in \mathcal{C}(\Delta'_C\Delta_C)$, then*

$$C \supseteq \Delta'_C(W). \tag{4.11}$$

Proof. By definition (see (1.29)), we have $C \in \mathcal{C}(\Delta'_C\Delta_C)$ if and only if

$$\forall x \in X \setminus C,\ \exists w \in W,\ C \subseteq \Delta'_C(\{w\}),\ x \in X \setminus \Delta'_C(\{w\}).$$

Hence, in particular, in this case

$$\forall x \in X \setminus C,\ \exists w \in W,\ x \in X \setminus \Delta'_C(\{w\}),$$

that is, we have

$$X \setminus C \subseteq \cup_{w \in W}[X \setminus \Delta'_C(\{w\})],$$

which, by lemma 8, is equivalent to (4.11). □

THEOREM 9. *If* $\Delta_C : 2^X \to 2^W$ *is a duality satisfying (4.11), then*

$$\beta^r_{\Delta_C} \leq \inf f(X \setminus C) \leq \widetilde{\beta}^r_{\Delta_C}. \tag{4.12}$$

Proof. By (4.11) we have $X \setminus C \subseteq X \setminus \Delta'_C(W)$, whence, by (4.3) and (1.18), we obtain (4.12). □

Remark 13. Theorem 9 gives an upper and a lower estimate for $\inf f(X \setminus C)$. In general, we may have one of the weak duality equalities $\inf f(X \setminus C) = \beta^r_{\Delta_C}$ or $\inf f(X \setminus C) = \widetilde{\beta}^r_{\Delta_C}$ without having the other one.

COROLLARY 10. *If* $\Delta_C : 2^X \to 2^W$ *is a duality satisfying (4.8), then we have (4.12).*

Remark 14. For some conditions equivalent to (4.8), see remark 11(b) above.

COROLLARY 11. *If* $\Delta_C : 2^X \to 2^W$ *is a duality such that* $C \in \mathcal{C}(\Delta'_C \Delta_C)$, *then*

$$\beta^r_{\Delta_C} \leq \inf f(X \setminus C) = \widetilde{\beta}^r_{\Delta_C}. \tag{4.13}$$

Proof. By lemma 9, condition $C \in \mathcal{C}(\Delta'_C \Delta_C)$ implies (4.11). Thus, by theorem 9, we obtain (4.12). Moreover, by remark 10(b), condition $C \in \mathcal{C}(\Delta'_C \Delta_C)$ implies that $\inf f(X \setminus C) = \widetilde{\beta}^r_{\Delta_C}$. □

COROLLARY 12. *If we have (4.11) (e.g., if we have (4.8) or* $C \in \mathcal{C}(\Delta'_C \Delta_C)$ *) and* $\beta^r_{\Delta_C} = \widetilde{\beta}^r_{\Delta_C}$, *then there holds (4.7).*

Remark 15. Since

$$\beta^r_{\Delta_C} = \inf_{w \in W} \inf f(X \setminus \Delta'_C(\{w\})) =$$
$$= \min \{\widetilde{\beta}^r_{\Delta_C}, \inf_{w \in W \setminus \Delta_C(C)} \inf f(X \setminus \Delta'_C(\{w\}))\},$$

condition $\beta^r_{\Delta_C} = \widetilde{\beta}^r_{\Delta_C}$ is equivalent to

$$\widetilde{\beta}^r_{\Delta_C} = \inf_{w \in \Delta_C(C)} \inf f(X \setminus \Delta'_C(\{w\})) \leq \inf_{w \in W \setminus \Delta_C(C)} \inf f(X \setminus \Delta'_C(\{w\})). \tag{4.14}$$

Combining theorems 7 and 9, we obtain

COROLLARY 13. *If* $\Delta_C : 2^X \to 2^W$ *is a duality such that*

$$C = \Delta'_C(W), \tag{4.15}$$

then

$$\inf f(X \setminus C) = \beta^r_{\Delta_C} \le \widetilde{\beta}^r_{\Delta_C}. \tag{4.16}$$

Combining theorems 8 and 9, we obtain

COROLLARY 14. *If* X *is a topological space and* $\Delta_C : 2^X \to 2^W$ *is a duality such that*

$$\operatorname{int} C \subseteq \Delta'_C(W) \subseteq C, \tag{4.17}$$

and if f *is upper semi-continuous, then we have (4.16).*

5. Duality for primal infimization, using special families of dualities $\Delta_C : 2^X \to 2^W$

Let us first observe the following complement to lemma 1.

LEMMA 10. *Let* X *and* W *be two sets,* C *a subset of* X, *and* $\Gamma_{C,w} \subseteq X$ $(w \in W)$ *a family of subsets of* X. *Then for the duality* $\Delta_C : 2^X \to 2^W$ *defined by (1.22) we have*

$$\Delta'_C(W) = \cap_{w\in W}(X \setminus \Gamma_{C,w}) = X \setminus \cup_{w\in W} \Gamma_{C,w}, \tag{5.1}$$

$$\Delta'_C \Delta_C(C) = \bigcap_{\substack{w\in W \\ C\cap\Gamma_{C,w}=\emptyset}} (X \setminus \Gamma_{C,w}). \tag{5.2}$$

Furthermore, $\emptyset \in \mathcal{C}(\Delta'_C\Delta_C)$ *if and only if*

$$X = \cup_{w\in W}\Gamma_{C,w}. \tag{5.3}$$

Also, we have $C \in \mathcal{C}(\Delta'_C\Delta_C)$ *if and only if for each* $x \in X\setminus C$ *there exists* $w = w_x \in W$ *such that*

$$C \cap \Gamma_{C,w} = \emptyset,\ x \in \Gamma_{C,w}. \tag{5.4}$$

Proof. By (1.4) (applied to Δ'_C) and (1.11) (which holds by lemma 1), we have

$$\Delta'_C(W) = \cap_{w\in W}\Delta'_C(\{w\}) = \cap_{w\in W}(X \setminus \Gamma_{C,w}),$$

which proves (5.1). Also, by (1.4), (1.22) and (1.11), we obtain

$$\Delta'_C\Delta_C(C) = \cap_{w\in\Delta_C(C)}\Delta'_C(\{w\}) = \bigcap_{\substack{w\in W \\ C\cap\Gamma_{C,w}=\emptyset}} (X \setminus \Gamma_{C,w}),$$

that is, (5.2). Finally, the last two statements follow from (1.29) and (1.11) (alternatively, they follow also from (5.1), remark 11(b) and (5.2)). □

In the sequel we shall assume, without any special mention, that X is a set and W is as in (3.1), and, in some cases, even that X is a locally convex space and $W \subseteq X^*$, the conjugate space of X, or $W \subseteq X^* \setminus \{0\}$. For such X and W, and for $C \subseteq X$, we shall study the dualities $\Delta_C : 2^X \to 2^W$ associated (by lemma 1) to the families of sets $\Gamma_{C,w} \subseteq X$ $(w \in W)$ used in [6] and [7] (it will turn out that they will coincide with those introduced in Section 3), and we shall apply to the dual problems to (1.15) defined with the aid of these families the results of Section 4.

(1) Some additional properties of the duality Δ_C^2 defined by (3.21) are given in

LEMMA 11. *For any set $C \subseteq X$ we have*

$$(\Delta_C^2)'\Delta_C^2(C) = (\Delta_C^2)'(W) = \mathrm{co}_W C, \tag{5.5}$$

$$\beta^r_{\Delta_C^2} = \widetilde{\beta}^r_{\Delta_C^2}. \tag{5.6}$$

Also, we have the equivalence

$$C \in \mathcal{C}((\Delta_C^2)'\Delta_C^2) \Leftrightarrow C \in \mathcal{K}(W). \tag{5.7}$$

Proof. By (3.23), (5.1), (3.20) and the expression of $\mathrm{co}_W C$ given in [9], formula (2.131), we obtain

$$(\Delta_C^2)'\Delta_C^2(C) =$$
$$= (\Delta_C^2)'(W) = \{x \in X \mid w(x) \leq \sup w(C)\ (w \in W)\} = \mathrm{co}_W C\,,$$

which proves (5.5). Consequently, we have the equivalences

$$C \in \mathcal{C}((\Delta_C^2)'\Delta_C^2) \Leftrightarrow C = (\Delta_C^2)'\Delta_C^2(C) = \mathrm{co}_W C \Leftrightarrow C \in \mathcal{K}(W),$$

which yield (5.7). Finally, from (1.19) for Δ_C^2, (3.23), and (1.20) for Δ_C^2, we infer

$$\beta^r_{\Delta_C^2} = \inf_{w \in W} \inf f(X \setminus (\Delta_C^2)'(\{w\})) = \inf_{w \in \Delta_C^2(C)} \inf f(X \setminus (\Delta_C^2)'(\{w\})) =$$
$$= \widetilde{\beta}^r_{\Delta_C^2}\,.$$

□

THEOREM 10. *We have*

$$\inf f(X \setminus C) \leq \inf_{w \in W} \inf_{\substack{x \in X \\ w(x) > \sup w(C)}} f(x), \tag{5.8}$$

and if $C \in \mathcal{K}(W)$, *then*

$$\inf f(X \setminus C) = \inf_{w \in W} \inf_{\substack{x \in X \\ w(x) > \sup w(C)}} f(x). \tag{5.9}$$

Proof. For each $w \in W$ we have $\{x \in X | w(x) > \sup w(C)\} \subseteq X \setminus C$, whence (5.8) follows. Furthermore, if $C \in \mathcal{K}(W)$, then, by (5.5), $C = \mathrm{co}_W C = (\Delta_C^2)'(W)$, and hence, by corollary 13 for $\Delta_C = \Delta_C^2$, we obtain (5.9). □

Remark 16. (a) In the particular case when X is a locally convex space and $W \subseteq X^* \setminus \{0\}$, theorem 10 has been obtained in [6], theorem 5.1. An extension of this result has been proved in [7], theorem 2.1.

(b) As has been observed in [7], the $\inf_{w \in W}$ in (5.8) and (5.9) can be replaced by $\inf_{w \in C^b}$, where $C^b = \{w \in W | \sup w(x) < +\infty\}$ (the set C^b is the barrier cone of C, when X is a locally convex space and $W = X^*$).

(2) Some additional properties of the duality Δ_C^3 defined by (3.33) are given in

LEMMA 12. *(a) For any set* $C \subseteq X$ *we have*

$$(\Delta_C^3)'(W) = \{x \in X | w(x) \neq \sup w(C) \ (w \in W)\}. \tag{5.10}$$

(b) There holds

$$C \in \mathcal{C}((\Delta_C^3)'\Delta_C^3) \tag{5.11}$$

if and only if for each $x \in X \backslash C$ *there exists* $w = w_x \in W$ *such that*

$$w(c) < \sup w(C) = w(x) \qquad (c \in C). \tag{5.12}$$

Proof. (a) By (1.5) for $\Delta = \Delta_C^3$ of (3.33), we have (5.10) (or, alternatively, (5.1) and (3.32) yield (5.10)).

(b) By (3.42), condition (5.12) means that we have

$$C \subseteq (\Delta_C^3)'(\{w\}), \ x \in X \backslash (\Delta_C^3)'(\{w\}),$$

that is, (1.29) for $\Delta = \Delta_C^3$, which yields the conclusion. □

We recall (see [9], p. 84, definition 2.12) that if X is a set and $\emptyset \neq W \subseteq \overline{R}^X$, a subset C of X is called *W-convexlike*, if

$$C = \{x \in X \mid w(x) < \sup w(C)\ (w \in W)\}. \tag{5.13}$$

For example (see [9], p. 85, remark 2.31(a)), if X is a locally convex space, then every non-empty open convex subset C of X is $(X^*\backslash\{0\})$-convexlike and, in the converse direction, every $(X^*\backslash\{0\})$-convexlike set C is convex (but it need not be open).

COROLLARY 15. *If for a set $C \subseteq X$ we have (5.11) and*

$$\Delta_C^3(C) = W, \tag{5.14}$$

then C is W-convexlike.

Proof. By (3.33), we have (5.14) if and only if for each $w \in W$ there holds $w(c) < \sup w(C)$ $(c \in C)$, that is, if and only if

$$C \subseteq \{x \in X \mid w(x) < \sup w(C)\ (w \in W)\}. \tag{5.15}$$

If also (5.11) holds, then, by lemma 12(b), we have the opposite inclusion as well, and hence the equality (5.13). □

Let us recall now

LEMMA 13. (see e.g. [9], p. 85, lemma 2.1). *Let C be an open convex set in a locally convex space X. Then for each $w \in X^* \setminus \{0\}$ we have*

$$w(c) < \sup w(C) \qquad (c \in C). \tag{5.16}$$

THEOREM 11. *If X is a locally convex space, then for any open convex set $C \subseteq X$ and any upper semi-continuous function $f : X \to \overline{R}$ we have*

$$\inf f(X \setminus C) = \inf_{w \in X^* \setminus \{0\}} \inf_{\substack{x \in X \\ w(x) = \sup w(C)}} f(x). \tag{5.17}$$

Proof. Since C is an open convex set, by lemma 13 there holds (5.16). Also, by [4], corollary 2, for each $x \in X \backslash C$ there exists $w \in X^* \setminus \{0\}$ such that $w(x) = \sup w(C)$. Thus, taking $W = X^* \setminus \{0\}$, by lemma 12(b) we have $C \in \mathcal{C}(\Delta_C^3)'\Delta_C^3)$. Furthermore, by (5.16) and (5.10), there holds $C \subseteq (\Delta_C^3)'(W)$. Hence, by theorem 7 (for $\Delta_C = \Delta_C^3$), we obtain (5.17). □

Remark 17. Under the assumption that C is either a bounded convex set with non-empty interior or a bounded closed convex set, the equality (5.17) has been proved in [6], theorem 5.2. One can also see directly that there the boundedness of C is superfluous. Indeed, this assumption has been used in [6] only to ensure that $\sup w(C) < +\infty$, but, if $w \in X^* \setminus \{0\}$, $\sup w(C) = +\infty$, then $\inf f(\{x \in X | w(x) = \sup w(C)\}) = \inf \emptyset = +\infty$, so these w do not alter the second term of (5.17). Formula (5.17) (and a generalization of it) has been shown to hold also under somewhat different assumptions, in [7], theorem 3.1.

(3) Some additional properties of the duality Δ_C^1 defined by (3.3) are given in

LEMMA 14. *(a) For any set $C \subseteq X$ we have*

$$(\Delta_C^1)'(\{w\}) \subseteq (\Delta_C^3)'(\{w\}) \qquad (w \in W). \tag{5.18}$$

(b) There holds

$$C \in \mathcal{C}((\Delta_C^1)'\Delta_C^1) \tag{5.19}$$

if and only if for each $x \in X \setminus C$ there exists $w = w_x \in W$ such that

$$w(c) < \sup w(C) \leq w(x) \qquad (c \in C). \tag{5.20}$$

(c) We have the implication (5.11) $\Rightarrow$ (5.19), that is,

$$C \in \mathcal{C}((\Delta_C^3)'\Delta_C^3) \Rightarrow C \in \mathcal{C}((\Delta_C^1)'\Delta_C^1). \tag{5.21}$$

Proof. (a) By (1.5), (3.3) and (3.33), for each $w \in W$ we have

$$\begin{aligned}(\Delta_C^1)'(\{w\}) &= \{x \in X | w(x) < \sup w(C)\}\\ &\subseteq \{x \in X | w(x) \neq \sup w(C)\} = (\Delta_C^3)'(\{w\}).\end{aligned}$$

(b) We have (5.19) if and only if for each $x \in X \setminus C$ there exists $w = w_x$ satisfying (1.29) for $\Delta = \Delta_C^1$, that is, (5.20).

(c) By (5.18) and (3.36), we have

$$C \subseteq (\Delta_C^1)'\Delta_C^1(C) \subseteq (\Delta_C^3)'\Delta_C^1(C) = (\Delta_C^3)'\Delta_C^3(C), \tag{5.22}$$

whence (5.21) follows. □

COROLLARY 16. *For a set $C \subseteq X$, let us consider the following statements:*

1°. We have (5.11) and

$$\Delta_C^1(C) = W. \tag{5.23}$$

2°. We have

$$(\Delta_C^1)'(W) = C. \tag{5.24}$$

3°. C is W-convexlike.

4°. We have (5.19) and (5.23).

Then

$$1° \Rightarrow 2° \Leftrightarrow 3° \Rightarrow 4°. \tag{5.25}$$

Proof. 1° ⇒ 3°. By (3.36), formula (5.23) implies (5.14). Hence, if 1° holds, then, by corollary 15, C is W-convexlike.

2° ⇔ 3°. By (1.5) and (3.3), we have

$$(\Delta_C^1)'(W) = \{x \in X \,|\, w(x) < \sup w(C) \ (w \in W)\}, \tag{5.26}$$

whence, by the definition (5.13) of W-convexlike sets, we obtain the equivalence 2° ⇔ 3°.

3° ⇒ 4°. If (5.13) holds, then we have (5.15), and for each $x \in X \backslash C$ there exists $w = w_x \in W$ such that $w(x) \geq \sup w(C)$. Hence, by lemma 14(b), we have (5.19). Also, if (5.13) holds, then, by (3.3), we have (5.23). □

COROLLARY 17. *If X is a locally convex space and $W = X^* \setminus \{0\}$, then for any convex set $C \subseteq X$ we have*

$$\operatorname{int} C \subseteq (\Delta_C^1)'(W) \subseteq (\Delta_C^3)'(W). \tag{5.27}$$

Proof. By (5.18), we have the second inclusion in (5.27). Thus, if $\operatorname{int} C = \emptyset$, we are done. If $\operatorname{int} C \neq \emptyset$, then, since C is convex, for each $c \in \operatorname{int} C$ we have $w(c) < \sup w(C)$ $(w \in W)$ (by lemma 13). Hence, by (5.26), $\operatorname{int} C \subseteq (\Delta_C^1)'(W)$. □

THEOREM 12. *If X is a locally convex space, then for any open convex set $C \subseteq X$ and any upper semi-continuous function $f : X \to \overline{R}$ we have*

$$\inf f(X \setminus C) = \inf_{w \in X^* \setminus \{0\}} \ \inf_{\substack{x \in X \\ w(x) \geq \sup w(C)}} f(x). \tag{5.28}$$

Proof. Since C is an open convex set, for each $w \in X^* \setminus \{0\}$ we have, by lemma 13,

$$\{x \in X \,|\, w(x) = \sup w(C)\} \subseteq \{x \in X \,|\, w(x) \geq \sup w(C)\} \subseteq X \setminus C,$$

whence

$$\inf_{\substack{x \in X \\ w(x) = \sup w(C)}} f(x) \geq \inf_{\substack{x \in X \\ w(x) \geq \sup w(C)}} f(x) \geq \inf f(X \setminus C) \qquad (w \in X^* \setminus \{0\}). \tag{5.29}$$

Hence, by (5.29) and theorem 11, we obtain (5.28). □

Remark 18. (a) Alternatively, one can also give the following proof of theorem 12: Let $W = X^* \setminus \{0\}$. Then, by corollary 17, we have $C = \text{int } C \subseteq (\Delta_C^1)'(W)$. Furthermore, since C is an open convex set, by the separation theorem and lemmas 14(b) and 13, we have (5.19). Hence, by theorem 7 (for $\Delta_C = \Delta_C^1$), we obtain (5.28).

(b) Under the assumptions mentioned in remark 17, the equality (5.28) has been proved in [6], corollary 5.2. For some related results, see also [8], corollaries 5.6, 5.13 and theorem 5.4.

(4) Some additional properties of the duality Δ_C^4 defined by (3.47) are given in

LEMMA 15. *For any set $C \subseteq X$ the duality Δ_C^4 satisfies (1.11) and*

$$\Delta_C^4(G) \subseteq \Delta_C^2(G) \qquad (G \subseteq X), \tag{5.30}$$

$$(\Delta_C^4)'\Delta_C^4(C) = (\Delta_C^4)'(W) = \{x \in X \,|\, w(x) \in w(C) \ (w \in W)\}, \tag{5.31}$$

$$\beta^r_{\Delta_C^4} = \widetilde{\beta}^r_{\Delta_C^4}. \tag{5.32}$$

Also, we have the equivalence

$$C \in \mathcal{C}((\Delta_C^4)'\Delta_C^4(C)) \Leftrightarrow C \in \mathcal{ECA}(W). \tag{5.33}$$

Proof. Formula (5.30) is obvious from (3.47). Hence, by (3.49), (1.19) and (1.17), we obtain (5.32). Furthermore, by (3.49) and (1.5) (or, alternatively, by (3.49) and (5.1) for $\Gamma_{C,w}^4$ of (3.46)), we obtain (5.31). Finally, by (5.31), we have $C = (\Delta_C^4)'\Delta_C^4(C)$ if and only if for each $x \in X \setminus C$ there exists $w = w_x \in W$ such that $w(x) \notin w(C)$, i.e., if and only if $C \in \mathcal{ECA}(W)$, which proves (5.33). □

Remark 19. Applying [9], p. 53, formula (2.23), to the family $\mathcal{M}$ of all subsets of X of the form $M = M_{w,d} = \{x \in X \,|\, w(x) \neq d\}$ $(w \in W, d \in R)$, one can show that the right hand side of (5.31) coincides with $\text{eca}_W C$.

THEOREM 13. *We have*

$$\inf f(X \setminus C) \leq \inf_{w \in W} \inf_{\substack{x \in X \\ w(x) \notin w(C)}} f(x), \tag{5.34}$$

and if $C \in \mathcal{ECA}(W)$, then

$$\inf f(X \setminus C) = \inf_{w \in W} \inf_{\substack{x \in X \\ w(x) \notin w(C)}} f(x). \tag{5.35}$$

Proof. For each $w \in W$ we have $\{x \in X | w(x) \notin w(C)\} \subseteq X \setminus C$, whence (5.34) follows. Furthermore, if $C \in \mathcal{ECA}(W)$, then, by (5.33) and (5.31), we have $C = (\Delta_C^4)'\Delta_C^4(C) = (\Delta_C^4)'(W)$, and hence, by corollary 13 for $\Delta = \Delta_C^4$, we obtain (5.35). □

Remark 20. By (1.30), theorem 13 is an improvement of [6], corollary 5.1, according to which, if X is a linear space, $W \subseteq X^\# \setminus \{0\}$, where $X^\#$ denotes the algebraic dual of X, and $C \in \mathcal{K}(W)$, then (5.35) holds.

References

1. N. Bourbaki, Topologie générale. Ch. IV: Nombres réels. Hermann, Paris (1942).
2. J.-E. Martínez-Legaz and I. Singer, Some characterizations of surrogate dual problems. Optimization 24 (1992), 1-11.
3. I. Singer, Maximization of lower semi-continuous convex functionals on bounded subsets of locally convex spaces. I: Hyperplane theorems. Appl. Math. Optim. 5 (1979), 349-362.
4. I. Singer, Minimization of continuous convex functionals on complements of convex subsets of locally convex spaces. Math. Oper. Stat. Ser. Optim. 11 (1980), 235-248.
5. I. Singer, Generalizations of convex supremization duality. In: Nonlinear and convex analysis (B.-L. Lin and S. Simons, eds.). Lecture Notes in Pure Appl. Math. 107, Marcel Dekker, New York (1987), 253-270.
6. I. Singer, Optimization by level set methods. VI: Generalizations of surrogate type reverse convex duality. Optimization 18 (1987), 485-499.
7. I. Singer, Some further duality theorems for optimization problems with reverse convex constraint sets. J. Math. Anal. Appl. 171 (1992), 205-219.
8. I. Singer, Duality in quasi-convex supremization and reverse convex infimization via abstract convex analysis, and applications to approximation. Optimization 45 (1999), 255-307.
9. I. Singer, Abstract convex analysis. Wiley-Interscience, New York, 1997.
10. M. Volle, Conjugaison par tranches et dualité de Toland. Optimization 18 (1987), 633-642.

A Minimization Problem in Function Spaces

B. Uhrin
Department of Mathematics
University of Pécs
Ifjúság u. 6.
7624 Pécs, Hungary
e-mail: uhrin@sztaki.hu

Abstract. Let $L \subset R^n$ be a point-lattice of dimension n and $P \subset R^n$ a basic cell of L in R^n. Let $f : R^n \to R^1$ be a bounded real valued function vanishing outside of a bounded set and put $\text{supp}(f) := \{x \in R^n : f(x) \neq 0\}$. We call f *[almost] extendable to an L-periodic function* if there is a function $g : R^n \to R^1$ which is periodic(mod L) and such that $f(x) = g(x)$, [a.e.] $x \in \text{supp}(f)$. We call the set $D \subseteq P$ *an L-periodic part of f* if (i) there are at least two $u, v \in L$ such that $u = a - y, v = b - z$ for some $a, b \in \text{supp}(f), y, z \in D$ and (ii) for all pairs of $u, v \in L$ with the property (i) and for all $x \in D$ we have $f(x + u) = f(x + v) \neq 0$. In the paper a *new minimization problem in the space* $L^2(R^n)$ is formulated and it is proved that f is a solution of this problem exactly when the following two conditions are satisfied: (c1) f is almost extendable to an L-periodic function and (c2) f has no L-periodic parts of positive measure. This reflects an approach which is a "converse" to a usual optimization procedure in the following sense: we give some properties of a function (in this case some periodic properties) and after that we find an optimization problem which solution possesses the given properties.

1. Introduction

In what follows V means the volume (Lebesgue measure, in short measure) in R^n, $\int \cdot \, dx$ stands for the integral and "a.e." stands for almost everywhere, respectively, with respect to the V.

$A + B := \{ a + b : a \in A, \, b \in B \}$ is the algebraic (Minkowski) sum of $A, B \subseteq R^n$, in particular $A - B := A + (-B)$. $\theta \in R^n$ is the zero vector. If for all $x \in A + B$ there are unique $a \in A$ and $b \in B$ such that $x = a + b$, then we shall write $A + B$ as $A \oplus B$ (the direct algebraic sum of A and B).

$| \cdot |$ is either the cardinality of a set or the absolute value of a real or complex number (the meaning will be clear form the context).

Put $\langle y, z \rangle := \sum_{i=1}^{n} y_i z_i$ for the usual scalar product in R^n.

Given n linearly independent vectors $b_1, \ldots, b_n \in R^n$, the set

$$L := \left\{ \sum_{i=1}^{n} u_i b_i : u_i \text{ integers}, \; i = 1, \ldots, n \right\}$$

is the point-lattice generated by the basis (b_i). The set $P := \{\sum_{i=1}^{n} \lambda_i b_i : 0 \leq \lambda_i < 1, i = 1, \cdots, n\}$ is the basic cell of L defined by the basis (b_i)

F. Giannessi et al. (eds.), Optimization Theory, 255–278.

(as one can easily see, P is in a one-to-one correspondence with the quotient space R^n/L). The number $d(L) := V(P)$ is the determinant of L. P and L also give a direct decomposition of R^n, $R^n = P \oplus L$, i.e., any $x \in R^n$ can be written uniquely as

$$x = \varphi(x) + [x],\ \varphi(x) \in P,\ [x] \in L\ .$$

This defines two 'canonical' projections $\varphi : R^n \to P$, $[\ ] : R^n \to L$. The definition of φ shows that for any $y, z \in R^n$ we have

$$\varphi(y) = \varphi(z) \quad \Leftrightarrow \quad y - z \in L\ . \tag{1.0}$$

For any set $A \subseteq R^n$ write

$$\varphi(A) := \bigcup_{x\in A} \{\varphi(x)\},\ [A] := \bigcup_{x \in A} \{[x]\}.$$

The set

$$L^* := \{\sum_{i=1}^n u_i b_i^* : u_i \text{ integers }, i = 1, \dots, n\}$$

is the point lattice polar to L where (b_i^*) is the system of vectors orthonormal to the system (b_i). As one can see easily, $L^* = \{x \in R^n : \langle x, u\rangle \in Z\ \forall u \in L\}$. On point-lattices consult the classical books [2,3,6,12].

In what follows, if not specified otherwise, $f : R^n \to R^1$ *is any real valued function defined and bounded on* R^n. The set supp(f) defined in the Abstract is the *support* of f.

In the literature f is called *periodic* (mod L), *L-periodic* in short if

$$f(x) = f(x + u),\ x \in R^n,\ u \in L.$$

It is clear that if supp(f) is bounded, then f cannot be L-periodic, on the one hand. On the other hand, any such function defines an L-periodic function $\bar{f}$ as follows

$$\bar{f}(x) := \sum_{u\in L} f(x+u)\,, \quad x \in R^n. \tag{1.1}$$

The function $\bar{f}$ plays interesting roles both in multi-dimensional Fourier analysis and in the geometry of numbers.

Namely, if f is Lebesgue-measurable (in short, measurable) and has bounded support, then the Fourier expansion of $\bar{f}$ is equal to

$$\sum_{v\in L^*} \hat{f}(v)\ e^{2\pi i\langle v,x\rangle},\ \ x \in R^n, \tag{1.2}$$

where

$$(1.3) \qquad \hat{f}(z) := \frac{1}{d(L)} \int_{R^n} f(y) e^{-2\pi i \langle y, z \rangle} dy, \quad z \in R^n.$$

This statement is, in fact, essentially equivalent to the Poisson summation formula (see [13], p. 251, Theorem 2.4).

As to the role of $\bar{f}$ in the geometry of numbers, let us recall an identity involving $\bar{f}$ which is due to Bombieri [1]: if f is bounded, measurable, non-negative and has bounded support, then

$$(1.4) \qquad \int_{R^n} f(x)\bar{f}(x)dx = d(L) \cdot \sum_{v \in L^*} |\hat{f}(v)|^2 ,$$

where $\hat{f}$ is the function (1.3).

The latter identity is an important tool for some improvements of basic theorems in the geometry of numbers (see, [6], pp. 128-131). It is a simple consequence of the classical Parseval formula and the following another identity: for any complex valued function g defined on R^n and integrable there, we have

$$(1.5) \qquad \int_{R^n} g(x)dx = \int_P \left(\sum_{u \in L} g(y+u) \right) dy.$$

Let us note that the latter identity holds in a much more general setting, for functions defined on any topological group G and its discrete subgroup Γ s.t. the quotient group G/Γ is compact see, e.g., [24]. For all above results consult [6].

Two more L periodic functions generated by f have been introduced in [15] which, together with $\bar{f}$, also are quite useful in the geometry of numbers, see [15] or [6], pp. 128-131.

In the results just mentioned different L-periodic functions *generated* by f rather than the periodic structure of f were important, while the main result of the paper stems out from our efforts to understand and describe the *inner periodic structure* of f itself.

The paper is divided into three more sections. Section 2 contains a detailed study of the two periodic properties of f defined in the Abstract. The results proved in Section 2 are not only needed for the proof of the main theorem, but they seem to be of independent interest as well. In Section 3 the formulation and the proof of the main theorem is given. While the results in Section 2 are true for general f, the main theorem is formulated for non-negative f, which is not a serious restriction, as it is shown at the end of Section 3. Finally, Section 4 contains some concluding remarks.

2. Two periodic properties of a function

In this section we study the two concepts concerning the periodic properties of a function defined in the Abstract. Let us remind all notations and definitions from the Abstract and the Introduction. So, $L \subset R^n$ is any point-lattice of dimension n and $f : R^n \to R^1$, if not stated otherwise, any bounded function such that $\text{supp}(f)$ is bounded. The two definitions in the Abstract read more formally as follows.

DEFINITION. *We call f [almost] extendable to an L-periodic function (in short, [almost] L-extendable) if there is a function $g : R^n \to R^1$ which is L-periodic (in the classical sense, see Introduction) and such that*

$$f(x) = g(x), \text{ [a.e.] } x \in \text{supp}(f). \tag{2.1}$$

DEFINITION. *We call $D \subseteq P$ a periodic part of f if*

$$\begin{cases} |(\text{supp}(f) - D) \cap L| \geq 2 \text{ and} \\ f(x+u) = f(x+v) \neq 0, \; x \in D, \; u, v \in (\text{supp}(f) - D) \cap L. \end{cases} \tag{2.2}$$

For any set $A \subseteq R^n$ denote

$$\mathcal{L}(A) := (A - A) \cap L . \tag{2.3}$$

Now, we have

THEOREM 2.1 *f is extendable to an L-periodic function and has no L-periodic parts if and only if*

$$\mathcal{L}(\text{supp}(f)) = \{\theta\}. \tag{2.4}$$

Proof. Assume first that f has got an L-periodic part $D \subseteq P$. This implies that $\exists\, x \in D$ and $u, v \in L$ such that $u \neq v$ and $x+u, x+v \in \text{supp}(f)$. This means that $\theta \neq u - v \in \text{supp}(f) - \text{supp}(f)$, a contradiction to (2.4).

Secondly, assume that f cannot be extended to any periodic function. This implies that $\bar{f}$, being a periodic function, cannot be an extension of f, hence $\exists\, x \in \text{supp}(f)$ with $f(x) \neq \bar{f}(x)$. This yields

$$\sum_{\theta \neq u \in L} f(x+u) \neq 0 ,$$

consequently, $\exists\, u \in L, u \neq \theta$, so that $x + u \in \text{supp}(f)$. This, together with $x \in \text{supp}(f)$ implies $\theta \neq u \in \text{supp}(f) - \text{supp}(f)$, a contradiction to (2.4).

These prove one direction of the theorem. To prove the converse direction, observe first that

$$\mathcal{L}(P) = \{\theta\} . \tag{2.5}$$

Assume (2.4) is not true, i.e, $\exists\ u = x - z \in L$ s.t. $u \neq \theta$ and $x, z \in \text{supp}(f)$. Using the projections φ and $[\cdot]$, this implies $u - [x] + [z] = \varphi(x) - \varphi(z) \in L$ which is by (2.5) possible only if $\varphi(x) = \varphi(z)$. The last two equalities together with $u \neq \theta$ also imply that $[x] \neq [z]$. Denote $y = \varphi(x) = \varphi(z)$. We have $x = y + [x], z = y + [z] \in \text{supp}(f)$. Now, if $f(x) = f(z)$, then either $D := \{y\}$ is a periodic part of f or if not, then there are $u, v \in (\text{supp}(f) - y) \cap L$ such that $f(u+y) \neq f(v+y)$. If $f(x) \neq f(z)$, then $f(x) = f(z+u) \neq f(z)$, where $z \in \text{supp}(f)$. In both cases we can see that f cannot be extended to any periodic function. By this, the converse direction and the whole theorem is proved. □

THEOREM 2.2 *f is extendable to an L-periodic function if and only if*
(2.6)
$$f(x) = f(x+u)\ \ \forall\ u \in \mathcal{L}(\text{supp}(f))\ \text{ and }\ \forall\ x \in \text{supp}(f) \cap (\text{supp}(f) - u).$$

Proof. Denote for short $A := \text{supp}(f)$. Assume that (2.6) is true. Define the function $g : R^n \to R^1$ as follows:

$$g(x) := \begin{cases} f(x)\,, & \text{if } x \in A; \\ f(y)\,, & \text{if } x \notin A,\ \exists\ y \in A \text{ s.t. } \varphi(x) = \varphi(y); \\ 0\,, & \text{if } x \notin A,\ \ \varphi(x) \notin \varphi(A); \end{cases} \tag{2.7}$$

This function is well defined, because

$$\{y, z \in A \text{ and } \varphi(y) = \varphi(z)\} \ \Rightarrow\ f(y) = f(z) . \tag{*}$$

Indeed, (1.0) implies that $y = z + v$ for some $v \in L$. The definition of $\mathcal{L}(A)$ and the fact that $y, z \in A$ in turn imply that $v \in \mathcal{L}(A)$ and $z \in A \cap (A - v)$. Finally, the property (2.6) implies $f(z) = f(z + v) = f(y)$, so (*) is true.

To prove that g is L-periodic, first, let $x \in A$. For any $u \in L$ we have $\varphi(x) = \varphi(x + u)$, hence if $x + u \notin A$, then we have $g(x + u) = f(x) = g(x)$, and if $x + u \in A$, then the property (2.6) implies $g(x) = f(x) = f(x + u) = g(x + u)$.

Secondly, let $x \notin A$. Then, for any $u \in L$ we have either $x+u \in A$ or $x+u \notin A$. In the first case $g(x) = f(x+u) = g(x+u)$. In the second case $g(x + u) = f(y)$ for some $y \in A$ with $\varphi(x + u) = \varphi(y)$ and $g(x) = f(z)$ for some $z \in A$ with $\varphi(x) = \varphi(z)$. But $\varphi(x+u) = \varphi(x) = \varphi(y) = \varphi(z)$, hence by (*) $f(y) = f(z)$, yielding $g(x) = f(z) = f(y) = g(x + u)$.

This proves one direction of the theorem. To prove the converse implication let h be an L-periodic function such that $f(x) = h(x)$, $x \in A$, and assume that (2.6) is not true. Then, there are $\theta \neq v \in \mathcal{L}(A)$ and $x \in A \cap (A - v)$ such that $f(x) \neq f(x+v)$. But clearly $x + v \in A$, hence $h(x+v) = f(x+v)$, consequently, $f(x) = h(x) \neq h(x+v)$, which contradicts the L-periodicity of h. By this, the theorem is proved. □

Remark 2.3 It is interesting to "check" both theorems for the "simplest" f, the constant functions, i.e., $f(x) = c$, $x \in A \subset R^n$, $f(x) = 0$, $x \in R^n \setminus A$. For these functions Theorem 2.2 is a triviality, because $g(x) = c$, $x \in R^n$, is trivially an L-periodic extension of f and f trivially satisfies the condition (2.6). This is in a sharp contrast with Theorem 2.1 where in this case the first part of the condition ("f is extendable to an L-periodic function") is trivially satisfied, consequently, it is the second part ("f has no L-periodic parts") that is equivalent to (2.4) and this is far from a triviality already, on the one hand. On the other hand, as one expects, the more special f is, the more sophisticated results can be proved. These deeper results for sets (for ~ characteristic functions of sets) have been proved in [23].

Theorems 2.1, 2.2 are too "stiff" in the sense that after changing the values of f at a few points, the conditions (2.4) and (2.6) (consequently, the properties of f equivalent to them) need not be true. To formulate theorems that are "insensitive" to such changes a new notion will be needed.

For any Lebesgue measurable (in short, measurable) set $A \subseteq R^n$ denote

$$\hat{\mathcal{L}}(A) := \{u \in L :\ V(A \cap (A+u)) > 0\}. \tag{2.8}$$

This set is a natural "restriction" of $\mathcal{L}(A)$ because, as we can easily check,

$$\mathcal{L}(A) = \{u \in L :\ A \cap (A+u) \neq \emptyset\}.$$

It is clear that if A is open, then $\hat{\mathcal{L}}(A) = \mathcal{L}(A)$ and that if, say A has got an open kernel A^o, i.e., there is an open set A^o such that $A^o \subseteq A$ and $V(A^o) = V(A)$, then

$$\mathcal{L}(A^o) = \hat{\mathcal{L}}(A). \tag{2.9}$$

By using (2.9) and Theorem 2.1 one immediately gets a result with $\hat{\mathcal{L}}(\mathrm{supp}(f))$ instead of $\mathcal{L}(\mathrm{supp}(f))$ for this special case, i.e. when $\mathrm{supp}(f)$ has got an open kernel. In the general case we have

THEOREM 2.4 *Let f be measurable and assume that* $V(\mathrm{supp}(f)) > 0$. *f is almost extendable to an L-periodic function and all L-periodic parts of f (if any) are measurable and have measure zero if and only if*

$$\hat{\mathcal{L}}(\mathrm{supp}(f)) = \{\theta\}. \tag{2.10}$$

Notice that (2.10) implies also the measurability of any L-periodic part of f, a strange thing enough.

As we shall see, the proof of this theorem is not so simple already, at least when compared with that of Theorem 2.1. Its reason is that, due to the invariance of everything upon changes of the function on sets of measures zero, we have to go more deeply into the "intersection structure" of the family $\{(\mathrm{supp}(f) - u) \cap P\}_{u \in L}$. This problem seems to be of independent interest, so let us make a small detour by proving some statements concerning it, which, however, will play important roles in the proofs of statements for measurable f.

Let $A \subseteq R^n$ be any (i.e., not necessarily measurable) bounded set. As one can see easily,

$$\varphi(A) = \{y \in P : (A - y) \cap L \neq \emptyset\} \tag{2.11}$$

and

$$\varphi(A) = \bigcup_{u \in L} ((A - u) \cap P) . \tag{2.12}$$

The results of [14], [15], [16], [17], [18] owed much, among other things, to a disjoint decomposition of the set $\varphi(A)$, based on the "identity" (2.11) where all points x of $\varphi(A)$ such that $|(A - x) \cap L| = k$ were collected to be one component and clearly such components for $k = 0, 1, \ldots,$ fill up $\varphi(A)$.

The "identity" (2.12) suggests another decomposition, which is much finer, and this will be the decomposition we shall work with.

Namely, considering the family of the sets in the union in (2.12) as a family *covering* the $\varphi(A)$, we take all the possible intersections of the members of the family and *distinguish the intersections according to the subsets of L from which they originated.*

This idea can be made precise by using either (2.11) or (2.12), resulting in two different descriptions of the same disjoint decomposition of $\varphi(A)$. Having (2.11) in mind, for any $x \in P$ define

$$B(x) := \{z \in P : (A - z) \cap L = (A - x) \cap L\}.$$

It is clear that $B(x) \neq \emptyset$ for all $x \in P$ (x is contained in $B(x)$) and one can see easily that for any two points $x, y \in P$ either $B(x) = B(y)$

or $B(x) \cap B(y) = \emptyset$ holds. Let $\mathcal{F}(A)$ denote the family of all *mutually disjoint* sets $B(x)$, $x \in P$ such that $(A-x)\cap L \neq \emptyset$, we get the following refinement of both (2.11) and (2.12)

$$\varphi(A) = \bigcup_{B\in\mathcal{F}(A)} B, \tag{2.13}$$

where the sets B in the latter union are mutually disjoint.

As to the description based on (2.12), let $H \subset L$ be any non-empty finite set and denote

$$B(H) := \Big(\bigcap_{u\in H}(A-u)\cap P\Big) \setminus \Big(\bigcup_{v\in L\setminus H}(A-v)\cap P\Big), \tag{2.14}$$

where $B(H)$ is empty by definition if the intersection in its definition is empty. Now, one can see easily that if H_1, H_2 are two finite subsets of L such that $H_1 \neq H_2$, then $B(H_1) \cap B(H_2) = \emptyset$. It is clear that the union of all non-empty $B(H)$s gives $\varphi(A)$, showing that any non-empty set $B(H)$ defined by (2.14) belongs to $\mathcal{F}(A)$ and conversely, for any $B \in \mathcal{F}(A)$ taking $H := (A-x)\cap L, x \in B$, we get that $B = B(H)$. In other words, (2.14) *characterizes the family* $\mathcal{F}(A)$.

In what follows, for $B \in \mathcal{F}(A)$ we shall denote $L(B) := (A-x) \cap L$, $x \in B$. It is easy to see that

$$A = \bigcup_{B\in\mathcal{F}(A)} (B + L(B)), \tag{2.15}$$

where the sets $B + L(B)$ in the latter union are mutually disjoint. By using (2.14) one can see that the members of the family $\mathcal{F}(A)$ have the following interesting "maximality" property.

Property 2.5 Any $B \in \mathcal{F}(A)$ is the maximal (w.r.t. the set inclusion) subset of P such that $L(B) = (A-x)\cap L, \ \forall\, x \in B$.

Let $\hat{\mathcal{F}}(A)$ denote the subfamily of $\mathcal{F}(A)$ consisting of all members B of $\mathcal{F}(A)$ such that $|L(B)| \geq 2$ (perhaps $\hat{\mathcal{F}}(A) = \emptyset$). For the proof of Theorem 2.4 we shall need the following purely technical lemma connecting $\hat{\mathcal{L}}(A)$ to $\hat{\mathcal{F}}(A)$.

LEMMA 2.6 *Let A be measurable. Then,*

$$\hat{\mathcal{L}}(A) = \{\theta\} \tag{2.16}$$

holds if and only if

$$\{\text{either } \hat{\mathcal{F}}(A) = \emptyset \ \text{ or } \ V(B) = 0 \quad \forall\, B \in \hat{\mathcal{F}}(A)\}. \tag{2.17}$$

Proof. By the definition of $\hat{\mathcal{L}}(A)$, the condition (2.16) is equivalent to the condition

$$V(A \cap (A+u)) = 0 \quad \forall\, u \in L\,, u \neq \theta\,. \tag{2.18}$$

By (2.14), it is clear that for any $B \in \mathcal{F}(A)$ we have

$$B = \bigcap_{u \in L(B)} (A-u) \cap P \setminus \bigcup_{v \in L \setminus L(B)} (A-v) \cap P\,. \tag{2.19}$$

The representation (2.19) shows immediately that the members of the family $\mathcal{F}(A)$ are measurable if A is. The sets $P+u$, $u \in L$, give a disjoint decomposition of the space R^n, hence for any measurable set $C \subseteq R^n$ we have

$$V(C) = \sum_{v \in L} V(C \cap (P+v))\,. \tag{2.20}$$

This implies that for all $u \in L$ we have

$$V(A \cap (A+u)) = \sum_{v \in L} V((A-v) \cap (A-v+u) \cap P)\,, \tag{2.21}$$

hence the condition (2.18) is equivalent to the condition

$$V((A-v) \cap (A-w) \cap P) = 0 \quad \forall\, v, w \in L,\ v \neq w\,, \tag{2.22}$$

which is in turn equivalent to

$$V\left(\bigcap_{v \in E} ((A-v) \cap P)\right) = 0 \quad \forall\, E \subset L,\ |E| \geq 2\,. \tag{2.23}$$

We can see that (2.23) is equivalent to (2.18), i.e., to (2.16). This proves the implication (2.16)⇒(2.17), because by (2.19) the condition (2.23) implies (2.17).

To prove the converse, assume that (2.23), (or (2.16), which is equivalent to it, as we have just seen), is not true, i.e., that $\exists\, E \subset L$, $|E| \geq 2$, such that

$$V\left(\bigcap_{v \in E} ((A-v) \cap P)\right) > 0\,. \tag{2.24}$$

Let H be the largest w.r.t. the set inclusion set containing E having this property, i.e.,

$$V\left(\bigcap_{v \in H} ((A-v) \cap P)\right) > 0\,, \quad E \subseteq H \subset L \tag{2.25}$$

and

$$V\left(\bigcap_{v\in H\cup\{w\}}((A-v)\cap P)\right) = 0, \quad \forall\, w\in L\setminus H. \tag{2.26}$$

We claim that there is $\tilde{B}\in\hat{\mathcal{F}}(A)$ such that

$$V(\tilde{B}) = V\left(\bigcap_{v\in H}((A-v)\cap P)\right). \tag{2.27}$$

Indeed, take

$$C := \bigcap_{v\in H}((A-v)\cap P) \tag{2.28}$$

and

$$\tilde{B} := C\setminus \bigcup_{w\in L\setminus H}((A-w)\cap P). \tag{2.29}$$

By the characterization (2.14) of elements of $\mathcal{F}(A)$, (2.28) and (2.29) imply that $\tilde{B}\in\hat{\mathcal{F}}(A)$. For this $\tilde{B}$ we have $V(\tilde{B})>0$, hence (2.17) is not true. This proves the implication (2.17)⇒(2.23) and by this the lemma is proved. □

The following lemma is an easy consequence of definitions.

LEMMA 2.7 *For any set $A\subset R^n$ the following three conditions are equivalent:*
(a) $\mathcal{L}(A)=\{\theta\}$;
(b) $\hat{\mathcal{F}}(A)=\emptyset$;
(c) $|(A-x)\cap L|\le 1 \quad \forall\, x\in P$.

The last lemma needed for the proof of Theorem 2.4 is

LEMMA 2.8 *For any (not necessarily measurable) f, if $D\subseteq P$ is an L-periodic part of f, then there is a $B\in\hat{\mathcal{F}}(\mathrm{supp}(f))$ such that $D\subseteq B$.*

Proof. The definition (2.2) of D implies that $u+D\subseteq \mathrm{supp}(f)$ for all $u\in(\mathrm{supp}(f)-D)\cap L$. This means that $(\mathrm{supp}(f)-x)\cap L = (\mathrm{supp}(f)-D)\cap L$ for all $x\in D$. This implies by the Property 2.5 that $\exists\, B\in\mathcal{F}(\mathrm{supp}(f))$ s.t. $D\subseteq B$. The definition (2.2) also shows that $|(\mathrm{supp}(f)-D)\cap L|\ge 2$, hence $B\in\hat{\mathcal{F}}(\mathrm{supp}(f))$. □

Proof of Theorem 2.4. By the Lemma 2.6, the condition (2.10) holds if and only if

$$\text{(2.30)}\quad \{\text{either } \hat{\mathcal{F}}(\operatorname{supp}(f)) = \emptyset \text{ or } V(B) = 0 \quad \forall\, B \in \hat{\mathcal{F}}(\operatorname{supp}(f))\}.$$

Assume now that (2.30), i.e., (2.10), is true. Then by the Lemma 2.8 all L-periodic parts of f (if any) are subsets of a set of measure zero, hence they are measurable and of measure zero.

To prove that (2.30) also implies that f is almost extendable to an L-periodic function, take the set

(2.31)

$$A := \begin{cases} \operatorname{supp}(f) & \text{if } \hat{\mathcal{F}}(\operatorname{supp}(f)) = \emptyset \\ \operatorname{supp}(f) \setminus \bigcup_{B \in \hat{\mathcal{F}}(\operatorname{supp}(f))} (B + L(B)) & \text{otherwise.} \end{cases}$$

(2.30) implies that $V(A) = V(\operatorname{supp}(f))$.

We claim that for this A necessarily

$$\text{(2.32)}\qquad |(A - x) \cap L| \ \leq\ 1 \quad \forall\, x \in P\,.$$

Indeed, assume first $\hat{\mathcal{F}}(\operatorname{supp}(f)) = \emptyset$. This implies by (2.31) that $A = \operatorname{supp}(f)$ and $\hat{\mathcal{F}}(A) = \emptyset$, which by the Lemma 2.7 yields (2.32).

Secondly, assume $\hat{\mathcal{F}}(\operatorname{supp}(f)) \neq \emptyset$ and that $\exists\, \tilde{x} \in P$ with $|(A - \tilde{x}) \cap L| \geq 2$. Then, clearly also $|(\operatorname{supp}(f) - \tilde{x}) \cap L| \geq 2$, hence $\exists\, B \in \hat{\mathcal{F}}(\operatorname{supp}(f))$ s.t. $\tilde{x} \in B$ and $L(B) = (\operatorname{supp}(f) - \tilde{x}) \cap L$. By (2.31) $A \subseteq \operatorname{supp}(f)$, hence $(A - \tilde{x}) \cap L \subseteq L(B)$. Now, put $u \in (A - \tilde{x}) \cap L \subseteq L(B)$, i.e., $u = a - \tilde{x}$ for some $a \in A$. But for this $a \in A$ we have also $a \in B + L(B)$, where $B \in \hat{\mathcal{F}}(\operatorname{supp}(f))$, a contradiction with the definition of A.

By Lemma 2.7, the condition (2.32) implies that

$$\text{(2.33)}\qquad \mathcal{L}(A) \ =\ \{\theta\}\,.$$

Applying Theorem 2.1 we can see that the restriction of f to the set A is extendable to an L-periodic function and this proves that f itself is almost extendable to an L-periodic function. By this one direction of the theorem is proved.

To prove the converse direction, assume that all L-periodic parts of f (if any) are measurable and of measures zero and that f is almost extendable to an L-periodic function, but (2.10) is not true. By the Lemma 2.6, the fact (2.10) is not true implies that either $\hat{\mathcal{F}}(\operatorname{supp}(f)) = \emptyset$ or there is $B \in \hat{\mathcal{F}}(\operatorname{supp}(f))$ with $V(B) > 0$.

If $\hat{\mathcal{F}}(\operatorname{supp}(f)) = \emptyset$, then Lemma 2.8 yields that f has no periodic parts. Let $\hat{\mathcal{F}}(\operatorname{supp}(f)) \neq \emptyset$. We claim that in this case for the latter B

there are $u, v \in L(B)$ such that the set

$$B(u,v) := \{x \in B : f(x+u) \neq f(x+v)\}$$

is of positive measure. Indeed, assume that $V(B(u,v)) = 0$ is true for all pairs $u, v \in L(B)$. This implies that, by denoting $D(u,v) := \{x \in B : f(x+u) = f(x+v)\}$, for all pairs $u, v \in L(B)$ we have $V(D(u,v)) = V(B)$. But one can check easily that $D := \bigcap_{u,v \in L(B)} D(u,v)$ is a periodic part of f, and this is a contradiction, because $V(D) = V(B) > 0$ and by the assumption all L-periodic parts have measures zero.

The fact $B(u,v) > 0$ implies that for any $A \subseteq \text{supp}(f)$ with $V(A) = V(\text{supp}(f))$ we have $V(B(u,v) \cap (A-u) \cap (A-v)) > 0$, hence $B(u,v) \cap (A-u) \cap (A-v)) \neq \emptyset$, consequently there are $a, b \in A$ such that $a = b + u - v$ and $f(b) \neq f(a) = f(b+u-v)$. By this we have just shown that there is no periodic function g that would be equal to f on A. As the above set A was arbitrary with $V(A) = V(\text{supp}(f))$, we see that f cannot be almost extendable to an L-periodic function. This proves the converse direction and by this the whole theorem is proved. □

While Theorem 2.4 is a "loosening" of Theorem 2.1, the following theorem "does the same" with Theorem 2.2.

THEOREM 2.9 *Let f be measurable and assume that $V(\text{supp}(f)) > 0$. f is almost extendable to an L-periodic function if and only if*

$$\begin{gathered} f(x) = f(x+u) \\ \forall u \in \hat{\mathcal{L}}(\text{supp}(f)) \text{ and for a.e. } x \in \text{supp}(f) \cap (\text{supp}(f) - u)\,. \end{gathered} \tag{2.34}$$

Proof. For $u \in L$ denote $B(u) := \text{supp}(f) \cap (\text{supp}(f) - u)$. Assume that f has the property (2.34) and for $u \in \hat{\mathcal{L}}(\text{supp}(f))$ let $\hat{B}(u)$ be the subset of $B(u)$ guaranteed by (2.34) such that $V(B(u)) = V(\hat{B}(u))$ and $f(x) = f(x+u) \;\; \forall \;\; x \in \hat{B}(u)$. Take all sets $B(u), u \notin \hat{\mathcal{L}}(\text{supp}(f))$ and all sets $B(u) \setminus \hat{B}(u), u \in \hat{\mathcal{L}}(\text{supp}(f))$ out of $\text{supp}(f)$, and denote the resulting subset of $\text{supp}(f)$ by A. It is clear that A has the same measure as $\text{supp}(f)$ and that $\mathcal{L}(A) = \hat{\mathcal{L}}(\text{supp}(f))$.

Now, define the function $g : R^n \to R^1$ analogously as in (2.7):

$$g(x) := \begin{cases} f(x)\,, & \text{if } x \in A; \\ f(y)\,, & \text{if } x \notin A,\ \exists\, y \in A \text{ s.t. } \varphi(x) = \varphi(y); \\ 0\,, & \text{if } x \notin A,\ \ \varphi(x) \notin \varphi(A); \end{cases} \tag{2.35}$$

This function is again well defined, because the implication

$$\{\ y, z \in A \ \text{ and } \ \varphi(y) = \varphi(z)\ \} \ \Rightarrow \ f(y) = f(z) \tag{$*$}$$

holds again. Indeed, by (1.0) we can see that $y = z + v$ for some $v \in L$. The definitions of $\hat{\mathcal{L}}(\mathrm{supp}(f))$, A and the fact $y, z \in A$ in turn imply that $v \in \hat{\mathcal{L}}(\mathrm{supp}(f))$ and $z \in \hat{B}(v)$. Finally, the property (2.34) implies $f(z) = f(z+v) = f(y)$, so (*) is true. The proof that the function g is L-periodic goes analogously as for (2.7). This proves one direction of the theorem.

To prove the converse implication let $C \subseteq \mathrm{supp}(f)$ with $V(C) = V(\mathrm{supp}(f)$ and let h be a periodic function such that $f(x) = h(x)$, $x \in C$. Assume (2.34) is not true. Then there is $\theta \neq v \in \hat{\mathcal{L}}(\mathrm{supp}(f))$ and $B \subseteq \hat{B}(v)$ with $V(B) > 0$ such that $f(x) \neq f(x+v)$ $\forall\ x \in B$. For this B we have $B, B+v \subseteq \mathrm{supp}(f)$,$V(B), V(B+v) > 0$, consequently, $B \cap (C-v) \cap C \neq \emptyset$. Let $y = z+v \in C$ with $z \in B \cap C$. For these y, z we have $f(y) = h(y)$. and $f(z) = h(z)$. On the other hand, $f(z) \neq f(z+v)$ because $z \in B \subseteq \hat{B}(v)$, which implies $h(z) \neq h(z+v)$, a contradiction with the periodicity of h. By this the theorem is proved. □

3. A minimization problem and its solution

Let us fix again that L is an n-dimensional point-lattice in R^n, and put $N(R^n)$ for the space of non-negative bounded measurable functions having bounded supports. Recall what the two periodic properties of f: f has no L-periodic parts of positive measure and f is almost extendable to an L-periodic function, respectively, mean.

Put $\|f\|_2$ for the L_2-norm of f, i.e.,

$$\|f\|_2 := \left(\int_{R^n} f^2(x)\, dx \right)^{1/2}$$

and put $\|\hat{f}\|_2$ for the square-root of the right-hand side expression in (1.4) ($\sim$ l_2-norm of the function $\hat{f}$ when restricted to L^*).

Now, we have

PROBLEM. *Given L and $a > 0$, find the number*

$$i(a) := \inf\{\|\hat{f}\|_2 : f \in N(R^n),\ \|f\|_2 = a\} \tag{3.1}$$

and characterize $f \in N(R^n)$ such that

$$i(a) = \|\hat{f}\|_2\,. \tag{3.2}$$

THEOREM 3.1 (The first solution of the Problem). *We have*

$$i(a) = a \tag{3.3}$$

and $i(a) = \|\hat{f}\|_2$ *holds if and only if*

(3.4) $$f(x) \cdot f(x+v) = 0 \quad \text{for a.e. } x \in R^n \text{ and } \forall\, \theta \neq v \in L.$$

Proof. The theorem is an immediate consequence of the Bombieri identity (1.4). □

THEOREM 3.2 (The second solution of the Problem). *The condition (3.3) is true. The condition (3.2) holds if and only if* f *is almost extendable to an L-periodic function and* f *has no periodic parts of positive measure.*

Proof. (3.3) follows from (1.4). As one can see easily, the condition (3.4) is equivalent to

(3.5) $$V(\mathrm{supp}(f) \cap (\mathrm{supp}(f) - v)) = 0, \quad \forall\, \theta \neq v \in L.$$

Now, taking the definition of $\hat{\mathcal{L}}(\mathrm{supp}(f)$ into account, the condition (3.5) is equivalent to the condition (2.10), so Theorem 2.4 gives the result. □

THEOREM 3.3 (A finer solution of the problem). *For any* $f \in N(R^n)$ *we have*

(3.6)
$$\int_{R^n} f^2(x)\,dx \leq^a$$
$$\leq^a \frac{1}{d(L)} \sum_{v \in L^*} \int_{R^n}\int_{R^n} \cos(2\pi\langle v, x-t\rangle) \min\{f^2(x), f^2(t)\}\,dx\,dt \leq^b \|\hat{f}\|_2^2\,.$$

$\leq^a$ *is equality if and only if* f *is almost extendable to an L-periodic function and has no L-periodic parts of positive measure.*

$\leq^b$ *is equality if and only if* f *is almost extendable to an L-periodic function.*

Theorem 3.3 is a consequence of the following two lemmas.

LEMMA 3.4 *For any* $f \in N(R^n)$ *we have*

(3.7)
$$\int_{R^n} f^2(x)\,dx \;\leq^1$$
$$\leq^1 \sum_{u \in L} \int_{R^n} \min\{f^2(x), f^2(x+u)\}\,dx \leq^2 \sum_{u \in L} \int_{R^n} f(x) \cdot f(x+u)\,dx\,.$$

$\leq^1$ *is equality if and only if* f *is almost extendable to an* L*-periodic function and has no* L*-periodic parts of positive measures.*

$\leq^2$ *is equality if and only if* f *is almost extendable to an* L*-periodic function.*

LEMMA 3.5 *For any* $f \in N(R^n)$ *the following identity holds*

$$\sum_{u\in L}\int_{R^n}\min\{f(x),f(x+u)\}\,dx = \tag{3.8}$$

$$= \frac{1}{d(L)}\sum_{v\in L^*}\int_{R^n}\int_{R^n}\cos(2\pi\langle v,x-t\rangle)\min\{f(x),f(t)\}\,dx\,dt\,.$$

Proof of Lemma 3.4. If $V(\mathrm{supp}(f))=0$, then the lemma trivially holds, so assume $V(\mathrm{supp}(f))>0$. The inequalities are clear. $\leq^1$ is clearly equality if and only if

$$\int_{R^n}\min\{f^2(x),f^2(x+u)\}\,dx \;=\; 0,\quad \forall\, u\in L\,,\; u\neq\theta\,. \tag{3.9}$$

The non-negativity of f ensures that the condition (3.9) is equivalent to (3.4) and by Theorem 3.2, we get the assertion on the equality in $\leq^1$.

$\leq^2$ is equality if and only

$$\min\{f^2(x),f^2(x+u)\} \;=\; f(x)\cdot f(x+u) \tag{3.10}$$

for all $u\in\hat{\mathcal{L}}(\mathrm{supp}(f))$ and for almost all $x\in\mathrm{supp}(f)\cap(\mathrm{supp}(f)-u)$. The condition (3.10) is equivalent to the condition

$$f(x)=f(x+u)$$

for all $u\in\hat{\mathcal{L}}(\mathrm{supp}(f))$ and for almost all $x\in\mathrm{supp}(f)\cap(\mathrm{supp}(f)-u)$. Now, using Theorem 2.9, we get the exact condition of equality in $\leq^2$.

□

We can see that Lemma 3.4 is a simple consequence of Theorem 3.2 (i.e., Theorem 2.4) and Theorem 2.9. As we shall see, the proof of the Lemma 3.5 is much more complicated. For it, we need the following three propositions.

The first is the following classical identity (the generalized Parseval 'formula' for real functions, see, e.g., [9], [13]).

PROPOSITION 3.6 *For two* f_1,f_2 *square-integrable non-negative functions defined on* P *we have*

$$\int_P f_1(y)f_2(y)\,dy \;=\; \frac{1}{d(\Lambda)}\sum_{v\in\Lambda^*}\int_P\int_P\cos(2\pi\langle v,y-z\rangle)f_1(y)f_2(z)\,dy\,dz\,. \tag{3.11}$$

For the second proposition, let us introduce the $n+1$-dimensional point-lattice $W \subset R^{n+1}$ as

$$W := L \oplus \{ k(\theta,1) : k \in Z^1 \}, \tag{3.12}$$

where θ is the zero vector of R^n and Z^1 means the set of integers. For any set $E \subset R^{n+1}$ the sets $\mathcal{W}(E)$ and $\hat{\mathcal{W}}(E)$ are defined by taking the lattice W instead of L in (2.3) and (2.8), respectively.

The last special notation we shall use is the so-called "hypograph of a function", defined for non-negative functions f as :

$$\text{hyp}(f) := \{ (x,\xi) \in R^n \times R^1 : 0 \le \xi \le f(x),\ x \in \text{supp}(f) \} . \tag{3.13}$$

For this set the following quite interesting identity involving supp(f) holds.

PROPOSITION 3.7 *For any $f \in N(R^n)$ and for any $(x,0) \in R^{n+1} = R^n \times R^1$ we have*

$$\bar{V}(\text{hyp}(f) \cap (\text{hyp}(f) - (x,0))) = \int_{\text{supp}(f)\cap(\text{supp}(f)-x)} \min\{f(y), f(y+x)\}\, dy\ , \tag{3.14}$$

where $\bar{V}$ is the Lebesgue measure in R^{n+1}.

Proof. First observe that for any set $E \subset R^{n+1}$ and any $z \in R^{n+1}$ we have

$$\chi_{E\cap(E-z)}(y) = \chi_E(y) \cdot \chi_E(y+z)\,, \quad y \in R^{n+1}\,, \tag{3.15}$$

where χ means the characteristic function. This implies

$$\bar{V}(E \cap (E - z)) = \int_{R^{n+1}} \chi_E(y) \cdot \chi_E(y+z)\, dy\,. \tag{3.16}$$

Now, for $x \in R^n$ let G and F be the sets of elements $(y,\eta) \in R^{n+1}$ such that $y \in \text{supp}(f) \cap (\text{supp}(f) - x)$, $0 \le \eta \le f(y) \le f(y+x)$ and $0 \le \eta \le f(y+x) < f(y)$, respectively. One of the sets may be empty and it is clear that both are empty if and only if supp$(f) - x$ does not intersect supp(f). It is also clear that G, F are measurable and give a disjoint decomposition of the set supp$(f) \cap (\text{supp}(f) - x)$. Taking $E :=$ hyp(f) in (3.16), after some calculations, we get

$$\bar{V}(\text{hyp}(f) \cap (\text{hyp}(f) - (x,0))) = \int_G f(y)\, dy \ + \ \int_F f(y+x)\, dy\ . \tag{3.17}$$

This implies that the function $\min\{f(y), f(y+x)\}$ (as a function of y) is measurable and that (3.14) holds. □

Also, we shall need the following consequence of (3.14).

PROPOSITION 3.8 *If $f \in R^n$ is such that*

$$0 < f(x) < 1\,, \quad x \in \mathrm{supp}(f)\,, \tag{3.18}$$

then

$$\mathcal{W}(\mathrm{hyp}(f)) = \mathcal{L}(\mathrm{supp}(f)) \tag{3.19}$$

and

$$\hat{\mathcal{W}}(\mathrm{hyp}(f)) = \hat{\mathcal{L}}(\mathrm{supp}(f))\,. \tag{3.20}$$

Proof. The definition (3.13) of hyp(f) and the assumption (3.18) imply that for $(x_1, \xi_1), (x_2, \xi_2) \in \mathrm{hyp}(f)$, the condition

$$(x_1, \xi_1) - (x_2, \xi_2) \in W$$

holds if and only if $\xi_1 = \xi_2$ and $x_1 - x_2 \in L$, that gives (3.19).

To prove (3.20) one first observes that

$$\hat{\mathcal{W}}(\mathrm{hyp}(f)) \subseteq \mathcal{L}(\mathrm{supp}(f)), \tag{3.21}$$

as a consequence of (3.19). Secondly, Proposition 3.7 shows that for any $(x, 0) \in R^{n+1}$, the condition

$$\bar{V}(\mathrm{hyp}(f) \cap (\mathrm{hyp}(f) - (x, 0))) > 0 \tag{3.22}$$

holds if and only if

$$V(\mathrm{supp}(f) \cap (\mathrm{supp}(f) - x)) > 0\,. \tag{3.23}$$

Using (3.21) this implies (3.20). □

We have prepared everything to prove the Lemma 3.5.

Proof of Lemma 3.5. We can assume w.t.l.g. that (3.18) holds, because both sides of (3.8) are invariant upon the multiplication of f by any positive constant. Let h denote the characteristic function of the set hyp(f). Let $S := P \oplus \{(\theta, \xi) : 0 \le \xi < 1\}$ be the basic cell of W in R^{n+1}, where P is the basic cell of L in R^n.

Applying (3.16) to $E := \mathrm{hyp}(f)$ we can see that

(3.24)

$$\sum_{w \in W} \bar{V}(\mathrm{hyp}(f) \cap (\mathrm{hyp}(f) - w)) \;=\; \int_{R^{n+1}} \left(h(y) \cdot \sum_{w \in W} h(y + w) \right) dy\,.$$

It is clear that (1.5) is valid also in R^{n+1} and using it to the right-hand side integral of (3.24) (for W, S instead of L, P), we get that

$$\sum_{w\in W} \bar{V}(\mathrm{hyp}(f)\cap(\mathrm{hyp}(f)-w)) = \int_S \left(\sum_{w\in W} h(z+w)\right)^2 dz \; . \tag{3.25}$$

On the other hand, (3.20) and (3.14) imply that

$$\sum_{w\in W} \bar{V}(\mathrm{hyp}(f)\cap(\mathrm{hyp}(f)-w)) = \sum_{u\in L}\int_{R^n} \min\{f(x), f(x+u)\}\, dx \; , \tag{3.26}$$

which yields by (3.25) that

$$\sum_{u\in L}\int_{R^n} \min\{f(x), f(x+u)\}\, dx = \int_S \left(\sum_{w\in W} h(z+w)\right)^2 dz \; . \tag{3.27}$$

Applying the generalized Parseval formula (3.11) for R^{n+1} to the right-hand side of (3.27) with $f_1(\cdot) = f_2(\cdot) = \sum_{w\in W} h(\cdot + w)$, $P := S$, $L := W$, using again (1.5) with W, S and the fact $\langle q, w\rangle$ is integer for $q \in W^*$, $w \in W$, we get

$$\begin{aligned} &\sum_{u\in L}\int_{R^n} \min\{f(x), f(x+u)\}\, dx = \\ &= \frac{1}{d(L)}\sum_{q\in W^*}\int_{R^{n+1}}\int_{R^{n+1}} \cos(2\pi\langle q, z-y\rangle) h(z)h(y)\, dz\, dy \; . \end{aligned} \tag{3.28}$$

The scalar product $\langle\cdot,\cdot\rangle$ within the cosine in (3.28) is now understood in the space R^{n+1}. Here we also used that $d(W) = d(L)$, which follows from the definition (3.12) of W.

To prove that the right-hand sides of (3.8) and (3.28) are the same, one needs a careful analysis, the main steps of which are as follows. (We recommend to the reader to check the validity of each step.)

(a) $W^* = L^* \oplus Z^1$, hence $\langle q, z-y\rangle$ in (3.28) is equal to $\langle v, x-t\rangle + k(\xi-\tau)$, where $v \in L^*$, $x, t \in R^n$, $k \in Z^1$, $\xi, \tau \in R^1$.

(b) For all $x, t \in R^n$ we have

$$\sum_{k\in Z^1}\int_{R^1}\int_{R^1} \sin(2\pi k(\xi-\tau)) h(x,\xi)h(t,\tau)\, d\xi\, d\tau = 0 \; ,$$

(because $\sin\alpha = -\sin(-\alpha)$).

(c) Using the formula $\cos(\alpha+\beta) = \cos\alpha \cdot \cos\beta - \sin\alpha \cdot \sin\beta$, (a) and (b) together imply that the right-hand side of (3.28) is equal to

$$\begin{aligned}&\frac{1}{d(L)} \sum_{v \in L^*} \int_{R^n} \int_{R^n} \cos(2\pi \langle v, x-t\rangle) \cdot \\ &\left(\sum_{k \in Z^1} \int_{R^1} \int_{R^1} \cos(2\pi k(\xi - \tau)) h(x,\xi) h(t,\tau)\, d\xi\, d\tau \right) dx\, dt\,.\end{aligned} \tag{3.29}$$

(d) Taking the definition of h into account, the inner sum in (3.29) boils down to

$$\sum_{k \in Z^1} \int_0^{f(t)} \int_0^{f(x)} \cos(2\pi k(\xi-\tau))\, d\xi\, d\tau\,. \tag{3.30}$$

(e) Performing the two integrations, (3.30) turns to:

$$\begin{aligned}&f(x) \cdot f(t) + \frac{1}{2\pi^2} \left(\sum_{k \geq 1} \frac{\cos(2\pi k(f(x) - f(t)))}{k^2} + \sum_{k \geq 1} \frac{1}{k^2} - \right.\\ &\left. - \sum_{k \geq 1} \frac{\cos(2\pi k f(x))}{k^2} - \sum_{k \geq 1} \frac{\cos(2\pi k f(t))}{k^2} \right).\end{aligned} \tag{3.31}$$

(f) For the calculations of four sums in (3.31) we use the well-known identity (see, e.g., [5]): for $-1 \leq \eta \leq +1$ we have

$$\sum_{k \geq 1} \frac{\cos(2\pi k \eta)}{k^2} = \frac{\pi^2}{6} - \pi^2 |\eta| + \pi^2 \eta^2\,. \tag{3.32}$$

(g): Finally, using (3.32) we can see that (3.31) is equal to

$$-\frac{1}{2}|f(x) - f(t)| + \frac{1}{2}(f(x) + f(t)) = \min\{f(x), f(t)\}. \tag{3.33}$$

Putting (a)-(g) together, we can see that the right-hand side of (3.28) is equal to that of (3.8). □

Remark 3.9 The identity (3.8) seems to be something completely new in Fourier analysis. As we can see, its proof (not counting the generalized Parseval formula and the identity (1.5)) is elementary, but not easy. It can be even generalized to point lattices of lower dimension, see remarks in Section 4.

Comparing the two identities (1.4) and (3.8), at first glance the latter seems to be of a different flavour than (1.4). An interesting feature of (3.8) is that the same sort of a "two variable" function created by f, the function $\min\{f(\cdot), f(\cdot)\}$, occurs on both sides of (3.8). Nevertheless, this feature can be observed in (1.4) as well, after transforming the right-hand side of (1.4) appropriately. Namely, we have

PROPOSITION 3.10 *For $f \in N(R^n)$ the following identity holds.*

$$\sum_{u\in L} \int_{R^n} f(x)f(x+u)\,dx = \tag{3.34}$$
$$\frac{1}{d(L)} \sum_{v\in L^*} \int_{R^n}\int_{R^n} \cos(2\pi\langle v, x-t\rangle) f(x) f(t)\,dx\,dt\,.$$

Proof. (3.34) is a simple consequence of the generalized Parseval formula (3.11) applied to the functions $f, \bar{f}$ and the relation (1.5). Namely, using (1.5) we can see easily that the left-hand side of (3.34) is equal to

$$\int_P (\bar{f}(y))^2\,dy\,.$$

Applying (3.11), we get

$$\int_P (\bar{f}(y))^2\,dy = \tag{3.35}$$
$$\frac{1}{d(L)} \sum_{v\in L^*} \int_P\int_P \cos(2\pi\langle v, y-z\rangle) \bar{f}(y)\bar{f}(z)\,dy\,dz\,.$$

Using again (1.5) and the fact that $\langle v, u\rangle$ is integer for $v \in L^*$ and $u \in L$, we easily derive that

$$\int_P\int_P \cos(2\pi\langle v, y-z\rangle) \bar{f}(y)\bar{f}(z)\,dy\,dz = \tag{3.36}$$
$$\int_{R^n}\int_{R^n} \cos(2\pi\langle v, x-t\rangle) f(x) f(t)\,dx\,dt\,.$$

This proves (3.34). □

Remark 3.11 Proposition 3.10 shows that the right-hand sides of (1.4) and (3.34) should be equal. This fact can be proved directly by simple calculations as well, hence (1.4) implies (3.34) and vice versa. This also implies that replacing in (3.6) the right-hand side of (3.34) instead of $\|\hat{f}\|_2^2$, one get another form of the inequality $\leq^b$, where now the functions $\min\{f^2(x), f^2(t)\}$ and $f(x) \cdot f(t)$ occur. These two forms are in fact two special cases of a continuous hierarchy of the so-called α-means and one can proof an inequality analogous to (3.7) by using the latter means (see Section 4).

A natural question is: can the results of this section be extended to any, not necessarily non-negative f? The answer is partly yes, as it is seen from the following theorems.

THEOREM 3.12 *Let f be a bounded square-integrable function having bounded support. Then the inequalities* (3.6, $\leq^a$) *and* (3.7, $\leq^1$) *are true with $|f|$ instead of f. The conditions of equalities in these new inequalities are the same as those in Theorem 3.3 and Lemma 3.4, respectively.*

Proof. According to Theorem 2.4, the conditions of equalities in inequalities in question depend only on $\text{supp}(f)$ and clearly $\text{supp}(f) = \text{supp}(|f|)$. □

For the remaining cases only the following weaker, a little "asymmetric" theorem holds.

THEOREM 3.13 *Let f be as in the Theorem 3.12. Then, the inequalities* (3.6, $\leq^b$) *and* (3.7, $\leq^2$) *are true with $|f|$ instead of f. If f is almost extendable to an L-periodic function, then both new inequalities turn to equalities.*

Proof. Theorem 2.9 shows that if f is almost extendable to an L-periodic function, then the same is true for $|f|$. □

Remark 3.14 It is clear that the converse statement in Theorem 3.13 is in general not true: $|f|$ can be almost extendable to an L-periodic function, while f need not. This is also a consequence of Theorem 2.9, because $|f(x)| = |f(x+u)|$ may mean that $f(x) = -f(x+u)$.

4. Concluding Remarks

4.1. As Theorem 2.1 shows, the two periodic properties of a function studied in the paper are closely related to the classical basic question of the geometry of numbers: how many lattice points are there in the difference set of a given set? For a comprehensive account of results concerning this question see the basic monographs [2, 3, 6, 12]. We have proved both improvements and extensions of some fundamental results concerning this question. For example, the papers [14, 15] contain improvements of the basic Theorem of Minkowski-Blichfeldt-V.d. Corput and its generalization due to Rado and Cassels, respectively (also see, [6], pp. 128-131). It turned out that the basic result of [14] also improves upon another fundamental theorem, the Siegel-Bombieri-formula, see, [16] (for the latter formula, see, [1, 11]). We have succeeded in extending the results to any discrete subgroup of R^n ($\sim$ point lattice of any dimension), see, e.g., [16, 17, 18], to closed subgroups of R^n, [19], as well as to some discrete subgroups of some topological groups, [20] (see subsection 4.4 below).

4.2. The inequalities proved in Section 3 are through the two periodicity properties of f directly connected, via the theorems of Section 2, to the problematics of finding some consequences of the fact that $A - A$ contains no non-zero lattice points. One of the classical results of this sort is due to Siegel, [11], which has been further developed by Bombieri, [1] ($\sim$ the Siegel-Bombieri-formula mentioned above). In [16], among other things, a method how to extend this formula to lower dimensional point lattices as well as some improvements of the formula has been also showed. However, the main emphasis in [16] was an application of a new lower estimation for $|H-H|$ proved in [4] into geometry of numbers (where H is any finite subset of R^n) and analytic exploration of the results achieved (also see, [17, 18]). A method how to extend Fourier-analytic techniques to lower dimensional point lattices has been worked out in [16]. Using this method, the basic results of the paper can be extended to point lattices in R^n of any dimension, see, [21, 22].

4.3. In fact, as it was mentioned in Remark 3.11, the inequality (3.6,$\leq^b$) is a special case of a whole "continuous" hierarchy of inequalities. Namely, in [21] and [22] the following theorem has been proved. First, introduce the following notation. Let a, b, α, λ be real numbers such that $a, b \geq 0$, $0 < \lambda < 1$ and $-\infty < \alpha < \infty$, $\alpha \neq 0$. Then, we define $M_\alpha^{(\lambda)}(a,b)$ to be equal to zero if one of a, b is zero, and to the number $(\lambda a^\alpha + (1-\lambda)b^\alpha)^{1/\alpha}$ otherwise. Taking the limits in α one arrives at three more numbers: $M_{-\infty}^{(\lambda)}(a,b) = \min\{a,b\}$; $M_{+\infty}^{(\lambda)}(a,b) = \max\{a,b\}$ if $a, b > 0$ and $= 0$ if $a \cdot b = 0$; $M_0^{(\lambda)}(a,b) = a^\lambda \cdot b^{(1-\lambda)}$. (These numbers might be called "extended means" of a, b. On 'usual' means $(\lambda a^\alpha + (1-\lambda)b^\alpha)^{1/\alpha}$, $-\infty \leq \alpha \leq +\infty$, see, e.g., [8].) The properties of usual means (see [8]) show that $M_\alpha^{(\lambda)}(a,b)$ is for positive $a, b, a \neq b$, a strictly increasing continuous function of α on $[-\infty, +\infty]$. Now, we have

THEOREM 4.1 ([21], [22]). *Let* $-\infty \leq \alpha \leq \beta \leq +\infty$, $f \in N(R^n)$, *and assume that* $V(\mathrm{supp}(f)) > 0$. *Then,*
(4.1)

$$\int_{R^n} f(x)\,dx \leq^1$$

$$\sum_{u\in L}\int_{R^n} M_\alpha^{(\lambda)}(f(x), f(x+u))\,dx \leq^2 \sum_{u\in L}\int_{R^n} M_\beta^{(\lambda)}(f(x), f(x+u))\,dx\,.$$

$\leq^1$ *is equality if and only if* f *is almost extendable to an* L*-periodic function and* f *has no periodic parts of positive measure.*

For $\alpha < \beta$, $\leq^2$ *is equality if and only if* f *is almost extendable to an* L*-periodic function.*

4.4. As it is mentioned above (subsections 4.1, 4.2), all the results of the paper (including Theorem 4.1) can be proved (by using the extension of Fourier techniques proposed in [16]) when L is any discrete subgroup of R^n, [21], [22].

Also, it is mentioned in subsection 4.1 that the basic result of [14] holds not only for any L, but also in any Abelian locally compact topological group G for any so-called "sufficiently large" discrete subgroup Γ of G (see [20] for more details). This hints a very interesting question: to what extent can the results proved for (R^n, L), say those mentioned above or those proved in the paper, be extended to (G, Γ)? As our techniques work in many situations also for general G, Γ, one has a feeling that some of the results might be extended to these more general structures.

An interesting "extension of the geometry of numbers" has been introduced and studied in [19]: one takes instead of L an arbitrary closed subgroup of R^n. It is sure that some basic theorems of geometry of numbers can, to some extent, be extended to this more general situation, [19]. What about the "periodic w.r.t. the closed subgroup" structure of a function?

References

1. E. Bombieri, Sulla dimostracione di C. L. Siegel del teorema fondamentale di Minkowski nella geometria dei numeri, *Boll. U. Mat. Ital.*, Ser. III, **17**(1962), 283-288.
2. J. W. S. Cassels, *An Introduction to the Geometry of Numbers*, Springer, Berlin-Göttingen-Heidelberg. 1959.
3. P. Erdős, P. M. Gruber, J. Hammer, *Lattice Points*, Longman Sci., London, 1989.
4. G. Freiman, A. Heppes, B. Uhrin, A lower estimation for the cardinality of finite difference sets in R^n, In: K. Győry, G. Halász, Eds., *Number Theory, Budapest,*

1987 , Coll. Math. Soc. J. Bolyai **51**, North-Holland, Amsterdam-New York, 1989, 125-139.
5. I. S. Gradstein, I. M. Ryzhik, *Tables of integrals, sums, series and products*, Fizmatgiz, Moscow, 1963, (in Russian).
6. P. M. Gruber, C. G. Lekkerkerker, *Geometry of Numbers*, 2nd Ed., North-Holland, Amsterdam-New York, 1987.
7. H. Hadwiger, Überdeckung des Raumes durch translationsgleiche Punktmengen und Nachbarnzahl, *Monatsh. f. Math.*, **73**(1969), 213-217.
8. G. H. Hardy, J. E. Littlewood, G. Pólya, *Inequalities*, Cambridge Univ. Press, Cambridge, 1951.
9. E. Hewitt, K. A. Ross, *Abstract Harmonic Analysis,* Vol. I, II, Springer, Berlin-Heidelberg-New York, 1963, 1970.
10. I. Z. Ruzsa, Sums of sets in several dimensions, *Combinatorica*, to appear.
11. C. L. Siegel, Über gitterpunkte in konvexen Körpern und ein damit zusammenhängendes Extremalproblem, *Acta Math.*, **68** (1935), 307-323.
12. C. L. Siegel, *Lectures on Geometry of Numbers*, Springer, Berlin-Heidelberg-New York, 1967.
13. E. M. Stein, G. Weiss, *Introduction to Fourier Analysis on Euclidean Spaces*, Princeton Univ. Press, Princeton, N.J., 1971.
14. B. Uhrin, Some useful estimations in geometry of numbers, *Period. Math. Hungar.*, **11**(1980), 95-103.
15. B. Uhrin, On a generalization of Minkowski convex body theorem, *J. of Number Th.*, **13**(1981), 192-209.
16. B. Uhrin, New lower bounds for the number of lattice points in a difference set, *Acta Sci. Math.(Szeged)*, **62** (1996), 81-99.
17. B. Uhrin, A description of the segment $[1, T]$, where T is the meeting number of a set-lattice, *Period. Math. Hungar.*, **26**(1993), 139-156.
18. B. Uhrin, The index of a point-lattice in a set, *J. Number Th.*, **54** (1995), 232-247.
19. B. Uhrin, The measure of covering the Euclidean space by group translates of a set, In: G. M. Rassias, Ed., *The Mathematical Heritage of C. F. Gauss*, World Scientific, Singapore-New Jersey-London, 1991, 785-805.
20. B. Uhrin, Some remarks about the lattice points in difference sets, In: J. Szabados, K. Tandori, Eds., *Proc. of the A. Haar Memorial Conference, Budapest, 1985,* Coll. Math. Soc. J. Bolyai **49**, North-Holland, Amsterdam-New York, 1987, 929-937.
21. B. Uhrin, The inner periodic structure of a function, Research Report, IRL-3-1995, Comp.and Automat.Inst, HAS, Informatics Research Laboratory, Budapest, pp. 1-21 (submitted for publication).
22. B. Uhrin, On the periodic structure of a function, Research Report, IRL-1-1995, Comp. and Automat. Inst., HAS, Informatics Res. Laboratory, pp. 1-23.
23. B. Uhrin, Inner aperiodicities and partitions of sets, *Linear Algebra and Appl.*, **241-243** (1996), 851-876.
24. A. Weil, *Basic Number Theory*, Springer, Berlin-Heidelberg-New York, 1967.

Applied Optimization

36. G. Di Pillo and F. Giannessi (eds.): *Nonlinear Optimization and Related Topics.* 2000
ISBN 0-7923-6109-1
37. V. Tsurkov: *Hierarchical Optimization and Mathematical Physics.* 2000
ISBN 0-7923-6175-X
38. C. Zopounidis and M. Doumpos: *Intelligent Decision Aiding Systems Based on Multiple Criteria for Financial Engineering.* 2000 ISBN 0-7923-6273-X
39. X. Yang, A.I. Mees, M. Fisher and L.Jennings (eds.): *Progress in Optimization.* Contributions from Australasia. 2000 ISBN 0-7923-6286-1
40. D. Butnariu and A.N. Iusem: *Totally Convex Functions for Fixed Points Computation and Infinite Dimensional Optimization.* 2000 ISBN 0-7923-6287-X
41. J. Mockus: *A Set of Examples of Global and Discrete Optimization.* Applications of Bayesian Heuristic Approach. 2000 ISBN 0-7923-6359-0
42. H. Neunzert and A.H. Siddiqi: *Topics in Industrial Mathematics.* Case Studies and Related Mathematical Methods. 2000 ISBN 0-7923-6417-1
43. K. Kogan and E. Khmelnitsky: *Scheduling: Control-Based Theory and Polynomial-Time Algorithms.* 2000 ISBN 0-7923-6486-4
44. E. Triantaphyllou: *Multi-Criteria Decision Making Methods.* A Comparative Study. 2000 ISBN 0-7923-6607-7
45. S.H. Zanakis, G. Doukidis and C. Zopounidis (eds.): *Decision Making: Recent Developments and Worldwide Applications.* 2000 ISBN 0-7923-6621-2
46. G.E. Stavroulakis: *Inverse and Crack Identification Problems in Engineering Mechanics.* 2000 ISBN 0-7923-6690-5
47. A. Rubinov and B. Glover (eds.): *Optimization and Related Topics.* 2001
ISBN 0-7923-6732-4
48. M. Pursula and J. Niittymäki (eds.): *Mathematical Methods on Optimization in Transportation Systems.* 2000 ISBN 0-7923-6774-X
49. E. Cascetta: *Transportation Systems Engineering: Theory and Methods.* 2001
ISBN 0-7923-6792-8
50. M.C. Ferris, O.L. Mangasarian and J.-S. Pang (eds.): *Complementarity: Applications, Algorithms and Extensions.* 2001 ISBN 0-7923-6816-9
51. V. Tsurkov: *Large-scale Optimization – Problems and Methods.* 2001
ISBN 0-7923-6817-7
52. X. Yang, K.L. Teo and L. Caccetta (eds.): *Optimization Methods and Applications.* 2001 ISBN 0-7923-6866-5
53. S.M. Stefanov: *Separable Programming Theory and Methods.* 2001
ISBN 0-7923-6882-7

Applied Optimization

54. S.P. Uryasev and P.M. Pardalos (eds.): *Stochastic Optimization: Algorithms and Applications.* 2001 ISBN 0-7923-6951-3
55. J. Gil-Aluja (ed.): *Handbook of Management under Uncertainty.* 2001 ISBN 0-7923-7025-2
56. B.-N. Vo, A. Cantoni and K.L. Teo: *Filter Design with Time Domain Mask Constraints: Theory and Applications.* 2001 ISBN 0-7923-7138-0
57. S. Zlobec: *Stable Parametric Programming.* 2001 ISBN 0-7923-7139-9
58. M.G. Nicholls, S. Clarke and B. Lehaney (eds.): *Mixed-Mode Modelling: Mixing Methodologies for Organisational Intervention.* 2001 ISBN 0-7923-7151-8

KLUWER ACADEMIC PUBLISHERS – DORDRECHT / BOSTON / LONDON